Sex Determination in Plants

EXPERIMENTAL BIOLOGY REVIEWS

Series advisors:

D.W. Lawlor
AFRC Institute of Arable Crops Research, Rothamsted Experimental Station, Harpenden, Hertfordshire AL5 2JQ, UK

M. Thorndyke
School of Biological Sciences, Royal Holloway, University of London, Egham, Surrey TW20 0EX, UK

Environmental Stress and Gene Regulation
Sex Determination in Plants

Forthcoming titles include:

Plant Carbohydrate Biochemistry
Biomechanics in Animal Behaviour
Cell Death in Health and Disease
Cambium: the biology of wood formation

Related titles from BIOS in the Environmental Plant Biology *series:*

Abscisic Acid: physiology and biochemistry
Biological Rhythms and Photoperiodism in Plants
Carbon Partitioning: within and between organisms
Embryogenesis: the generation of a plant
Environment and Plant Metabolism: flexibility and acclimation
Pests and Pathogens: plant responses to foliar attack
Photoinhibition of Photosynthesis: from molecular mechanisms to the field
Plant Cuticles: an integrated functional approach
Stable Isotopes: the integration of biological, ecological and geochemical processes
Water Deficits: plant responses from cell to community

C.C. AINSWORTH
Plant Molecular Biology Laboratory, Department of Biological Sciences, Wye College, University of London, Wye, Ashford, Kent TN25 5AH, UK

Sex Determination in Plants

Oxford • Washington DC

First published in 1999

A CIP catalogue record for this book is available from the British Library.

ISBN 1 85996 042 1

BIOS Scientific Publishers Ltd
9 Newtec Place, Magdalen Road, Oxford OX4 1RE, UK.
Tel. +44 (0) 1865 726286. Fax. +44 (0) 1865 246823
World Wide Web home page: http://www.bios.co.uk/

Production Editor: Jonathan Gunning.
Typeset by Saxon Graphics Ltd, Derby, UK.
Printed by Biddles Ltd, Guildford, UK.

Contents

Contributors

Aghmir, A., Plantengenetica, Institute of Molecular Biology, B-1640 St Genesius Rode, Belgium

Ainsworth, C.C., Plant Molecular Biology Laboratory, Department of Biological Sciences, Wye College, University of London, Wye, Ashford, Kent TN25 5AH, UK

Caporali, E., Università di Milano, Dipartimento di Biologia, Sezione di Botanica Generale, Via Celoria 26, I-20133 Milano, Italy

Causier, B., Centre for Plant Sciences, Leeds Institute for Plant Biotechnology and Agriculture, University of Leeds, Leeds LS2 9JT, UK

Charlesworth, D., Institute of Cell, Animal and Population Biology, University of Edinburgh, Ashworth Lab., King's Buildings, W. Mains Road, Edinburgh EH9 3JT, UK

Cipriani, G., Dipartimento di Produzione vegetale e tecnologie agrarie, University of Udine, Via delle scienze 208, I-33100 Udine, Italy

Darby, P., Department of Hop Research, Horticulture Research International, Wye, Ashford, Kent TN25 5AH, UK

Davies, B., Centre for Plant Sciences, Leeds Institute for Plant Biotechnology and Agriculture, University of Leeds, Leeds LS2 9JT, UK

Donnison, I.S., Institute of Grassland and Environmental Research, Plas Gogerddan, Aberystwyth, Ceredigion, Wales SY23 3EB

Georgiev, S., Reproduction et Développement des Plantes, UMR 9938 CNRS/ INRA/ENS, Ecole Normale Supérieure, 46 Allée d'Italie, F-69364 Lyon Cedex 07, France

Gilmartin, P.M., Centre for Plant Sciences, Leeds Institute for Plant Biotechnology and Agriculture, University of Leeds, Leeds LS2 9JT, UK

Grant, S.G., Coker Hall CB#3280, Department of Biology, University of North Carolina, Chapel Hill, NC 27599-3280, USA

Guttman, D.S., Department of Molecular Genetics and Cell Biology, University of Chicago, 1103 East 57th Street, Chicago, IL 60637, USA

Irish, E.E., Department of Biological Sciences, University of Iowa, Iowa City, IA 52242, USA

Kahl, G., Plant Molecular Biology, University of Frankfurt, Marie-Curie-Str. 9, D-60439, Frankfurt am Main, Germany

Kamisugi, Y., Centre for Plant Sciences, Leeds Institute for Plant Biotechnology and Agriculture, University of Leeds, Leeds LS2 9JT, UK

Lardon, A., Reproduction et Développement des Plantes, UMR 9938 CNRS/INRA/ENS, Ecole Normale Supérieure, 46 Allée d'Italie, F-69364 Lyon Cedex 07, France

Li, Y., The Plant Laboratory, University of York, PO Box 373, York YO10 5YW, UK

Lu, J., Plant Molecular Biology Laboratory, Department of Biological Sciences, Wye College, University of London, Wye, Ashford, Kent TN25 5AH, UK

Martin, H., Centre for Plant Sciences, Leeds Institute for Plant Biotechnology and Agriculture, University of Leeds, Leeds LS2 9JT, UK

Marziani, G., Università di Milano, Dipartimento di Biologia, Sezione di Botanica Generale, Via Celoria 26, I-20133 Milano, Italy

Messina, R., Dipartimento di Produzione vegetale e tecnologie agrarie, University of Udine, Via delle scienze 208, I-33100 Udine, Italy

Monéger, F., Reproduction et Développement des Plantes, UMR 9938 CNRS/INRA/ENS, Ecole Normale Supérieure, 46 Allée d'Italie, F-69364 Lyon Cedex 07, France

Negrutiu, I., Reproduction et Développement des Plantes, UMR 9938 CNRS/INRA/ENS, Ecole Normale Supérieure, 46 Allée d'Italie, F-69364 Lyon Cedex 07, France

Parker, J.S., University Botanic Garden, Cory Lodge, Bateman Street, Cambridge CB2 1JF, UK

Perl-Treves, R., Faculty of Life Sciences, Bar-Ilan University, Ramat Gan 52900, Israel

Robertson, S.E., Centre for Plant Sciences, Leeds Institute for Plant Biotechnology and Agriculture, University of Leeds, Leeds LS2 9JT, UK

Scutt, C.P., Reproduction et Développement des Plantes, Ecole Normale Supérieure de Lyon, 46 Allée d'Italie, F-69364 Lyon Cedex 07, France

Shenton, M.R., Plant Molecular Science, Division of Biochemistry and Molecular Biology, University of Glasgow, Glasgow G12 8QQ, UK

Shephard, H., Plant Molecular Biology Laboratory, Department of Biological Sciences, Wye College, University of London, Wye, Ashford, Kent TN25 5AH, UK

Smith, R.H., Department of Botany and Plant Pathology, Purdue University, West Lafayette, IN 47907–1155, USA

Spada, A., Università di Milano, Dipartimento di Biologia, Sezione di Botanica Generale, Via Celoria 26, I-20133 Milano, Italy

Terauchi, R., Plant Molecular Biology, University of Frankfurt, Marie-Curie-Str. 9, D-60439, Frankfurt am Main, Germany

Testolin, R., Dipartimento di Produzione vegetale e tecnologie agrarie, University of Udine, Via della scienze 208, I-33100 Udine, Italy

Vyskot, B., Institute of Biophysics, Czech Academy of Sciences, CZ-612 65 Brno, Czech Republic

Weir, I., Centre for Plant Sciences, Leeds Institute for Plant Biotechnology and Agriculture, University of Leeds, Leeds LS2 9JT, UK

Willis, M.E., Centre for Plant Sciences, Leeds Institute for Plant Biotechnology and Agriculture, University of Leeds, Leeds LS2 9JT, UK

Winfield, M., University Botanic Garden, Cory Lodge, Bateman Street, Cambridge CB2 1JF, UK

Abbreviations

ABA	abscisic acid
ACC	1-aminocyclopropane-1-carboxylate
ACO	ACC oxidase
ACS	ACC synthase
AFLP	amplified fragment length polymorphism
AOA	α-aminoxyacetic acid
AVG	aminoethoxyvinyl glycine
5-azaC	5-azacytidine
BAC	bacterial artificial chromosome
bp	base pair
DAPI	4′,6-diamidino-2-phenylindole
DOP-PCR	degenerate oligonucleotide primer polymerase chain reaction
EMS	ethane methane sulphate
FBP	floral binding protein
FDD	fluorescent differential display
FISH	fluorescent *in situ* hybridization
GA	gibberellic acid
GISH	genomic *in situ* hybridization
GSF	gynoecium suppression function
IAA	indole acetic acid
ITS	internal transcribed spacer
LD	long day
LINE	long interspersed nuclear element
LOD	Log-Odds
LTR	long-terminal-repeat
5-mC	5-methylcytosine
MDH	malate dehydrogenase
MEF2	myocyte enhancer factor
NAA	naphthalene acetic acid
NILS	nearly isogenic lines
PCIB	α-(*p*-chlorophenoxy)isobutyric acid
PCR	polymerase chain reaction
PGI	phosphoglucose isomerase

PMC	pollen mother cell
PPO	polyphenol oxidase
Prep-ISH	preparative *in situ* hybridization
PX	peroxidase
QTL	quantitative trait loci
RACE	rapid amplification of cDNA ends
RAPD	random amplified polymorphic DNA
RDA	representational difference analysis
RFLP	restriction fragment length polymorphism
SCAR	sequence characterized amplified region
SD	short day
SEM	scanning electron microscopy
SPF	stamen-promoting function
SRF	serum response factor
SSCP	single-strand conformation polymorphism
TCF	ternary complex factor

Preface

Most animal species are unisexual, with male and female gametes produced directly by meiosis in different individuals. By contrast, the great majority of flowering plants are hermaphrodite, producing flowers which are 'perfect' and which contain the 'male' organs of a flower (the stamens) and the 'female' organs (the carpels or, collectively, the gynoecium). A minority of plant species (precise estimates are difficult, but probably about 10%) produce flowers which are unisexual. This can be as dioecy, where male and female flowers are carried on separate individuals, or monoecy, where male and female flowers are produced on the same individual. The determination of sex in plants is thus a key developmental decision which leads to the suppression of the programme of development of either the male or female organs. This intriguing group of plants has gone about unisexuality in many different ways with the developmental blocks occurring across the spectrum of organ initiation and development. For this reason, unisexual plants offer us the opportunity of investigating plant development in a way which is complementary to the study of flower development in hermaphrodite species. Indeed, we may learn a considerable amount about development in hermaphrodites from the study of unisexual plants.

Classical genetic studies of dioecious plant species, and the first identification of sex chromosomes in plants, date back to the early decades of this century. The advent of new molecular biology techniques and the ability to express genes in transgenic plants has led to a resurgence of research on sex determination in monoecious and dioecious plants. In addition to the possibility of isolating the genes which are responsible for the unisexual conditions, these techniques are allowing new insights to be gained in the nature and evolution of plant sex chromosomes. This volume is a collection of papers based on the 'Sex Determination in Plants' symposium held at the 1997 Society for Experimental Biology meeting in York and presents the state of research in the major groups of unisexual plants, both dioecious and monoecious.

The plants discussed here include those where sex development is underpinned by well characterized sex-chromosome systems (*Silene latifolia*, *Humulus* species, *Rumex acetosa*) and plants where sex chromosomes are either absent or whose existence has not been proven. The evolution of unisexuality and sex-chromosome systems is given specific attention. The range of species discussed encompasses those in which the inappropriate organs are not initiated (*Humulus* species), those in which the inappropriate organs are initiated but which do not develop significantly (maize, *S. latifolia*, *R. acetosa*, *Cucumis sativus*), and those in which the arrest in the development of the inappropriate sex organs occurs very late so that male and female flowers are indistinguishable from each other and from perfect flowers (kiwi fruit, *Asparagus officinalis*).

Charles Ainsworth

1

MADS-box factors in hermaphrodite flower development

Barry Causier, Irene Weir and Brendan Davies

1. Introduction

MADS-box proteins are a family of transcription factors found in vertebrates, insects, plants and fungi, which regulate diverse processes (for review see Shore and Sharrocks, 1995). MADS-box factors share a highly conserved domain of 57 amino acids, known as the MADS-box (Schwarz-Sommer *et al.*, 1990), which binds DNA. However, there are significantly more MADS-box genes in plants than are found in members of the other kingdoms. Currently, 36 MADS-box genes have been described from *Arabidopsis* (Liljegren *et al.*, 1998a), 20 from *Antirrhinum* (Sommer and Davies, unpublished results) and 24 from *Petunia* (Angenent, personal communication). Plant MADS-box factors are expressed in the roots and aerial parts, implicating them in both vegetative and reproductive plant development. In spite of this diversity of expression, most of the known MADS-box genes are expressed in the flower and those involved in flower development have been studied most extensively. MADS-box transcription factors mediate floral meristem identity, floral organ identity and various other aspects of flower development. The regulation of floral organ identity might suggest that factors of this type play a role, directly or indirectly, in sex determination. The factors controlling sex determination and the resultant expression of the floral organ identity genes, are beyond the scope of this review, but are covered in other chapters of this volume.

The study of floral homeotic mutants of *Antirrhinum* and *Arabidopsis* led directly to the isolation of the first members of the plant MADS-box gene family (Sommer *et al.*, 1990; Yanofsky *et al.*, 1990). Further studies on such mutants and on related members of the MADS-box gene family resulted in the establishment of a more or less universal model for the specification of floral organ identity within the hermaphrodite flower. This model, known as the ABC model, relied on the activities of three classes of genes, most of which were identified as MADS-box genes, to determine the four

Sex Determination in Plants, edited by C.C. Ainsworth.

floral organ types. The ABC model and the genes which constitute the A, B and C functions have been reviewed at length elsewhere (Coen and Meyerowitz, 1991; Davies and Schwarz-Sommer, 1994; Haughn *et al.*, 1995; Ma, 1994; Riechmann and Meyerowitz, 1997a; Weigel and Meyerowitz, 1994; Yanofsky, 1995) and will not be the subject of this review. The purpose of this chapter is to review the plant MADS-box transcription factors for which a biological function has been assigned and to give a brief account of the current biochemical and molecular biological evidence for how plant MADS-box transcription factors act to regulate development. We will demonstrate the importance in flower development of protein–protein interactions and MADS-box genes other than those strictly designated as A, B and C genes. In this way it is hoped that we will identify genes and mechanisms which could usefully be studied in plants which are subject to sex determination as well as pointing out some other roles of this important family of plant regulators.

2. Functions of the plant MADS-box factors

Recent evidence suggests that a typical dicotyledon contains 40 or more different members of the MADS-box family, each of which is subject to a precise temporal and spatial regulation. Although MADS-factors are best characterized as central regulators of floral organ identity, the biological function of the majority currently remains unclear. The use of genetic, reverse genetic and transgenic approaches is enabling functions to be assigned to increasing numbers of MADS-box factors in a diverse range of developmental processes (summarized in *Table 1*). Although some MADS-box genes have expression patterns which suggest that they are not involved in flower development (see below), the majority are expressed within the flower and their continued study is likely to provide further insights into flower development. Similarities in sequence and expression patterns amongst the known MADS-box genes might indicate a degree of functional redundancy which could be addressed by testing combinations of MADS-box gene mutants for genetic interaction. Examples are given below of redundancy and partial redundancy.

2.1 *The control of meristem identity*

The MADS-box genes involved in floral meristem identity in *Antirrhinum* (*SQUAMOSA*, *SQUA*) and *Arabidopsis* (*APETALA1*, *AP1* and *CAULIFLOWER*, *CAL*) have been described and reviewed at length elsewhere (Yanofsky, 1995; see *Table 1*). Mutants affecting *SQUA* or *AP1* result in a partial failure to make the transition from inflorescence to floral meristem. However, important differences between the two mutant phenotypes suggest different regulatory mechanisms are at work in these two species (Motte *et al.*, 1998).

The *CAL* gene was isolated when the *ap1* mutation was crossed into a phenotypically wild-type background which, fortuitously, carried a *cal* mutation. Although *cal* alone appears to have no phenotype, the floral meristem of the *ap1/cal* double mutant produces a mass of inflorescence tissue instead of flowers (Bowman *et al.*, 1993). This phenotype is more familiar as the garden cauliflower *Brassica oleracea var. botrytis.* Mutations resulting in loss of function in the cauliflower *CAL* gene have been demonstrated although its relationship with other factors such as cauliflower *AP1* remains to be determined (Kempin *et al.*, 1995).

2.2 *The organ identity genes*

As with the meristem identity genes, organ identity MADS-box genes have been associated with homeotic mutants in several species (*Table 1*). Once again, these genes have been reviewed extensively (Davies and Schwarz-Sommer, 1994; Haughn *et al.*, 1995; Riechmann and Meyerowitz, 1997a). MADS-box genes affecting organ identity have been divided into the subgroups A, B and C, based on the organ identities they control. Class A genes are considered to control the identity of sepals and petals, class B genes control the identity of petals and stamens and class C genes control the identity of the reproductive organs (Riechmann and Meyerowitz, 1997a; see *Table 1*).

Although there is remarkable similarity between the sequences, expression patterns and functions of these genes in different species, there are also sufficient differences to suggest that part of the diversity of flower structure might be related to subtle changes in these genes. The variability of the C-function in different species serves to illustrate this point. In *Arabidopsis* a single C-function gene *AGAMOUS (AG)* has been identified which is responsible for the determinate growth of the flower and the identity of the reproductive organs (Yanofsky *et al.*, 1990). However, two C-function genes have been reported in other species including maize, *Petunia*, *Antirrhinum*, tobacco and cucumber (Davies *et al.*, 1996a; Kater *et al.*, 1998; Kempin *et al.*, 1993; Mena *et al.*, 1996; Tsuchimoto *et al.*, 1993). In maize, the two genes were named *ZAG1* and *ZMM2*. A transposon insertion into the maize gene *ZAG1* led to a loss of determinate growth within the carpel, but reproductive organ identity was otherwise unaffected. Both genes are expressed in both reproductive tissues; *ZAG1* is more abundant in carpels and *ZMM2* in stamens. The identification of a *ZMM2* mutant will allow the relative contribution of both genes to the C-function to be assessed. The two *Antirrhinum* C-function genes are called *PLENA* (*PLE*) and *FARINELLI* (*FAR*). Flowers on *ple* mutant plants show the typical C-function phenotype of loss of reproductive organs and indeterminacy (Bradley *et al.*, 1993). Reverse genetic selection for transposon insertions into the *FAR* gene resulted in the isolation of two independent mutants showing partial male sterility with no obvious homeotic effects. Interestingly, the *far/ple* double mutant shows increased petal identity in the fourth whorl indicating an additional role for *FAR* in the negative regulation of petal identity in the inner whorls (Davies, unpublished). These results show that *FAR* and *PLE* are partially redundant with respect to negative regulation of the B-function and determination of the identity of the reproductive organs. Differences are also observed when the B-functions and, particularly the A-functions, are compared between species. Comparative studies of these functions in divergent plant species, including those subject to sex determination mechanisms, will be of great interest.

2.3 *Intermediate genes*

Intermediate MADS-box genes are first expressed after the meristem identity genes but before the onset of organ identity gene expression. Co-suppression of a *Petunia* intermediate gene, *FBP2*, results in homeotic changes to organ identity in the inner three whorls including: reduced green petals, stamen to petal conversions and production of leaf like carpels (see *Table 1*). *FBP2*-co-suppressed plants also produce inflorescence tissue in the centre of the flower (Angenent *et al.*, 1994). Antisense inhibition of a tomato intermediate gene, *TM5*, also alters organ identity in the inner whorls and affects determinacy

Table 1. *Known functions of MADS-box genes* [a]

Gene	Species	Phenotype of loss of function mutant	Proposed role in development	References
AP1 *APETALA1*	*Arabidopsis*	Carpelloidy of the first whorl, absence of the second whorl. Mutant flowers within the first whorl	A function, meristem identity and regulation of organ identity	Mandel *et al.*, 1992
CAL *CAULIFLOWER*	*Arabidopsis*	In combination with ap1, proliferation of inflorescence meristems	Floral meristem identity and regulation of organ identity genes	Kempin *et al.*, 1995
SQUA *SQUAMOSA*	*Antirrhinum*	Flowers usually replaced by secondary inflorescence shoots	Floral meristem identity	Huijser *et al.*, 1992
AP3 *APETALA 3*	*Arabidopsis*	Petal to sepal and stamen to carpel conversions	B-function	Jack *et al.*, 1992
DEF *DEFICIENS*	*Antirrhinum*	Sepaloid petals and carpelloid stamens, absence of fourth whorl	B-function. Prevention of premature termination of floral whorl production	Sommer *et al.*, 1990; Schwarz-Sommer *et al.*, 1992
GP *GREEN PETALS* (also termed *pMADS1)*	*Petunia*	Sepaloid petals	B-function. Regulation of petal development	van der Krol *et al.*, 1993
FBP1	*Petunia*	Sepaloid petals and carpelloid stamens	B-function	Angenent *et al.*, 1993
GLO *GLOBOSA*	*Antirrhinum*	As for *def*	B-function	Tröbner *et al.*, 1992
PI *PISTILLATA*	*Arabidopsis*	As for *ap3*	B-function	Goto and Meyerowitz, 1994
PLE *PLENA*	*Antirrhinum*	Stamens become petaloid and sepaloid/petaloid organs replace carpels. Reiteration of flowers	C-function. Promotion of floral determinacy	Bradley *et al.*, 1993
AG *AGAMOUS*	*Arabidopsis*	Stamen to petal conversions, replacement of carpels by a reiteration of the mutant flower	C-function. Promotion of floral determinacy	Yanofsky *et al.*, 1990
FAR *FARINELLI*	*Antirrhinum*	Male sterile but otherwise normal	C-function required for male gametogenesis. Negative regulation of B function	Davies, unpublished

ZAG 1	Maize	Loss of determinate growth of the carpel	C-function. Carpel development and control of determinacy	Mena *et al.*, 1996
TAG 1	Tomato	Stamen to petal conversions, replacement of carpels with a reiteration of mutant flowers	C-function and control of determinacy	Pneuli *et al.*, 1994b
FBP2	*Petunia*	Green petals, stamen to petal conversions and the formation of secondary inflorescence in the centre of the flower	Intermediate between meristem and organ identity. Control of determinacy	Angenent *et al.*, 1994
TM5	Tomato	Green petals and stamens, stamens unfused, gynoecium variable. Indeterminant	Intermediate between meristem and organ identity. Control of determinacy	Pneuli *et al.*, 1994a
FBP11/FBP7	*Petunia*	Conversion of ovules to carpelloid structures. Aberrant seed coat and endosperm formation	Induction of ovule development and correct development of the seed coat	Angenent *et al.*, 1995
FUL FRUITFULL	*Arabidopsis*	Failure of the carpel valve cells to elongate correctly	Regulation of fruit development	Gu *et al.*, 1998
ANR1	*Arabidopsis*	Failure of lateral roots to elongate in the presence of localized pockets of nitrate	Control of root architecture	Zhang and Forde, 1998

[a] Those MADS-box genes for which a function has currently been determined, either by mutant analysis or by transgenic means, are listed. A brief description of the effect of loss of gene function and the implied role in development is given.

(Pnueli *et al.*, 1994a). In *Antirrhinum*, there are at least three genes which fall into this category on the basis of sequence similarity and expression pattern. Importantly, the *Antirrhinum* intermediate MADS-box factors have been shown to interact directly with the C-function organ identity factors (Davies *et al.*, 1996b) and with the meristem identity factor SQUA (see below). Evidence suggests that the intermediate genes may be required for at least some of the organ identity genes to function (see below). The interactions and timing of expression of the intermediate genes suggest that they also play other roles in flower development. It is likely that future experiments will reveal the true importance of this class of MADS-box genes.

2.4 *The next generation: ovules and seeds*

The ability to produce viable seed is of obvious importance to any angiosperm but comparatively little is known about the specification of ovule identity, seed maturation and fruit formation. Recent work in *Petunia* has identified two MADS-box factors expressed within the ovule. Co-suppression of *FBP11/FBP7* resulted in a homeotic conversion of ovules to carpeloid structures (Angenent *et al.*, 1995). Ectopic expression of *FBP11* caused ovule formation on sepals and to a lesser extent on petals (Colombo *et al.*, 1995). *FBP7* and *FBP11* continue to be expressed following fertilization in tissue that contributes to the seed coat. Co-suppressed plants show degeneration of the inner epidermis of the seed coat (endothelium) and a shrunken endosperm within the seed. It is thought that the effect on endosperm development results from an inability of nutrients to pass through the degenerated endothelium to the endosperm within the seed (Colombo *et al.*, 1997).

A MADS-box factor with a role in fruit development has recently been identified in *Arabidopsis*. Reverse genetic screens for insertions into a previously identified MADS-box gene, *AGL8*, resulted in a mutation which affects silique elongation, causing overcrowding and premature bursting of the silique (Gu *et al.*, 1998). The *agl8* mutant has been named *fruitfull* (*ful*). Other phenotypes observed in *fruitfull* mutants include: localized absence of stomata, poor development of vascular tissue, increased number of smaller epidermal cells produced within the inner carpel valve tissue and reduction in seed weight. Furthermore, an enhancement of the *ap1/cal* phenotype is seen in *ful/ap1/cal* triple mutants (Liljegren *et al.*, 1998a) indicating that *FUL* is also partially redundant.

Another example of redundancy amongst MADS-box genes is provided by the *Arabidopsis* genes *AGL1* and *AGL5*. Recently, *AGL5* was disrupted by homologous recombination (Kempin *et al.*, 1997), but the *agl5* mutant had no observable phenotype. Reverse genetic screens were used to identify a loss of function mutant of an *AGL5*-like gene, *AGL1* (Liljegren *et al.*, 1998b). This mutant also showed no detectable alteration in phenotype. However, the *agl1/agl5* double mutant did show a phenotype; the silique failed to dehisce. This once again illustrates that although expression patterns and sequence similarity can serve to suggest possible functions, it is only when loss-of-function mutants are examined alone and in combination that the true functions will be revealed.

2.5 *MADS-box factors in vegetative development*

MADS-box factors are not confined to floral tissues; several have been identified which are expressed in vegetative tissues, but an associated mutant phenotype for these factors has yet to be assigned (Heard *et al.*, 1997; Rounsley *et al.*, 1995). Recently, a root specific factor from *Arabidopsis* has been found to be important for root architecture (Zhang and Forde, 1998). The *ANR1* gene promotes lateral root growth in the presence of localized pockets of nitrate ions. Detailed examination showed that this was not due to root initiation but root elongation. This ability enables the plant to optimize root growth in the presence of an important mineral nutrient. It is proposed that lateral root elongation is repressed when nitrate levels within the shoot are high, but that inhibition is lifted by the localized induction of *ANR1* in response to the presence of nitrate.

Further work will be required to determine all the functions of the MADS-box family of genes. The genetic approach is very powerful where a clear mutant phenotype can be observed. However, where this is not the case, the use of reverse genetics is proving increasingly appropriate. Polymerase chain reaction (PCR) based screening for transposon or T-DNA insertions into cloned genes has already successfully identified MADS-box mutants in maize, *Arabidopsis* and *Antirrhinum*. The recent reports of targeted gene disruption in *Arabidopsis* (Kempin *et al.*, 1997) will extend our ability to generate stable recessive mutations without the variability associated with other transgenic approaches. The generation of recessive loss-of-function mutants will provide the opportunity to combine mutants to look for genetic interaction and also to initiate new screens for enhancers and suppressers of phenotypes. As more MADS-box factors are characterized it should be possible to identify conserved functions and species-specific diversity in an increasing number of developmental situations.

3. Properties of plant MADS-box factors

The similarity between plant and animal MADS-box factors is confined to the MADS-box. Plant MADS-box factors comprise four distinct domains; MADS (M), Intervening (I), K-domain (K) and C-terminal domain (C) (*Figure 1*). The plant MADS-domains, like that of the myocyte enhancer factor 2 (MEF2), are found at or near the amino terminus of the protein. In a few cases, there are a variable number of

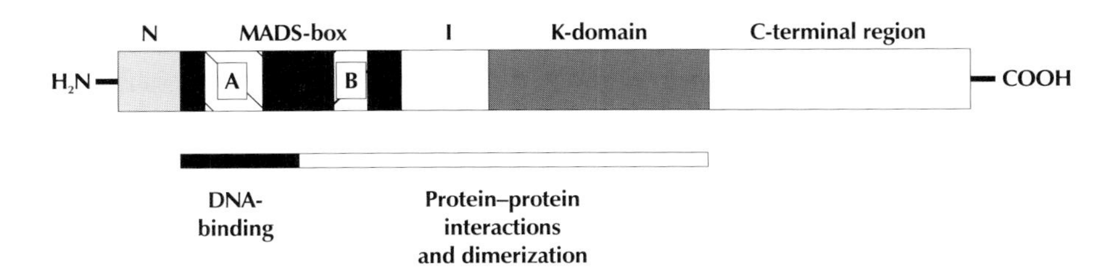

Figure 1. *Schematic representation of a plant MADS-box protein. Plant MADS-box transcription factors possess a number of distinct domains with overlapping functions. At, or near, the amino terminus of the protein is the MADS-box. This domain is highly conserved between all MADS-box proteins from animals, plants and fungi and is involved in both DNA-binding and dimerization. The N-terminal half of the MADS-box is largely hydrophilic and contains a small region (consensus sequence = ----R--T-KR-; labelled A in figure) predicted to form an α-helix which interacts directly with DNA. The remainder of the protein is involved in protein–protein interactions with other MADS-box factors (either through the MADS and I-domains or through the K-domain) and with any potential accessory factors. The dimerization between MADS-box proteins, at least in the case of animal and yeast factors, is mediated to some extent by a hexameric patch of hydrophobic amino acid residues present in the C-terminal half of the MADS-box (labelled B in the figure).*

amino acids (typically about 15) just before the MADS-box. Genes of this type are always related to the C-function class of the MADS-box gene and the function of this N-terminal extension is unknown. The MADS-domain of human serum response factor (SRF) is found more centrally, while that of ARG80 (a yeast MADS-box protein involved in arginine metabolism) is found towards the C-terminus (reviewed in Shore and Sharrocks, 1995). The intervening domain (or I-domain) lies immediately C-terminal to the MADS-box and like the K-domain and C-terminal region of the protein, is poorly conserved between the MADS-box proteins. The predicted roles for each of these domains will be discussed in the following sections.

MADS-box transcription factors bind DNA as either homo- or heterodimers. Generally speaking, the MADS domain is responsible for DNA-binding while the remainder of the protein is involved in protein–protein interactions. Such interactions may occur between two MADS-box proteins or between MADS-box proteins and other, unrelated, proteins.

3.1 *The MADS and I-domains are required for DNA-binding*

In vitro studies with a number of MADS-box proteins, primarily from yeast and vertebrates, have demonstrated that the MADS domain and an extension of approximately 30 amino acids at its C-terminus, constitute the minimal DNA-binding domains. Similarly, in the case of plant MADS-box proteins, *in vitro* DNA-binding studies utilizing various truncated MADS-box proteins have suggested that in a number of cases, the MADS-box and the I-domain form the minimal DNA-binding domain (Huang *et al.*, 1996; Mizukami *et al.*, 1996; Riechmann *et al.*, 1996b; West and Sharrocks, personal communication; see *Table 2*).

The N-terminal half of the MADS-box contains a high proportion of basic residues while the C-terminal half is largely hydrophobic. Studies with SRF have shown that the basic N-terminus of the MADS-box forms the site-specific DNA-binding domain. More precisely, residues within the N-terminal half of the MADS domain, predicted to form an α-helix, bind DNA, and three basic residues within the helix are critical for DNA-binding. Arginine164 contacts the DNA via the phosphate backbone and also appears to be involved in dimerization. Lysine163 makes contact with the two guanine residues in the CArG box (see below). Finally, the third residue (arginine157), although it does not contact DNA directly, plays a role in ensuring that various other residues within the MADS domain are in the correct orientation (reviewed in Davies and Schwarz-Sommer, 1994; Shore and Sharrocks, 1995; West *et al.*, 1997). These three residues are conserved in almost all of the plant MADS-box proteins suggesting that the mechanism of DNA-binding is conserved amongst all the MADS-box factors of the different kingdoms. Indeed, chimaeric AP1, AP3, PI and AG proteins in which the N-terminal half of the MADS domain was swapped with the same regions from SRF and MEF2A, displayed SRF and MEF2A DNA-binding specificities *in vitro*, suggesting that the determinants of DNA-binding specificities amongst the plant MADS-box factors are also within the N-terminal half of the MADS domain. However, in the case of some plant MADS-box factors, residues within the C-terminal half of the MADS domain may also be involved in determining DNA-binding specificity (Riechmann and Meyerowitz, 1997b).

MADS-box proteins bind DNA sequences closely related to the SRF consensus binding site ($CC(A/T)_6GG$; the CArG-box) or the MEF2 consensus binding site

(CTA(A/T)$_4$TAG) (see Shore and Sharrocks, 1995; West *et al.*, 1997 and references therein). For simplicity, both binding sites will be referred to as the CArG-box. Plant MADS-box proteins also bind CArG-like sequences *in vitro* either as homodimers (e.g. AG and AP1) or heterodimers (e.g. AP3/PI and DEF/GLO; see *Table 1*). CArG-box-like sequences have been found associated with several putative MADS-box factor target genes. Both *DEF* and *GLO* contain CArG box-like sequences within their promoters to which it is proposed the DEF/GLO heterodimer binds to regulate transcription of each gene (reviewed by Davies and Schwarz-Sommer, 1994). Similarly, three CArG-box sequences were found within the promoter of *AP3* and were shown to play different roles in the regulation of *AP3* transcription, possibly by binding the AP3/PI heterodimer (Hill *et al.*, 1998; Tilly *et al.*, 1998). Other potential target genes have been isolated by differential screening (Nacken, 1991). Ito *et al.* (1997) isolated DNA encoding a serine/threonine kinase from genomic DNA fragments bound to AG. AG was shown to bind a divergent CArG box-like sequence (CC(A/T)$_6$CC) in the 3′ untranslated sequence of the kinase gene, and since the expression pattern of the kinase gene was antagonistic to that of AG, it was suggested that AG negatively regulated expression of the kinase gene. Savidge *et al.* (1995) suggested that *AGL5* may also be a target of AG and showed that AG was capable of binding a CArG-box (CC(A)$_6$GG) present in the promoter region of *AGL5*. Although the genes mentioned above are suggested to be targets of MADS-box factors, the studies suffer from the inability to distinguish direct from indirect target genes. Proof of *in vitro* DNA binding does not constitute proof of direct binding and regulation *in vivo*. The *Arabidopsis* gene *NAP*, which shows some sequence homology with the *Petunia* gene *NO APICAL MERISTEM*, is one of the best candidates for an immediate target of AP3/PI (Sablowski and Meyerowitz, 1998). *NAP* was cloned, along with two other genes, using differential display in an inducible system in the absence of protein synthesis.

Interestingly, the association of MADS-box factors with DNA induces a contortion of the DNA which results in DNA bending towards the minor groove. Several transcription factors induce DNA bending upon binding. Sharrocks and Shore (1995) reported that core^{SRF} (the minimal DNA-binding domain of SRF) binds the SRE as a homodimer and induces DNA bending. In addition, the crystal structure of SRF bound to DNA showed the DNA to be bent around the protein (Pellegrini *et al.*, 1995). It was proposed that the SRF-induced bending of DNA may be partly responsible for its DNA-binding specificity. Later studies, which showed that many MADS-box proteins show differential induction of DNA bending (West *et al.*, 1997; West and Sharrocks, personal communication), support this hypothesis since many of these proteins also display differential DNA-binding specificities. The role of transcription factor-induced DNA bending is currently unclear. However, there is evidence to suggest that it may play a role in ensuring that the promoter region of the target gene is in the correct conformation to bring the appropriate transcription machinery together to regulate gene expression. In the case of the yeast MADS-box factor MCM1, DNA bending appears to be sequence-dependent since alterations to the sequence affects DNA bending and transcriptional activation by MCM1. In contrast, DNA-binding affinity is largely unaffected by these alterations suggesting that the degree of DNA bending is an important determinant in the regulation of gene transcription by MCM1 (Acton *et al.*, 1997). Plant MADS-box factors also induce DNA bending (Riechmann *et al.*, 1996b; West and Sharrocks, personal communication). Furthermore, many plant

Table 2. *Identification of the functions of the plant MADS-box protein domains*[a]

MADS	I	K	C	Phenotype	DNA-binding specificity	Conclusions
AG	AP3	AP3	AP3	Ectopic *AP3* (carpels to stamens), complements *ap3-3* but not *ag-3*	Binds DNA with PI but not AP1, AP3, AG or as a homodimer	The various phenotypes suggest that the functional specificity of AP3 is independent of the MADS-box, but requires the I- and K-domains
AG	AP3	AP3	AG	Ectopic *AP3*, complements *ap3-3*	NR	
AG	AG	AP3	AP3	Wild-type	Binds DNA as homodimer but not with PI	The *in vitro* DNA-binding partner specificities are in broad agreement with this notion
AP3	AP3	AG	AG	Curled leaves, inflorescence terminates early	No binding detected with AP1, AP3, PI, AG or as homodimer	(Krizek and Meyerowitz, 1996; Riechmann *et al.*, 1996a)
	AP3	AP3	AP3	Wild-type	NR	
AG	AP3	AP3		Wild-type	Binds DNA with PI	
AG	PI	PI	PI	Ectopic *PI* (sepals to petals)	NR	The functional specificity of PI is mediated by the I- and K-domains
AG	PI	PI	AG	Ectopic *PI* complements *pi-1*	Binds DNA with AP3 and as a homodimer	(Krizek and Meyerowitz, 1996; Riechmann *et al.*, 1996a)
AG	AG	AG	AP3	Ectopic *AG* (sepals to carpels, petals to stamens, inflorescence terminates early)	NR	The key determinants of the functional specificities of AG and AP1 lie within the MADS and I-domains
AG	AG	AP1	AP1	Ectopic *AG*, partially complements *ag-3* but not *ap1-1*	NR	However, the MADS and I-domains alone are not sufficient to specify the function of AG and AP1, the K- and C-domains are also required
AG	AP1	AP1	AP1	Lateral inflorescence to terminal flower	NR	

AP1 AP1 AG AP1	Ectoptic *AP1* (apical and lateral inflorescences terminate in floral structure), complements *ap1-1* but not *ag-3*	NR	(Krizek and Meyerowitz, 1996)
AP1 AG AG AP1	Curled leaves, inflorescence terminates early	NR	
AG AG	Wild-type	NR	
AP1 AP1	Wild-type	NR	
AP1 AP1 AP1 AP1		Binds DNA as a homodimer	The MADS-box and the majority of the I-domain are required for DNA-binding *in vitro*
AP1 AP1(3)	NR	No binding detected	
AP1 AP1(15)	NR	Binds DNA weakly	(Riechmann *et al*., 1996b)
AP1 AP1(29)	NR	Binds DNA strongly	
AG AG AG AG[b]		Binds DNA as a homodimer	MADS-box and I-domain necessary and sufficient for DNA-binding *in vitro*
AG	NR	No binding detected	(Mizukami *et al*., 1996; Riechmann *et al*., 1996b)
AG AG(22)	NR	Binds DNA	
AG(43) AG AG AG	NR	No binding detected	
AG AG AG	NR	No binding detected	

Table 2. *Continued*

MADS	I	K	C	Phenotype	DNA-binding specificity	Conclusions
AGL2	AGL2	AGL2	AGL2		Binds DNA as a homodimer	*In vitro* DNA-binding is mediated by the MADS-box and N-terminal portion of the I-domain
AGL2	AGL2(10)			NR	Binds DNA as a homodimer	
						(Huang *et al.*, 1996)
AP3	AP3	AP3	AP3		Binds DNA with PI	*In vitro* DNA-binding by B-function transcription factors requires the MADS-box, I-domain and part of the K-domain
AP3	AP3	AP3(12)		NR	No binding detected with PI	
AP3	AP3	AP3(31)		NR	Binds DNA with PI	
DEF	DEF	DEF	DEF		Binds DNA with GLO	The K-domain of DEF is predicted to have two amphipathic helices. Only the most amino-terminal of these is required for DNA-binding with GLO
DEF	DEF	DEF(10)		NR	No binding detected with GLO	
DEF	DEF	DEF(25)		NR	Binds DNA with GLO	The K-domain of GLO has three amphipathic helices. Removal of the second and third helices prevents DNA-binding with DEF
PI	PI	PI	PI		Binds DNA with AP3	
PI	PI			NR	No binding detected with AP3	
PI	PI	PI(16)		NR	Binds DNA with truncated AP3[c]	(Riechmann *et al.*, 1996b; Zachgo *et al.*, 1995)
GLO	GLO	GLO	GLO		Binds DNA with DEF	
GLO	GLO	GLO(26)		NR	No binding detected with DEF	
GLO	GLO	GLO(58)		NR	Binds DNA with DEF	

SRF/MEF AG AG AG AG	Ectopic *AG*, partially complements *ag-3*, activates *GUS* from *AGL5* promoter[d]	SRF or MEF2A specificity	The amino-terminal half of the plant MADS-box determines DNA-binding specificity
SRF/MEF AP3 AP3 AP3 AP3	Ectopic *AP3*, partially complements *ap3-3*	SRF or MEF2A specificity	Ectopic expression of the hybrid proteins gives the same phenotype as that of plants ectopically expressing wild-type *AG*, *AP3*, *PI* or *AP1*. Similarly, each hybrid gene was able to rescue the appropriate mutant phenotype. This suggests that the control of floral organ identity by AP1, AG, AP3 and PI is independent of, or loosely related to, their DNA-binding specificity
SRF/MEF PI PI PI PI	Ectopic *PI*, partially complements *pi-1*	SRF or MEF2A specificity	
SRF/MEF AP1 AP1 AP1 AP1	Ectopic *AP1*, partially complements *ap1-1*	SRF or MEF2A specificity	(Riechmann and Meyerowitz, 1997b)

[a]The table describes the phenotype of transgenic *Arabidopsis* expressing chimaeric or truncated MADS-box transcription factors from the 35S promoter. The expression of the altered proteins was analysed in either a wild-type background or in the appropriate mutant background detailed in the table. Where an altered gene complements a mutant phenotype, the degree of rescue is similar to that obtained with the appropriate wild-type gene. *In vitro* DNA-binding experiments tested the ability of the altered proteins to bind to CArG-box DNA in band shift assays. The numbers in parentheses listed next to a truncated protein indicate the number of amino acid residues remaining in the truncated domain. While this table is not exhaustive, it does summarize the major findings. For a more detailed review of the experiments summarized in the table, see the appropriate reference or refer to the text.

[b]The wild-type AG protein has an N-terminal region preceding the MADS-box (see *Figure 1*). To aid clarity, experiments involving AG plus N-terminal region are omitted from this table. However, it should be noted that no DNA-binding was detected for AG with 50 amino acid residues N-terminal to the MADS domain. The same is also true for AG truncations bearing this N-terminal domain. AG was only able to bind DNA if this N-terminal region was completely or partially (i.e. 17 residues remaining N-terminal to the MADS-box) removed.

[c]DNA-binding was only detected between the truncated PI protein and a truncated AP3 protein (consisting of the MADS- and I-domains plus 31 amino acid residues C-terminal to I) but not with full-length AP3.

[d]Wild-type AG activates transcription of *GUS* driven from the *AGL5* promoter. A similar pattern of GUS expression occurs in *AGL5::GUS* plants crossed with 35S-*AG*, 35S-*SRF-AG* or 35S-*MEF-AG* plants.

NR = not reported.

MADS-box proteins do not possess an obvious transcriptional activation domain. The function of plant MADS-box factors may simply be one of altering promoter architecture to allow either activation or repression of target gene transcription.

The plant MADS-box transcription factors regulate different developmental processes and differential DNA-binding specificities exhibited by these proteins was thought to reflect these different biological roles. However, the involvement of DNA-binding specificity in the biological function of the plant MADS-box factors has recently been called into question. Studies with chimaeric MADS-box genes demonstrated that the MADS-boxes of AP1, AP3, PI and AG were functionally interchangeable. For example, ectopic expression of an *AP3* gene in which the MADS-box encoding sequence was swapped with that of *AG*, gave a phenotype similar to that obtained in transgenic *Arabidopsis* ectopically expressing the wild-type *AP3* gene. Furthermore, the *AG/AP3* chimaera was also able to complement the *ap3* mutation. Taken together, the findings suggest that the biological specificity of *AP3* and *PI* is not determined by the MADS-box and that function is not dependent upon the DNA-binding specificities of the MADS-box proteins (Krizek and Meyerowitz, 1996; Riechmann and Meyerowitz, 1997b; summarized in *Table 2*). More recent studies in which the N-terminal half of the MADS domain from AP1, AP3, PI and AG was replaced with those from SRF and MEF2A also suggest that the biological function of the MADS-box proteins is independent of DNA-binding specificities. While the chimaeric proteins demonstrated SRF and MEF2A DNA-binding specificities *in vitro*, ectopic expression and complementation experiments indicated that the chimaeric MADS-box proteins had a similar biological activity to the wild-type AP1, AP3, PI and AG proteins (Riechmann and Meyerowitz, 1997b; summarized in *Table 2*). However, since residues within the C-terminal half of the plant MADS domain also play a role in DNA-binding specificity, it is conceivable that although the chimaeric proteins show SRF and MEF2A binding specificities *in vitro*, *in planta* the residues mediating DNA-binding specificity located in the C-terminal half of the MADS domain, coupled with potential ternary complex factors (discussed in Section 3.4) which may also play a critical role in DNA-binding, ensure that the chimaeric proteins have a biological function equivalent to that of the wild-type proteins. In addition, one of the two MADS domains of a dimer is sufficient to mediate DNA-binding specificity. In the transgenic experiments it is possible that the chimaeric proteins formed heterodimers with wild-type MADS-box factors. In this situation, the wild-type partner may be the key determinant for DNA-binding specificity (Riechmann and Meyerowitz, 1997b). Finally, in the domain swap experiments a constitutive viral promoter was used to drive the transgene and the resulting changes in temporal, spatial and quantitative expression might be sufficient to influence a developmental decision based on relative affinities and complex interactions.

3.2 *The K-domain is required for dimerization*

Most plant MADS-box factors possess a domain which is not found in animal or yeast MADS-box proteins. Termed the K-domain, due to its predicted structural similarity to the coiled-coil domain of keratin (Ma *et al.*, 1991), it is situated immediately C-terminal to the I-domain (see *Figure 1*). While the K-domain is poorly conserved at the sequence level, in all cases it is predicted to form α-helices in which hydrophobic residues are located on one face of the helix. Amphipathic helices of this type are predicted to mediate protein–protein interactions – one helix interacting with a similar helix of a separate polypeptide. Temperature-sensitive mutants of *DEF* and *AP3*, where substitutions have occurred within the K-domain, are consistent with a disruption to protein–protein interactions (reviewed by Davies and Schwarz-Sommer, 1994).

Dimerization between MADS-box proteins is central to their role as transcription factors. In floral tissues a number of different MADS-box factors may be expressed at any given time. The formation of numerous, but specific, combinations of homo- and heterodimers amongst these factors, their subsequent interaction with ternary factors and the binding of these complexes to the regulatory sequences of their target genes, may provide a means for regulation of gene transcription in these tissues.

In vitro experiments have shown that, in many cases, only the MADS and I-domains are required for DNA-binding (see *Table 2* and Section 3.1). Since MADS-box proteins bind DNA as either homo- or heterodimers, it was suggested that the K-domain may only be required to mediate interactions with other, non-MADS-box, proteins (Huang *et al*., 1996; Shore and Sharrocks, 1995). However, more recent studies using yeast two-hybrid assays, which provides a sensitive system for studying the interactions between eukaryotic proteins *in vivo*, suggested that the K-domain is required for dimerization. Davies *et al*. (1996b) demonstrated that certain truncated versions of DEF were able to interact with the full-length GLO protein. Amongst these was a DEF polypeptide consisting of only the I- and K-domains. Furthermore, no interactions were detected between GLO and DEF polypeptides lacking some, or all, of the K-domain (but maintaining the MADS-box and I-domains). Later studies, using two-hybrid technology and immunoprecipitation techniques, demonstrated that the K-domain of AG alone interacted with AGL2, AGL4, AGL6 and AGL9 (Fan *et al*., 1997). Recent results in our laboratory also show that the interaction between MADS-box proteins does not require the MADS domain. A yeast two-hybrid screen was conducted to identify the interacting partners of DEFH72, a protein encoded by an intermediate gene (see Section 2.3), which was previously isolated as an interacting partner of PLE (Davies *et al*., 1996b). From the library screen six positives were isolated. Each one was a truncated form of the protein encoded by the meristem identity gene *SQUA*, lacking the entire MADS-box, approximately 90% of the I-domain and approximately 63% from the end of the C-terminal region of the protein (Causier and Davies, unpublished data).

The importance of the K-domain in the interactions between MADS-box factors is further demonstrated on comparison of the sequences of the K-domain from the various proteins currently identified. Such analyses reveal a significantly higher conservation within the K-domains of factors which interact with the same partners than in those which interact with different partners (Davies *et al*., 1996b).

A number of studies have attempted to define the influence that the various domains of the floral organ identity MADS-box factors (Section 2.2) have on their biological function, and these are summarized in *Table 2*. Chimaeric *AP3* genes in which the I- and K-domains were replaced with the equivalent regions from *AG*, were ectopically expressed in *Arabidopsis*. The phenotype consistent with ectopic wild-type *AP3* expression (i.e. partial homeotic conversion of fourth whorl carpels to stamens) was not observed, suggesting that the I- and K-domains are required for AP3 functional specificity, presumably through interaction with *PI* (Krizek and Meyerowitz, 1996). Immunoprecipitation experiments with various combinations of the AP1,

AP3, PI and AG proteins suggested that all of these proteins were capable of interacting with one another. However, *in vitro* experiments demonstrated that only certain dimer pairs were able to bind DNA (the AG homodimer, the AP1 homodimer and the AP3/PI heterodimer). In addition, analysis of the DNA-binding of proteins encoded by chimaeric MADS-box genes suggested the I-domain played a key role in determining the specificity of the interaction between the MADS-box proteins (Riechmann *et al.*, 1996a). In *Antirrhinum*, two-hybrid experiments with DEF, GLO and PLE indicate that the B-function proteins DEF and GLO interact exclusively with one another. In contrast, the C-function protein PLE interacts with numerous MADS-box factors, but not the B-function proteins (Davies *et al.*, 1996b). This suggests that, at least in *Antirrhinum*, the combinatorial model for organ identity does not rely on simple direct interactions between the B- and C-functions.

3.3 *Floral development and the interactions between floral organ identity MADS-box factors*

In the previous sections we reviewed the interactions detected between the *Antirrhinum* and *Arabidopsis* floral organ identity MADS-box factors, in yeast two-hybrid assays and *in vitro* DNA-binding studies. Here we will discuss the implications of these findings for the *Antirrhinum* proteins. A summary of the expression patterns of, and the protein–protein interactions between a number of the *Antirrhinum* MADS-box factors, is presented in *Figure 2*. The ABC model of floral development predicts the concomitant expression of *DEF*, *GLO* and *PLE* to specify stamen identity and the expression of *DEF*, *GLO* and A-function genes to specify petals. However, since interactions between DEF or GLO and PLE, and DEF or GLO and SQUA (the *Antirrhinum* MADS-box factor most similar to the *Arabidopsis* A-function factor AP1) were not detected, other mechanisms such as interaction with accessory factors, competition for DNA binding sites, more complex direct interactions or

---

Figure 2. *The temporal and spatial expression patterns, and dimerization partner specificity of selected* Antirrhinum *MADS-box factors. (a) The figure summarizes the temporal and spatial expression of* SQUA, DEFH72, DEF, GLO, PLE *and* DEFH49. SQUA *is expressed early in the development of the* Antirrhinum *flower and is detectable in emerging flower primordia. It is expressed in the cells which give rise to the floral organs but becomes excluded from stamen primordia.* SQUA *expression in the fourth whorl is restricted to specific cells and overall expression in the carpels is weak (Huijser* et al., *1992). The expression of the floral organ identity genes* DEF, GLO *and* PLE *is detectable as the appropriate primordia emerge. Generally,* DEF *and* GLO *expression is restricted to the second and third whorls while* PLE *is restricted to the third and fourth whorls. Once expression of the meristem identity genes has been activated and before the floral organ identity genes are expressed, a group of intermediate genes become active. In the case of* Antirrhinum, *such intermediate genes include* DEFH72 *and* DEFH200 *which share a similar expression pattern. They are expressed in all four whorls, although weakly in the sepals. Finally, the expression of late-acting genes occurs after that of the organ identity genes. In* Antirrhinum, DEFH49 *represents a gene of this temporal class and its expression is restricted mainly to the fourth whorl, although weak expression is detectable in whorls 2 and 3 (Davies* et al., *1996b). (b) The figure represents the protein–protein interactions detected amongst the different temporal classes of floral MADS-box factors. Yeast two-hybrid studies have demonstrated that PLE is capable of forming heterodimers with MADS-box factors from a number of temporal classes (i.e. SQUA, DEFH72 and DEFH49) and that SQUA interacts with DEFH72 (Davies* et al., *1996b; Causier and Davies, unpublished data). In contrast, the B-function proteins DEF and GLO interact exclusively with one another (Davies* et al., *1996b).*

interactions between downstream target genes, may explain how these proteins act in combination to specify the floral organ identities (Davies *et al.*, 1996b). Further investigation of the protein–protein interactions, expression patterns and mutant phenotypes of the MADS-box factors will begin to suggest how these proteins act to regulate floral development.

The C-function genes from *Arabidopsis* (*AG*) and *Antirrhinum* (*PLE*) play a role in both floral meristem identity and floral organ identity. The phenotypes of *ag* and *ple* mutant flowers are very similar. In both cases the mutants exhibit conversion of third whorl stamens to petals (representing altered organ identity) and a loss of determinacy within the third or fourth whorls. Furthermore, ectopic expression of *AG* or *PLE* also points to the roles of these C-function genes since, in both cases, the flowers show partial homeotic conversion of first whorl sepals to carpels and second whorl

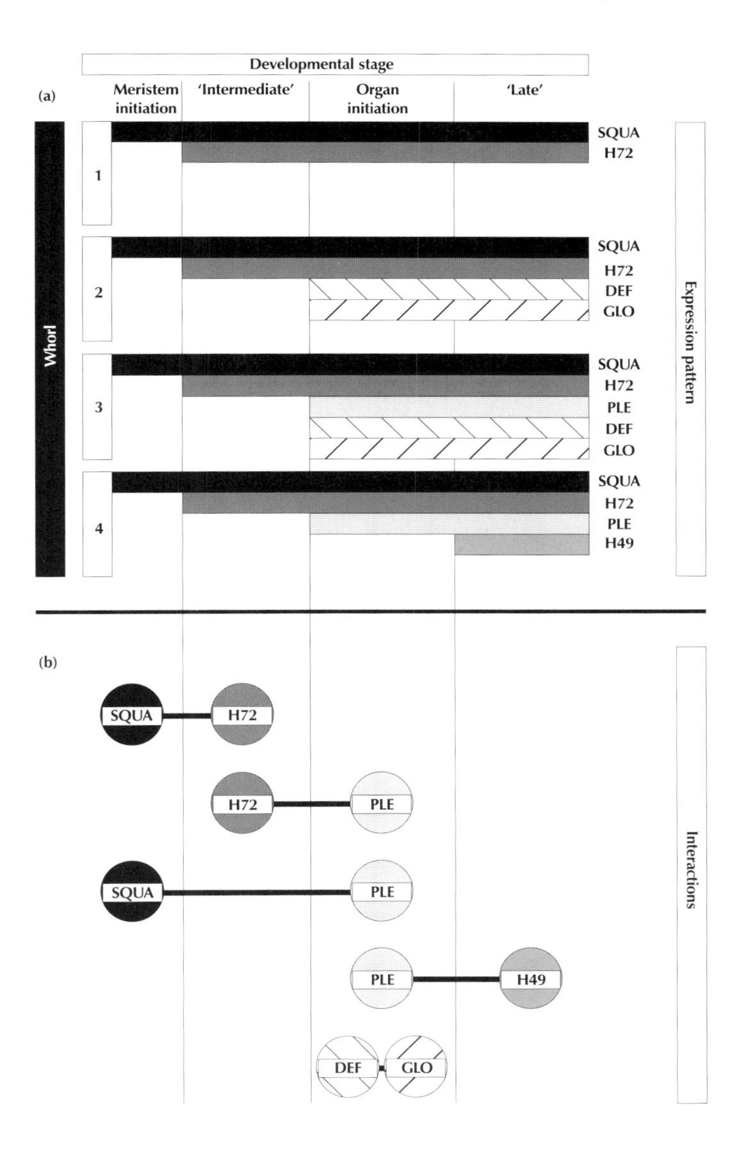

petals to stamens (Davies *et al.*, 1996b; Mizukami and Ma, 1992). In addition, terminal floral structures were formed in inflorescence apices of 35S-*AG* plants, suggesting a role for *AG* in floral determinacy and consequently in floral meristem identity (Mizukami and Ma, 1992; Mizukami *et al.*, 1996). The organ identity and meristem identity functions of *AG* are separable. For example, in strong *ag* mutants, defects in both floral organ identity (stamens converted to petals) and determinacy (abnormal flowers developing within abnormal flowers) are seen. In weaker *ag* phenotypes, the flowers are indeterminate but floral organ identity is affected to a much lesser extent (Sieburth *et al.*, 1995). It is possible that the dual role of *AG* and *PLE* may be mediated through their interactions with the meristem identity MADS-factors on one hand, and the intermediate MADS-factors, on the other (Causier and Davies, unpublished; Davies *et al.*, 1996b; Fan *et al.*, 1997).

Analysis of transgenic tomato and *Petunia* plants in which specific intermediate genes (*TM5* and *FBP2*, respectively) were down-regulated, indicate that these genes are required for reproductive organ identity (Angenent *et al.*, 1994; Pnueli *et al.*, 1994a). That the correct operation of the C-function is dependent upon the intermediate genes was demonstrated by crossing antisense *TM5* plants with 35S-*TAG* (*TAG* is the tomato C-function gene) plants. In 35S-*TAG* plants, the ectopic expression of *TAG* in the second whorl caused a partial homeotic conversion of petals to stamens. However, in the antisense *TM5* × 35S-*TAG* plants, petals were again found in the second whorl (Lifschitz, 1996). Thus, the intermediate MADS genes are required for the C-function to confer reproductive organ identity. The intermediate genes were also predicted to play a role in floral meristem identity since a reduction in *TM5* or *FBP2* expression caused a loss of determinacy in the inner whorls of mutant flowers (Angenent *et al.*, 1994; Pnueli *et al.*, 1994a). The interaction between SQUA and DEFH72 (Causier and Davies, unpublished data) and AP1 and AGL9 (reported by Fan *et al.*, 1997) is consistent with this predicted role.

3.4 *MADS-box factors and ternary complex factors (TCFs)*

Animal and yeast MADS-box transcription factors interact with unrelated DNA-binding proteins which are required for the appropriate regulation of target gene transcription (reviewed by Shore and Sharrocks, 1995). MCM1 and SRF both interact with accessory factors which bind to DNA sequences adjacent to the MADS-box factor binding sites within the promoters of their respective target genes. In addition, the interactions between accessory factor and MADS-box protein appear to be mediated, at least in part, by $core^{MCM1}$ and $core^{SRF}$ (reviewed by Shore and Sharrocks, 1995). The formation of ternary complexes between a MADS-box dimer, accessory factor and the appropriate promoter element may also be necessary for the regulation of plant genes. To date, no ternary complex factor has been identified for the plant MADS-box factors, although PLE was shown to interact with a non-MADS-box protein in two-hybrid assays (Davies *et al.*, 1996b). However, since the plant MADS-box transcription factors do not possess an obvious transcriptional activation domain, it is tempting to speculate that plant MADS-box factors would form a ternary complex at the target gene promoter with accessory factors capable of activating gene transcription. Another mechanism by which plant MADS-factor/accessory factor complexes may act to regulate gene transcription is by influencing the architecture of the target gene promoter. In the case of SRF, interaction with the ternary complex factor Elk-1 at the SRE, alters the degree of DNA bending (Sharrocks and Shore, 1995). Finally, TCFs may function to mediate the DNA-binding specificity of a particular MADS-box dimer. A number of recent reports have indicated that the biological and DNA-binding specificities of MADS-box dimers are independent of one another and, in addition, only one MADS domain of the dimer is sufficient for DNA-binding specificity (Krizek and Meyerowitz, 1996; Riechmann and Meyerowitz, 1997b; see Section 3.1). This suggests that the numerous possible combinations of different MADS-box factors which can form DNA-binding dimers, may not offer new DNA-binding specificities. However, the different combinations of MADS-box factors may provide for new TCF specificities which may influence the DNA-binding specificity, and consequently the biological specificity, of the MADS-box dimer.

The interaction between a plant MADS-box protein and a potential accessory factor may be achieved in one of two ways. Firstly, the interaction may occur as it does between animal MADS-box proteins and TCFs, via the dimerization region (the MADS domain plus C-terminal extension). However, in the case of plant MADS-box proteins, dimerization may involve two separate mechanisms. The first of these may be a DNA-binding dependent mechanism (*in vitro* studies indicate that only the MADS-box and I-domains are required for DNA-binding and dimerization of some plant MADS-factors), and the second may be mediated through the K-domain (as suggested by *in vivo* studies). It is possible then that the K-domain may mediate direct interactions between two MADS-box proteins, or between separate dimers (forming multimeric MADS-box proteins) (Fan *et al.*, 1997), or between a MADS-box dimer and an accessory factor. Yeast two-hybrid experiments may allow the isolation of genes encoding putative ternary complex factors which interact with plant MADS-box transcription factors. However, since MADS-box proteins bind DNA as dimers, and in many cases, as heterodimers, the two-hybrid approach may be limited. A modification of the two-hybrid system, termed the three-hybrid system (Zhang and Lautar, 1996), in which a library can be screened for proteins that interact with a protein dimer, may prove a more appropriate approach.

4. Future directions

Much progress has been made since the first plant MADS-box gene was cloned in 1990. Within the flower, plants make use of the MADS-box family of transcription factors to determine organ identity in much the same way as animals use homeobox factors. However, as the plant MADS-box gene family has begun to expand and more functions are identified it has become clear that the MADS-box family is neither confined to establishment of organ identity nor to the flower. A remaining challenge is to identify all the members of this class of plant regulators and establish their individual and combined roles in development. It will also become important to analyse these genes in different plant species in order to discover the common conserved mechanisms and species-specific differences which might relate to the variety of plant form which exists. From an evolutionary point of view it would be very interesting to discover the function of MADS-box genes in primitive, non-flowering, plants (Hasebe *et al.*, 1998; Münster *et al.*, 1997).

Several questions remain outstanding at the molecular level. Further work needs to be done to establish the possible network of protein–protein interactions both between MADS-box factors and between MADS-box factors and ternary complex factors. It is still not clear how MADS-factors exert their influence over developmental processes and the isolation and study of target genes and the interaction of MADS-box complexes with the promoters of target genes will help in this aim. It is clear that certain MADS-box factors influence, directly or indirectly, the progression of the cell cycle and links must be found between these two elements. Furthermore, although mutants are known in several species which affect the expression of MADS-box genes, we are still ignorant of the mechanisms, both transcriptional and post-transcriptional, which determine the intricate spatial and temporal expression patterns of these genes. Understanding the biochemistry of MADS-box factor interaction, function and regulation will help to explain the different flower and plant types that occur naturally and will also allow defined changes to be made to manipulate plant development artificially.

Acknowledgements

We regret that space constraints and the necessity to avoid too much detail have resulted in us being unable to cite all the literature in full. The authors would like to thank Phil Gilmartin, Karen Hansen, Martin Kieffer, Andy Sharrocks and Zsuzsanna Schwarz-Sommer for helpful suggestions and critical comments during the preparation of this chapter. We would also like to thank Andrew Sharrocks, Adam West and Gerco Angenent for allowing us to report results prior to publication. BC is supported by a grant from the Leverhulme Trust and IW is supported by a BBSRC studentship.

References

Acton, T.B., Zhong, H. and Vershon, A.K. (1997) DNA-binding specificity of Mcm1: operator mutations that alter DNA-bending and transcriptional activities by a MADS-box protein. *Mol. Cell. Biol.* **17**: 1881–1889.

Angenent, G.C., Franken, J., Busscher, M., Colombo, L. and van Tunen, A.J. (1993) Petal and stamen formation in petunia is regulated by the homeotic gene *fbp1*. *Plant J.* **4**: 101–112.

Angenent, G.C., Franken, J., Busscher, M., Weiss, D. and van Tunen, A.J. (1994) Co-suppression of the petunia homeotic gene *fbp2* affects the identity of the generative meristem. *Plant J.* **5**: 33–44.

Angenent, G.C., Franken, J., Busscher, M., van Dijken, A., van Went, J.L., Dons, H. and van Tunen, A.J. (1995) A novel class of MADS box genes is involved in ovule development in petunia. *Plant Cell* **7**: 1569–1582.

Bowman, J.L., Alvarez, J., Weigel, D., Meyerowitz, E. and Smyth, D.R. (1993) Control of flower development in *Arabidopsis thaliana* by *APETALA1* and interacting genes. *Development* **119**: 721–743.

Bradley, D., Carpenter, R., Sommer, H., Hartley, N. and Coen, E. (1993) Complementary floral homeotic phenotypes result from opposite orientations of a transposon at the *plena* locus of *Antirrhinum*. *Cell* **72**: 85–95.

Coen, E.S. and Meyerowitz, E.M. (1991) The war of the whorls: genetic interactions controlling flower development. *Nature* **353**: 31–37.

Colombo, L., Franken, J., Koetje, E., Vanwent, J., Dons, H., Angenent, G.C. and van Tunen, A.J. (1995) The petunia MADS box gene *FBP11* determines ovule identity. *Plant Cell* **7**: 1859–1868.

Colombo, L., Franken, J., van der Krol, A.R., Wittich, P.E., Dons, H. and Angenent, G.C. (1997) Downregulation of ovule-specific MADS box genes from petunia results in maternally controlled defects in seed development. *Plant Cell* **9**: 703–715.

Davies, B. and Schwarz-Sommer, Zs. (1994) Control of floral organ identity by homeotic MADS-box transcription factors. *Results and Problems in Cell Differentiation* **20**: 235–258

Davies, B., DiRosa, A., Eneva, T., Saedler, H. and Sommer, H. (1996a) Alteration of tobacco floral organ identity by expression of combinations of Antirrhinum MADS-box genes. *Plant J.* **10**: 663–677.

Davies, B., Egea-Cortines, M., de Andrade Silva, E., Saedler, H. and Sommer, H. (1996b) Multiple interactions amongst floral homeotic proteins. *EMBO J.* **15**: 4330–4343.

Fan, H.Y., Hu, Y., Tudor, M. and Ma, H. (1997) Specific interactions between the K domains of AG and AGLs, members of the MADS domain family of DNA binding proteins. *Plant J.* **12**:999–1010.

Goto, K. and Meyerowitz, E.M. (1994) Function and regulation of the *Arabidopsis* floral homeotic gene *PISTILLATA*. *Genes Develop.* **8**: 1548–1560.

Gu, Q., Ferrándiz, C., Yanofsky, M.F. and Martienssen, R. (1998) The *FRUITFULL* MADS-box gene mediates cell differentiation during *Arabidopsis* fruit development. *Development* **125**: 1509–1517.

Hasebe, M., Wen, C.K., Kato, M. and Banks, J.A. (1998) Characterization of MADS homeotic genes in the fern *Ceratopteris richardii*. *Proc. Natl Acad. Sci. USA* **95**: 6222–6227.

Haughn, G.W, Schultz, E.A. and Martinez-Zapater, J.M. (1995) The regulation of flowering in *Arabidopsis thaliana* – meristems, morphogenesis, and mutants. *Can. J. Bot.* **73**: 959–981.

Heard, J., Caspi, M. and Dunn, K. (1997) Evolutionary diversity of symbiotically induced nodule MADS box genes: characterization of *nmhC5*, a member of a novel subfamily. *Mol. Plant Microbe Interact.* **10**: 665–676.

Hill, T.A., Day, C.D., Zondlo, S.C., Thackeray, A.G. and Irish, V.F. (1998). Discrete spatial and temporal *cis*-acting elements regulate transcription of the *Arabidopsis* floral homeotic gene *APETALA3*. *Development* **125**: 1711–1721.

Huang, H., Tudor, M., Su, T., Zhang, Y., Hu, Y. and Ma, H. (1996) DNA binding properties of two *Arabidopsis* MADS domain proteins: binding consensus and dimer formation. *Plant Cell* **8**: 81–94.

Huijser, P., Klein, J., Lönnig, W.-E., Meijer, H., Saedler, H. and Sommer, H. (1992) Bractomania, an inflorescence anomaly, is caused by the loss of function of the MADS-box gene *squamosa* in *Antirrhinum majus*. *EMBO J.* **11**: 1239–1249.

Ito, T., Takahashi, N., Shimura, Y. and Okada, K. (1997) A serine/threonine protein kinase gene isolated by an *in vivo* binding procedure using the *Arabidopsis* floral homeotic gene product, *AGAMOUS*. *Plant Cell Physiol.* **38**: 248–258.

Jack, T., Brockman, L.L. and Meyerowitz, E.M. (1992) The homeotic gene *APETALA3* of *Arabidopsis thaliana* encodes a MADS box and is expressed in petals and stamens. *Cell* **68**: 683–697.

Kater, M.M., Colombo, L., Franken, J., Busscher, M., Masiero, S., Campagne, M. and Angenent, G.C. (1998) Multiple *AGAMOUS* homologs from cucumber and petunia differ in their ability to induce reproductive organ fate. *Plant Cell* **10**: 171–182.

Kempin, S.A., Mandel, M.A. and Yanofsky, M.F. (1993) Conversion of perianth into reproductive organs by ectopic expression of the tobacco floral homeotic gene *NAG1*. *Plant Physiol.* **103**: 1041–1046.

Kempin, S.A., Savidge, B. and Yanofsky, M.F. (1995) Molecular basis of the *cauliflower* phenotype in *Arabidopsis*. *Science* **267**: 522–525.

Kempin, S.A., Liljegren, S.J., Block, L.M., Rounsley, S.D., Yanofsky, M.F. and Lam, E. (1997) Targeted disruption in *Arabidopsis*. *Nature* **389**: 802–803.

Krizek, B.A. and Meyerowitz, E.M. (1996) Mapping the protein regions responsible for the functional specificities of the *Arabidopsis* mads domain organ-identity proteins. *Proc. Natl Acad. Sci. USA* **93**: 4063–4070.

Lifschitz, E. (1996) Flowers, leaves and inflorescences: an integrated approach. *Flowering Newsletter* **21**: 28–33.

Liljegren, S.J. and Yanofsky, M.F. (1998b) Towards targeted transformation in plants response: targeting *Arabidopsis. Trends in Plant Sci.* **3**: 79–80.

Liljegren, S.J., Ferrándiz, C., Alvarez-Buylla, E.R., Pelaz, S. and Yanofsky, M.F. (1998a) *Arabidopsis* MADS-box genes involved in fruit dehiscence. *Flowering Newsletter* **25**:9–19.

Ma, H. (1994) The unfolding drama of flower development: recent results from genetic and molecular analyses. *Genes Develop.* **8**: 745–756.

Ma, H., Yanofsky, M.F. and Meyerowitz, E.M. (1991) *AGL1-AGL6*, an *Arabidopsis* gene family with similarity to floral homeotic and transcription factor genes. *Genes Develop.* **5**: 484–495.

Mandel, M.A., Gustafson-Brown, C., Savidge, B. and Yanofsky, M.F. (1992) Molecular characterization of the *Arabidopsis* floral homeotic gene *APETALA1. Nature* **360**: 273–277.

Mena, M., Ambrose, B.A., Meeley, R.B., Briggs, S.P., Yanofsky, M.F. and Schmidt, R.J. (1996) Diversification of C-function activity in maize flower development. *Science* **274**: 1537–1540.

Mizukami, Y. and Ma, H. (1992) Ectopic expression of the floral homeotic gene *AGAMOUS* in transgenic *Arabidopsis* plants alters floral organ identity. *Cell* **71**: 119–131.

Mizukami, Y., Huang, H., Tudor, M., Hu, Y. and Ma, H. (1996) Functional domains of the floral regulator AGAMOUS: characterization of the DNA-binding domain and analysis of dominant-negative mutations. *Plant Cell* **8**: 831–845.

Motte, P., Wilkinson, M. and Schwarz-Sommer, Zs. (1998) Floral meristem identity and the A function in *Antirrhinum. Flowering Newsletter* **25**: 41–43.

Münster, T., Pahnke, J., Di Rosa, A., Kim, J.T., Martin, W., Saedler, H. and Theissen, G. (1997) Floral homeotic genes were recruited from homologous MADS-box genes preexisting in the common ancestor of ferns and seed plants. *Proc. Natl Acad. Sci. USA* **94**: 2415–2420.

Nacken, W.K., Huijser, P., Beltrán, J.P., Saedler, H. and Sommer, H. (1991) Molecular characterization of two stamen-specific genes, *tap1* and *fil1*, that are expressed in the wild type, but not in the *deficiens* mutant of *Antirrhinum majus. Mol. Gen. Genet.* **229**: 129–136.

Pellegrini, L., Tan, S. and Richmond, T.J. (1995) Structure of serum response factor core bound to DNA. *Nature* **376**: 490–498.

Pnueli, L., Hareven, D., Broday, L., Hurwitz, C. and Lifschitz, E. (1994a) The *TM5* MADS-box gene mediates organ differentiation in the three inner whorls of tomato flowers: analysis of transgenic plants expressing antisense RNA. *Plant Cell* **6**: 175–186.

Pnueli, L., Hareven, D., Rounsley, S.D., Yanofsky, M.F. and Lifschitz, E. (1994b) Isolation of the tomato *AGAMOUS* gene, *TAG1*, and analysis of its homeotic role in transgenic plants. *Plant Cell* **6**: 163–173.

Riechmann, J.L. and Meyerowitz, E.M. (1997a) MADS domain proteins in plant development. *Biol. Chem.* **378**: 1079–1101.

Riechmann, J.L. and Meyerowitz, E.M. (1997b) Determination of floral organ identity by *Arabidopsis* MADS domain homeotic proteins AP1, AP3, PI, and AG is independent of their DNA-binding specificity. *Mol. Biol. Cell* **8**: 1243–1259.

Riechmann, J.L., Krizek, B.A. and Meyerowitz, E.M. (1996a) Dimerization specificity of *Arabidopsis* MADS domain homeotic proteins APETALA1, APETALA3, PISTILLATA, and AGAMOUS. *Proc. Natl Acad. Sci. USA* **93**: 4793–4798.

Riechmann, J.L., Wang, M.Q. and Meyerowitz, E.M. (1996b) DNA-binding properties of *Arabidopsis* MADS domain homeotic proteins APETALA1, APETALA3, PISTILLATA and AGAMOUS. *Nucleic Acids Res.* **24**: 3134–3141.

Rounsley, S.D., Ditta, G.S. and Yanofsky, M.F. (1995) Diverse roles for MADS box genes in *Arabidopsis* development. *Plant Cell* **7**: 1259–1269.

Sablowski, R.W.M. and Meyerowitz, E.M. (1998) A homolog of *NO APICAL MERISTEM* is an immediate target of the floral homeotic genes *APETALA3/PISTILLATA. Cell* **92**: 93–103.

Savidge, B., Rounsley, S.D. and Yanofsky, M.F. (1995) Temporal relationship between the transcription of two *Arabidopsis* MADS-box genes and the floral organ identity genes. *Plant Cell* **7**: 721–733.

Schwarz-Sommer, Zs., Huijser, P., Nacken, W., Saedler, H. and Sommer, H. (1990) Genetic control of flower development by homeotic genes in *Antirrhinum majus. Science* **250**: 931–936.

Schwarz-Sommer, Zs., Hue, I., Huijser, P., Flor, P.J., Hansen, R., Tetens, F., Lönnig, W.-E., Saedler, H. and Sommer, H. (1992) Characterization of the *Antirrhinum* floral homeotic MADS-box gene *deficiens*: evidence for DNA binding and autoregulation of its persistent expression throughout flower development. *EMBO J.* **11**: 251–263.

Sharrocks, A.D. and Shore, P. (1995) DNA bending in the ternary nucleoprotein complex at the *c-fos* promoter. *Nucleic Acids Res.* **23**: 2442–2449.

Shore, P. and Sharrocks, A.D. (1995) The MADS-box family of transcription factors. *Eur. J. Biochem.* **229**: 1–13.

Sieburth, L.E., Running, M.P. and Meyerowitz, E.M. (1995) Genetic separation of 3rd and 4th whorl functions of *AGAMOUS. Plant Cell* **7**: 1249–1258.

Sommer, H., Beltrán, J.-P., Huijser, P., Pape, H., Lönnig, W.-E., Saedler, H. and Schwarz-Sommer, Z. (1990) *Deficiens*, a homeotic gene involved in the control of flower morphogenesis in *Antirrhinum majus*: the protein shows homology to transcription factors. *EMBO J.* **9**: 605–613.

Tilly, J.J., Allen, D.W. and Jack, T. (1998) The CArG boxes in the promoter of the *Arabidopsis* floral organ identity gene *APETALA3* mediate diverse regulatory effects. *Development* **125**: 1647–1657.

Tröbner, W., Ramirez, L., Motte, P., Hue, I., Huijser, P., Lönnig, W.-E., Saedler, H., Sommer, H. and Schwarz-Sommer, Zs. (1992) *GLOBOSA*: a homeotic gene which interacts with *DEFICIENS* in the control of *Antirrhinum* floral organogenesis. *EMBO J.* **11**: 4693–4704.

Tsuchimoto, S., van der Krol, A.R. and Chua, N.H. (1993) Ectopic expression of *pmads3* in transgenic petunia phenocopies the petunia *blind* mutant. *Plant Cell* **5**: 843–853.

van der Krol, A.R., Brunelle, A., Tsuchimoto, S. and Chua, N.H. (1993) Functional analysis of petunia floral homeotic MADS box gene *pMADS1. Genes Dev.* **7**: 1214–1228.

Weigel, D. and Meyerowitz, E.M. (1994) The ABCs of floral homeotic genes. *Cell* **78**: 203–209.

West, A.G., Shore, P. and Sharrocks, A.D. (1997) DNA binding by MADS-box transcription factors: a molecular mechanism for differential DNA bending. *Mol. Cell. Biol.* **17**: 2876–2887.

Yanofsky, M.F. (1995) Floral meristems to floral organs: genes controlling early events in *Arabidopsis* flower development. *Ann. Rev. Plant Physiol. Plant Mol. Biol.* **46**: 167–188.

Yanofsky, M.F., Ma, H., Bowman, J.L., Drews, G.N., Feldmann, K.A. and Meyerowitz, E.M. (1990) The protein encoded by the *Arabidopsis* homeotic gene *agamous* resembles transcription factors. *Nature* **346**: 35–39.

Zachgo, S., de Andrade Silva, E., Motte, P., Trobner, W., Saedler, H. and Schwarz-Sommer, Zs. (1995) Functional analysis of the *Antirrhinum* floral homeotic *DEFICIENS* gene *in vivo* and *in vitro* by using a temperature-sensitive mutant. *Development* **121**: 2861–2875.

Zhang, H.M. and Forde, B.G. (1998) An *Arabidopsis* MADS box gene that controls nutrient-induced changes in root architecture. *Science* **279**: 407–409.

Zhang, J. and Lautar, S. (1996) A yeast three-hybrid method to clone ternary protein complex components. *Analyt. Biochem.* **242**: 68–72.

2

The evolution of dioecy and plant sex chromosome systems

Deborah Charlesworth and David S. Guttman

1. Introduction: why are plant sex chromosomes of particular interest?

The genetic control of sex determination is becoming very well understood in several animal systems, particularly *Drosophila melanogaster*, *Caenorhabditis elegans* and mammals (Hodgkin, 1992). In plants, we can hope to investigate the developmental controls involved in sex determination systems, which should eventually form an important part of our understanding of plant developmental systems. In addition, plants are expected to contribute to our understanding of the evolution of separate sexes, and of sex chromosomes. The fact that in relatively recent evolutionary time there has been a major radiation of an important group of organisms, the flowering plants, is often overlooked by biologists. The statement that: 'Since then [the Cambrian], more than 500 years of wonderful stories, triumphs and tragedies, but not a single new phylum, or basic anatomical design, added to the Burgess complement' (Gould, 1989, p. 60) typifies the animal-centred attitude of much biological writing. In this chapter, we hope to illustrate by the example of the evolution of dioecy (separation of the sexes into different individuals) that the angiosperms have undergone evolutionary transformations as interesting as those studied by zoologists, and that study of the evolution of sex separation is important in a broad biological context. The evolutionary problems that arise in attempting to understand the evolution of sex chromosomes have relevance to many different questions in modern biology, and the study of these problems can integrate studies of plants with work on related issues in animal systems.

2. Sex and gender conditions in plants

Before discussing the evolution of dioecious plants, it is helpful to outline briefly the known facts about gender conditions in plants in general, and to make clear what the

Sex Determination in Plants, edited by C.C. Ainsworth.

evolutionary origins of dioecy must be. Here, we confine our attention to species and populations that are not asexually reproducing or apomictic. Sexually reproducing plants and animals can be classified into two groups by the fundamental distinction of whether all individuals are essentially alike in their gender condition ('sexually monomorphic', as in hermaphrodite populations), and those in which there are different kinds of individuals ('sexually polymorphic', as in dioecious species, with separate male and female individuals). The useful term 'cosexual' (Lloyd, 1984) is also employed by botanists to refer to the situation when the plant performs both sex functions, without specifying whether these are done within each flower (hermaphrodite), or by separate male and female flowers (monoecious). *Table 1* defines and lists the different systems seen in plants. The table gives some information on the frequencies of the different systems, but of course, the frequencies differ in different localities and circumstances. Floras of oceanic islands, for instance, appear to have higher frequencies of dioecious species than other floras (Bawa, 1980; Sakai *et al.*, 1995). Such correlations with particular circumstances can provide some evidence on the ecological circumstances that may promote the evolution of dioecy.

Table 1. *Sex and gender systems of sexually reproducing flowering plants*

Plant term	Definition of plant term	Occurrence in plants and examples
Sexually monomorphic		
Hermaphrodite	Flowers have both male and female functions	~90% of flowering plants (e.g. roses)
Monoecious	Separate sex flowers on the same individuals	~5% of flowering plants, often those with catkins (e.g. hazel), and many gymnosperms (e.g. pines)
Gynomonoecious	Individuals have both female and hermaphrodite flowers	e.g. daisies
Andromonoecious	Individuals have both male and hermaphrodite flowers	
Sexually polymorphic		
Dioecious	Separate sex individuals (male and female plants)	~5% of flowering plants (e.g. holly), and some gymnosperms (e.g. yew, cycads)
Gynodioecious	Individuals either female or hermaphrodite	e.g. ribwort plantain (*Plantago lanceolata*), bladder campion (*Silene vulgaris*)
Androdioecious	Individuals either male or hermaphrodite	very rare, see text for examples

3. Distribution of dioecy and sex chromosomes in flowering plants

The great majority of flowering plants are hermaphrodites, but a large proportion of angiosperm families have dioecious members. Extensive surveys of the literature have shown that dioecy is widely scattered taxonomically, being found in all six dicotyledonous, and all five monocotyledonous, subclasses (Renner and Ricklefs, 1995; Yampolsky and Yampolsky, 1922). Dioecious species occur in at least 7.1% of angiosperm genera (959 genera) and 38% of families (Renner and Ricklefs, 1995). *Table 2* shows some aspects of this distribution by listing species that have been studied genetically, showing the presence or absence of sex chromosome heteromorphism (see Westergaard, 1958). In addition to the species listed, dioecy is known from seven further orders of monocotyledonous plants, some of which may possibly have sex chromosomes (Westergaard, 1958). Sex chromosomes are known in every dicotyledonous subclass that has been subjected to study (though not certainly in the Asteridae), but in none of them are sex chromosomes found in all the species.

The much lower frequency of dioecy in genera than families implies origins within families. Many dioecious species are in genera most of whose members are hermaphroditic, and in many species with hermaphrodite relatives there are evident rudiments of opposite sex structures in flowers of plants of each sex: males may have substantial pistil rudiments, and females staminodes (Darwin, 1877). For examples, see Chapters 3, 9, 11 and 13. These observations suggest strongly that dioecy evolved quite recently, and this in turn implies that it has evolved repeatedly.

In plants, we may thus be able to estimate how long ago dioecy evolved in suitable taxonomic groups, and perhaps find out how much evolutionary time is necessary for morphologically distinct sex chromosomes to appear. In groups where there is good phylogenetic information, it is sometimes possible to estimate how many times dioecy has evolved. In the Hawaiian genus *Schiedia* (Caryophyllaceae), dioecy has probably evolved twice (Weller *et al.*, 1995, 1998). The best studied case at present is the large genus *Silene*. Most species in this genus are gynodioecious, though many are hermaphroditic, as are the vast majority of the approximately 2000 species in its family (Caryophyllaceae, with 80 genera). The dioecious *Silene* species are members of three sections of the genus, and include the close relatives *S. latifolia* and *S. dioica*, both which have heteromorphic sex chromosomes. Based on a phylogeny constructed from ITS sequences of nuclear ribosomal RNA genes of species sampled from the genus (Desfeux and Lejeune, 1996), dioecy has probably evolved at least twice in this genus also (Desfeux *et al.*, 1996; Zhang *et al.*, 1998), within a period estimated to be between 8 and 24 million years. It will be very interesting to have further sequence data from these species, and to attempt to refine this time estimate. It will also be worthwhile to study other taxonomic groups, to test the belief stated above that dioecy has most often evolved within the families in which it is seen, rather than having evolved earlier in the evolution of these taxa, with later loss of dioecy in some genera and species. If this belief is correct in general, as it appears to be in the Hawaiian flora (Sakai *et al.*, 1995), it suggests that separate sexes may have evolved more than 100 times in the flowering plants, to account for the current number of about 160 families with dioecious members. Such data should also tell us what fraction of such changes have led to the evolution of sex chromosomes, and provide rough estimates of the amount of time needed for them to evolve.

Table 2. *Sex systems of dioecious plants for which there have been genetic studies, arranged by subclass number and order in Cronquist's system.*

Subclass number, order, family	Species	Sex chromosome heteromorphism			Viability of genotype homozygous YY or WW	Sex determination	Close relatives hermaphrodite or monoecious
		Present or not	Female	Male			
Monocotyledons							
V, Liliales							
Liliaceae	*Asparagus officinalis*	No (males heterozygous)			Yes[a]	Male-determ. Y	Hermaphrodite
Dioscoreaceae	*Dioscorea* species	No (males heterozygous)			–	Male-determ. Y[b]	Hermaphrodite
Dicotyledons							
I, Ranunculales, Ranunculaceae	*Thalictrum* species	No (males heterozygous)			Yes		Hermaphrodite
II, Urticales, Cannabidaceae	*Cannabis sativa*	Yes	XX	XY	Yes?	X-autosome ?	Monoecious
	Humulus lupulus	Yes	XX	XY	Yes?	X-autosome ?	Monoecious
	Humulus japonicus	Yes	X_1X_1 X_2X_2	X_1Y_1 X_2Y_2	No	X-autosome ?	Monoecious
III, Polygonales							
Polygonaceae	*Rumex acetosa*	Yes	XX	XY_1Y_2		X-autosome ?	Hermaphrodite
	Rumex hastatulus	Yes	XX	XY or XY_1Y_2		X-autosome ?	Hermaphrodite
	Rumex acetosella					Male-determ. Y?	Hermaphrodite
III, Caryophyllales							
Caryophyllaceae	*Silene latifolia*, *S. dioica*	Yes	XX	XY	No	Male-determ. Y	Hermaphrodite
	Silene otites	No (uncertain)			–	–	Hermaphrodite
Chenopodiaceae	*Spinacia*	No (males heterozygous)			Yes	–	Monoecious ?
Amaranthaceae	*Acnida* species	No (males heterozygous)			–	Male-determ. Y	Monoecious

IV							
Theales, Actinidiaceae	*Actinidia chinensis*	(males heterozygous)			Yes[c]	Male-determ. Y	Hermaphrodite[d]
Violales, Cucurbitaceae	*Bryonia multiflora*	(males heterozygous)			–	–	Monoecious
	Ecballium elaterium	(males heterozygous)			–	Male-determ. Y	Monoecious
	Coccinia indica	Yes	XX	XY		Male-determ. Y	Monoecious
Violales, Caricaceae	*Carica papaya*	(males heterozygous)			No	Male-determ. Y	Monoecious
V							
Rosales, Rosaceae	*Fragaria* species	(females heterozygous)			–	–	Hermaphrodite
Rhamnales, Vitaceae	*Vitis*	Yes	XX	XY	Yes	Male-determ. Y	Hermaphrodite
Euphorbiales, Euphorbiaceae	*Mercurialis annua*	(males heterozygous)			Yes	Male-determ. Y	Monoecious
VI, Asterales							
Asteraceae	*Antennaria dioica*	Maybe (males heterozygous)			Probably no	–	

Most data are from Westergaard (1958), so no references are given. Some conclusions have not been established with certainty (particularly uncertain information is indicated by '?'). Species for which information is missing or not applicable are denoted by – in the relevant columns.

[a] Viable but male-sterile.

[b] Based on the fact that species are dioecious regardless of ploidy.

[c] Testolin *et al.*, 1995.

[d] Based on the presence of opposite sex organs in flowers of the two sexes.

4. Genetics of sex determination in dioecious angiosperms

Before considering how dioecy may have evolved, we briefly review data on the genetics of sex determination. Sex inheritance systems in plants show striking similarities to those in animals. The first similarity is that, as in most animal groups, the majority of plants studied have male heterogamety (XY males, XX females), though in plants, this term is not used exclusively to imply that the chromosomes are visibly different, but is sometimes used to include cases when males are the heterozygous sex. A second similarity is that visible chromosome differences between the sexes are found. This is true for perhaps half of plants that have separate sexes, though there may be more instances in species whose chromosomes are too small for definitive determination of heteromorphism, or in species which are polymorphic so that examination of different strains causes uncertainty (reviewed in Westergaard, 1958).

A third similarity to animal systems is that Y chromosomes appear to be genetically degenerate, or at least partially so. A difference from animals is that in many dioecious plants, hermaphrodites or 'inconstant males' occur, that is, males that produce occasional fruits (Lloyd, 1975b; Lloyd and Bawa, 1984). Self-fertilization of such plants is one major source of evidence for the conclusions just mentioned that males are usually the heterozygous sex, and that the Y is degenerated. Males are thought to have the genotype $Su^F S^M/Su^f S^m$, where Su^F is a dominant suppressor of femaleness, and S^m is a recessive male-sterility factor (i.e. females are homozygous S^m/S^m). On selfing inconstant males, a 3:1 ratio of males to females is expected if Su^F/Su^F is viable, but if the Su^F is on a Y chromosome that has degenerated, so that the YY genotype is inviable, the ratio would be 1:2. Both these ratios are observed, in different species (Testolin *et al.*, 1995; Westergaard, 1958).

Evidence that two loci are indeed involved in sex determination has been obtained in a number of dioecious plants in which crosses can be made to related monoecious or hermaphrodite species (see review in Westergaard, 1958). Cytological evidence from *S. dioica* and *S. latifolia* provides direct support for this. Y chromosomes of these species have distinct regions containing sex-determining genes with the expected properties and dominance relationships (Westergaard, 1958), that is, despite complete genetic linkage, the loci are separate. Reversion to hermaphrodite can occur by loss of the end of the non-pairing (p) arm of the Y, suggesting that this region of the normal Y contains a dominant suppressor of femaleness (Su^F). Deletions of two other Y regions yield sterile males, that is, the recessive male-sterility allele of females (S^m) is expressed. One male sterile deletion category (XY^3) has some hermaphrodite function. Heterozygotes for this deletion with the normal Y (which should not carry any male-sterility factors) should be normal males, and this is observed. By cytological studies, a total of three regions of the Y chromosome have thus been defined, the Su^F region, and two regions containing factors controlling early and late anther development (Grant *et al.*, 1994). This interpretation of the sex-determining 'locus' implies that recombination is absent or suppressed in the part of the chromosome on which these loci are carried, the differential segment, a final parallel with the situation in many animal sex chromosomes. In most animals, the Y chromosome is very nearly completely genetically inert and devoid of active gene loci, relative to what would be expected from the DNA content of the differential segment, though recently more loci have been discovered on the human Y chromosome (Lahn and Page, 1997). The genetic deterioration of this part of the Y chromosome has generally been attributed to its lack of recombination (e.g. Lewis, 1942; Westergaard, 1958), but the mechanism by which this occurs is still not fully understood. We shall return to this issue below, but will first discuss reasons why the sex-determining loci should be linked.

5. Why are sex-determining loci linked?

There is a well-developed theory that can account, in principle, for the evolution of the most fundamental feature of sex chromosomes: that they include a non-recombining region in which the sex-determining loci are found. The absence of recombination makes sense because it is evident that recombination between the sex-determining loci will produce maladaptive phenotypes, particularly neuter individuals. This is based on the implicit, but very reasonable, assumption that there must be multiple sex-determining loci (Lewis, 1942). For instance, in a species in which males are heterozygous and females the homozygous sex (as in species with XX/XY sex chromosome systems) it is reasonable to think that male-determining chromosomes carry female-sterility factors and female-determining chromosomes have male-sterility factors (see above). If these chromosomes recombine, genotypes could be produced with both male and female sterility, which would be neuter (see *Figure 2*, below). The reciprocal recombinants would have neither sterility factor, and would presumably be hermaphroditic (or cosexual).

It is sometimes thought that the linkage necessary to keep the two sexes separate, and to avoid production of neuters, evolved after establishment of male and female sterility genes, and that these loci were initially less close together but have been brought into proximity by inversions and/or translocations (Lewis, 1942). While this is a possibility, an alternative possibility is that quite close linkage may often be necessary for the evolution of separate sexes. If two loci are the minimum required, this becomes quite plausible (Charlesworth and Charlesworth, 1978). It is therefore worthwhile here to review in some detail the theory of how this evolutionary transition from cosexuality to dioecy could occur. Here, we will not discuss in detail the selective forces that might cause the genetic transitions that are believed to occur, but concentrate on the genetic changes themselves.

Figure 1 summarizes the possible genetic changes that could occur in the transition from cosexual to separate-sexed populations. Starting from an initially monomorphic hermaphroditic or monoecious cosexual state, the evolution of two sexes clearly requires a minimum of two genetic changes, one (male-sterility) creating females and the other (female-sterility) producing males. Although sex-determination systems can exist with both sexes under the control of a single locus (some genotypes being male while others are female, see, e.g. Dellaporta and Urrea, 1994), this situation cannot have been achieved in a single mutational step from an initial hermaphroditic state, except under the extremely improbable assumption that a mutation arises in a cosex whose heterozygotes have one sex, and homozygotes the other sex (e.g. *Aa* male and *aa* female). The two-step scenarios in *Figure 1* are much more plausible, given that mutations causing male or female sterility are well known in many plants (Kaul 1987). It must be borne in mind, however, that a minimum of two steps is required, and the process may sometimes have been more gradual, with partial sterility mutations occurring (Charlesworth, B. and Charlesworth, D., 1978; Lloyd, 1975a).

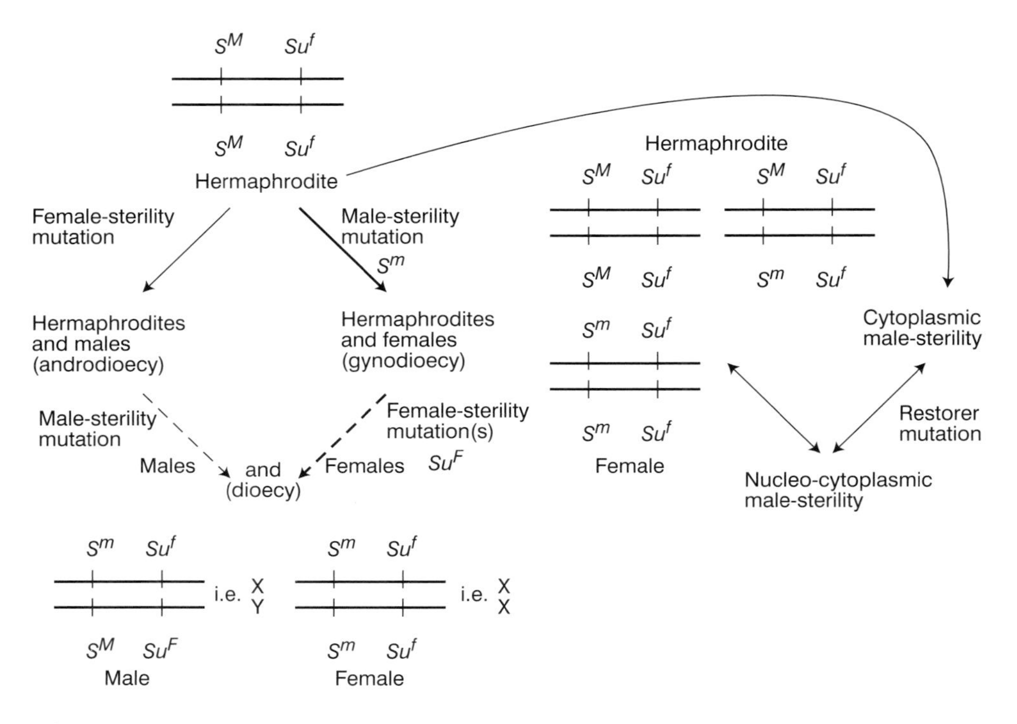

Figure 1. *The possible genetic changes that could occur in the transition from cosexual to separate-sexed populations.*

6. Pathways to dioecy: evidence for the gynodioecy pathway

6.1 *Relative frequencies of androdioecy and gynodioecy*

Of the two alternative pathways shown in *Figure 1*, there is both empirical and theoretical evidence that the pathway via gynodioecy is the more important. Firstly, the androdioecious condition is very rare compared with gynodioecy. It is difficult to estimate the frequency of gynodioecy, because a continuous distribution of femaleness exists between gynodioecious and subdioecious species, but gynodioecy is known in at least 36 angiosperm families, many of them containing several or many gynodioecious species (Delannay, 1978), whereas functional androdioecy is currently known or claimed in very few species (Charlesworth, 1984; Lepart and Dommée, 1992; Pannell, 1997; Rieseberg *et al.*, 1992; Traveset, 1994; Wolf *et al.*, 1997). The difference in frequencies of these two breeding systems must be due in part to the fact that gynodioecy can be caused by cytoplasmic male-sterility factors. Only a few gynodioecious species appear to have single-gene male-sterility systems (Eckhart, 1992; Kohn, 1988; Weller and Sakai, 1991), whereas cytoplasmic sterility factors are known, together with nuclear genes with alleles restoring male fertility, in most gynodioecious species whose genetics has been extensively studied (Belhassen *et al.*, 1991; Charlesworth and Laporte, 1998; Damme, 1983; DeHaan *et al.*, 1997; Kheyr-Pour, 1980; Koelewijn and Damme, 1995a, 1995b; Lewis, 1941; Sun, 1987).

Cytoplasmic male-sterility can invade populations if it confers even a slightly increased female fertility (Lewis, 1942), so it is not surprising that gynodioecy should occur and that cytoplasmic genes are involved. Gynodioecious systems with cytoplasmic inheritance are readily invaded by nuclear restorers of male fertility, sometimes leading to nucleocytoplasmic inheritance of male sterility (Charlesworth, 1981; Delannay *et al.*, 1981). Theoretically, it is possible for such systems to evolve into dioecy, though it is not clear how likely it is that this will happen (Delannay *et al.*, 1981; Maurice *et al.*, 1993, 1994; Nordborg, 1994; Schultz, 1994), and there is at present no empirical evidence that it has occurred.

6.2 *Taxonomic distributions of androdioecy and gynodioecy*

For those few species or populations that are androdioecious, most, if not all, of their closest relatives are found to be dioecious. This probably does not imply that the dioecious species have evolved via androdioecy, but is more likely to be because these populations evolved by breakdown of dioecy, that is, dioecious populations have been invaded by modified females having some degree of male function. The phylogenetic evidence for this is clear in the case of *Datisca glomerata* (Rieseberg *et al.*, 1992), and it appears highly likely for *Mercurialis annua* (Pannell, 1997), though further work on the relationships between populations of this species would be valuable.

We are aware of no case of androdioecy that may represent a population in an intermediate stage during the process of evolving dioecy. The remarkable rarity of androdioecy thus suggests either that it evolves rarely, or when it does, it evolves quickly into dioecy and is therefore not often found in contemporary populations. Gynodioecy is also taxonomically associated with dioecy. Several genera contain both breeding systems (Maurice *et al.*, 1993), which is a significant finding, given the low frequency of gynodioecy in general. But again, there are at best very few cases in which one can clearly conclude that a genus contains gynodioecious species and dioecious ones in an evolutionary progression from cosexuality to dioecy. Perhaps the best candidate for such a progression is the genus *Schiedia* (Weller *et al.*, 1995, 1998). A large set of potential examples includes several genera in which species have breeding systems ranging from gynodioecious to subdioecious, that is, where there are gynodioecious systems with females and cosexual plants, and the cosexes' female function ranges from similar to that of females to very low (Burrows, 1960; Sun and Ganders, 1986; Webb, 1979).

Another important issue with respect to the distribution of dioecy is whether it generally evolves from self-incompatible or highly outcrossing cosexes, or from partially inbreeding populations. Most close relatives of dioecious species that have been examined are not self-incompatible (Charlesworth, 1985), though their selfing rates have rarely been quantified. Furthermore, there appears to be a tendency for dioecy to be more frequent in genera that do not have self-incompatible species, compared with those where self-incompatibility is known. These findings suggest a potential role for inbreeding avoidance (see below).

6.3 *Theoretical conditions for the invasion of cosexual populations by males and females*

A third piece of evidence for the greater importance of the gynodioecy pathway, compared to that via androdioecy, is that polymorphisms for male-sterility factors (that is, gynodioecy) can be established under relatively plausible assumptions about the advantages of femaleness, whereas female sterility is always less advantageous, often greatly so; it follows that androdioecy is expected to evolve only very rarely from an initial cosexual state. This difference comes from differences in the way plants gain fitness via male and female functions. Female functioning, that is, seed production, is presumed to depend on resources available to the maternal plant. Male reproduction, however, is limited by the availability of unfertilized ovules, even though pollen output probably increases as resources available for reproduction increase. This implies that even a large increase in pollen output may result in a smaller increase in fitness via male functions (that is, seeds sired), because siring opportunities are limited by ovule availability.

The first examination of the conditions necessary for invasion of hermaphroditic populations by unisexual morphs caused by nuclear male-sterility genes (Lewis, 1941) showed that the fertility of females with nuclear-controlled male sterility must double in order for the allele to invade. This can be generalized to the conclusion that, for unisexual mutants, with one sex function remaining, this function must be doubled compared with the original cosex, for invasion to occur. Some increase in fertility seems plausible. Resources available at the time of reproduction are presumably finite. Thus, high pollen production must limit seed output, and *vice versa*; that is, there must be trade-offs between male and female reproduction. It follows that females should often be able to produce more seeds than cosexes. This is often referred to as the resource re-allocation advantage of unisexuality. There is some empirical evidence for resource re-allocation (Ashman, 1992; Delph, 1990).

Theoretical results on invasion conditions are derived from simple models based on the reasonable assumption that fitness via reproductive functions can be measured in terms of the numbers of genes contributed to the progeny produced. This explicitly accounts for contributions via both female and male reproduction, that is, via both ovules and pollen, and it ensures that the total progeny produced via pollen equals the total produced via ovules, so that the individuals in the population as a whole each have one male and one female parent, as they should. The models can be extended to include some self-fertilization, relaxing Lewis's assumption that the initial cosexual population consists of outcrossing plants. If some self-fertilization occurs in the initial cosexual population, unisexuals can gain additional fitness benefits from inbreeding avoidance, which is advantageous if there is inbreeding depression. The conditions for invasion by nuclear sterility factors given above are, therefore, somewhat modified (Charlesworth, B. and Charlesworth, D., 1978; Lloyd, 1974, 1979).

The conditions for invasion by females and males (that is, for rare unisexuals to have higher fitness than the initial cosexual) can be expressed as quite simple inequalities involving all the effects outlined here: the effect of the sterility in terms of one sex function on that of the remaining function (for females, the relative increase in fertility can be expressed as $k = (f_f - f_{cosex})/f_{cosex})$, where f_{cosex} and f_f are the female fertilities of cosexes and females, respectively; for males $K = (m_f - m_{cosex})/m_{cosex})$ where m_{cosex} and m_f are the male fertilities of cosexes and males). This results in the conditions:

$$\text{Females:} \qquad k > 1 - 2S\delta \qquad (1a)$$

$$\text{Males:} \qquad K > \frac{1 - 2S\delta + S}{1 - S} \qquad (1b)$$

where S is the cosexes' selfing rate and δ is the inbreeding depression (1 – the fitness of progeny produced by self-fertilization, relative to that of progeny produced by outcrossing to unrelated plants). If there were no inbreeding depression ($\delta = 0$), females could invade only if their fertility was doubled ($k > 1$), as outlined above. Alternatively, with no increase in fertility ($k = 0$), females can invade only if the product of the selfing rate and inbreeding depression exceeds 1/2. It seems unlikely that inbreeding depression would be as intense as this (for $S = 0.7$, for instance, δ would have to be above 0.7). More biologically plausible are situations in which both these advantages to females are present simultaneously, that is, both the quantity and quality of seeds are higher. A population of cosexes with selfing rate S and inbreeding depression 50% would require an increase in fertility of $1 - S$. This shows that, even if inbreeding depression is rather intense, a large increase in female fertility is always required unless the selfing rate is high. Similarly, males can invade only if their fertility is doubled ($K > 1$) compared with an outcrossing initial cosexual form. Other things being equal, Equation (1b) shows that the invasion of males into partially selfing populations requires even higher fertility differences (Charlesworth, B. and Charlesworth, D., 1978; Lloyd, 1974, 1979). Recent models that incorporate more biological realism confirm these conclusions (Pannell, in press; Seger and Eckhart, 1996).

Two conclusions can be drawn from these results. The first is that resource re-allocation and inbreeding avoidance are probably both involved whenever unisexuality evolves from cosexuality. Thus it is probably unhelpful to view these two advantages of unisexuality as alternatives. Second, it is more likely that females will invade populations than that males will. Hence the conclusion mentioned above that gynodioecy appears more likely than androdioecy, as a route to dioecy.

7. The transition to dioecy from gynodioecy: establishment of males in gynodioecious populations

Assuming that our conclusion above is correct (that the first step in the evolution of dioecy is establishment of unisexual females in a population), one might expect the cosexual morph to undergo further evolution to a new allocation of reproductive resources. This would be expected to be more male biased than before, since the availability of the females' ovules should lead to greater returns on investment in pollen output. It can be shown that this intuition is correct. Fully or partially female sterile phenotypes (caused by female-sterility factors or by modifier genes that change to more a male-biased allocation of reproductive resources) can invade gynodioecious populations under conditions that are less difficult to satisfy than those for invasion of the same phenotypes in the absence of females (Charlesworth, B. and Charlesworth, D., 1978). The observation of inconstant males (but not females) in dioecious species (e.g. Galli *et al.*, 1993; Testolin *et al.*, 1995) is completely consistent with this scenario of a major mutation leading to females, followed by selection for modifiers making the cosexes more male. In this second stage of the transition to dioecy, it turns out that each gene substituted must increase male fertility more than the proportional decrease it causes in female fertility (Charlesworth, B. and Charlesworth, D., 1978). This reinforces the conclusion already reached, that some reallocation of resources from female to male fertility is necessary in the evolution of dioecy via the gynodioecy pathway.

7.1 *Linkage between male and female suppressor loci*

The conclusion just given is only a partial one, however, because it considers only the direction of selection on the cosexual morph. If we make a genetic model and ask whether a modifier making cosexes more male-like can invade a gynodioecious population, the assumption of trade-offs clearly implies that modifier genes that make this morph more male in function should also reduce female fertility (*Figure 2*). Unless the modifiers are free from this trade-off (that is, the modifiers of allocation are sex-limited in their expression, so that they can act in cosexes alone, without affecting females), this produces a counter-selection that will hinder the spread of such factors. The invasion condition for a modifier producing a more male form thus depends on the linkage between the initial male sterility locus and the modifier locus. Linked modifiers can sometimes invade under conditions where unlinked modifiers would fail to do so (Charlesworth, B. and Charlesworth, D., 1978; Nordborg, 1994). As already outlined above, there may also be selection for tighter linkage between the male-sterility locus and modifier loci. The conclusion is that the evolution of dioecy will probably lead to a cluster of linked loci in a particular chromosomal region, in which recombination will be suppressed. After these evolutionary changes, one chromosome would carry the male-sterility allele that produces females, while the other would carry female-sterility alleles (and the wild-type allele at the male-sterility locus), and would be confined to males (*Figure 1*). In such a dioecious population, maleness would be controlled by a possession of a dominant male-determining chromosome. This is indeed the genetic system found in several plants that do not have heteromorphic sex chromosomes (see above).

This model also suggests that evolution of dioecy is not inevitable, since any step in the scenario described here may fail to occur. For instance, if linkage between male- and female-sterility loci is not very close, populations would become subdioecious, with females, males and also recombinant forms, that is, hermaphrodites (with neither male nor female sterility) and a low frequency of neuters, with both kinds of sterility factors. Such systems are known in some plants, for instance in the cactus *Pachycereus pringlei* (Fleming *et al*., 1994), though it is unknown whether the genetic situation underlying this species' 'trioecious' condition is as suggested here (Maurice and Fleming, 1995). A similar state might arise if sex-limited genes were involved, but in that case, no neuter plants should be present. Furthermore, given that recombination should sometimes produce hermaphrodites, it is not surprising that dioecy should

	Genotype at first locus	
	S^M/S^M or S^M/S^m	S^m/S^m
Genotype at modifier locus		
Su^f/Su^f	Hermaphrodite	Female
	↓	↓
Su^F/Su^f or Su^F/Su^F	Male	Neuter
Effects of change		
On fertility	Hermaphrodite male fertility increased	Female fertility decreased
On fitness	Increased	Decreased

Figure 2. *Effects of a female-sterility 'modifier' allele on hermaphrodites and females, in the absence of sex-limitation of its phenotypic effects. A trade-off between male and female functions is assumed, such that a gene increasing male fertility will often have the effect of reducing female fertility.*

sometimes revert to cosexuality. When this occurs, it could be due to a single gene difference in the sex-determining genes, which is in agreement with the few genetic data available for such cases (see above).

We have not discussed alternative evolutionary pathways to dioecy in any detail, but the model given above is just one particularly simple case, which we have argued, probably crudely represents the most frequent situation. Dioecy has probably often evolved from monoecy (Bawa, 1980; Darwin, 1877; Renner and Ricklef, 1995) and this might involve gradual specialization of individuals as one sex or the other (Charlesworth, B. and Charlesworth, D., 1978; Lloyd, 1975a), rather than the major changes assumed in the above model. It is, however, implausible that many mutational steps could all occur on the same chromosome. One might therefore argue that sex-limited modifiers would be of greater importance in such a gradual process. This, however, does not entirely solve the problem, because availability of many different genes with suitable effects on the proportions of flowers of each sex, and all able to act in a sex-limited manner, also seems implausible in organisms with very slight sex differences other than the sex organ differences themselves. Even when dioecy evolves from monoecy, it is therefore likely that some major factors may be necessary, and that they will need to be linked, on a single chromosome. The same is probably true for the final possibility, evolution from distyly. Distyly is a cosexual system with two mutually interfertilizing cosexual morphs (Barrett, 1992). In a few taxa, there are clear indications that one morph has become specialized as male and the other as female (Darwin, 1877; Lloyd, 1979; Muenchow and Grebus, 1987). The genetic details of how this may occur are still not well understood, and no study has yet examined the evolutionarily intermediate stages, but a very gradual process is difficult to envisage, as just explained with respect to monoecy. The only theoretical work has treated this as a problem of allocation of reproductive resources between the two sex functions (Casper and Charnov, 1982; Charlesworth, 1989), and this approach does not deal explicitly with the genetic steps involved.

8. Evolution of sex chromosomes

The theory outlined in the previous section shows how a rarely recombining chromosome region containing the sex-determining genes could evolve. Once this has happened, it could become the nucleus of an incipient sex chromosome system (see Charlesworth, 1991). In this section, we briefly review plant sex chromosomes in the light of hypotheses concerning how this evolutionary change might occur.

8.1 *Current knowledge about plant sex chromosomes*

The arguments and data reviewed above suggest that dioecy in plants is of relatively recent evolutionary origin. Intermediate stages are therefore likely to be found in extant plants, and this expectation appears to be correct. Classical and cytogenetic studies of angiosperm sex chromosomes and sex determination show that, as expected, some plants have well-differentiated sex chromosomes, while others do not. Species without heteromorphic sex chromosomes are not necessarily cases in which dioecy evolved only recently, as it is possible that there may be situations when chromosome heteromorphism does not evolve, but it appears likely that evolution of sex chromosomes very often follows once dioecy is established. Phylogenetic studies to estimate the ages of dioecious systems and their sex chromosomes should illuminate this process. If dioecy can evolve by gradual modification from cosexuality into two morphs with increasingly great emphasis on male and female functions, as has been suggested in the evolutionary pathway from monoecy (see above), sex chromosomes would not be expected to evolve. However, sex chromosomes are present as frequently in taxonomic groups in which dioecy probably evolved from monoecy, as among taxa in which dioecy appears to be derived from hermaphrodite species (*Table 2*).

It is also important to realize that plants and animals include many species with systems other than the simple dominant male-determining system just described (Bull, 1983). Systems such as those with X-autosome balance, and multiple sex chromosomes (e.g. XY1Y2, and X/0, see Smith, 1963, 1969; Westergaard, 1958), are most likely derived from systems with male-determining chromosomes. It is not at present known whether systems such as female heterogamety are also derived, or whether they evolve independently by a different route from the one outlined here. In some animal species, a sex-determining region has been replaced by a new gene taking over control of sex determination (Bull, 1983; Traut and Willhoeft, 1990). As might be expected, these new genes are not part of heteromorphic sex chromosome systems.

The Y chromosomes of several plant species are known to have terminal regions that pair with the X in male meiosis (Parker, 1990; Westergaard, 1958), while recombination is absent between other parts of the sex chromosomes. Genetic mapping is beginning to be feasible, and both X- and Y-linked markers have been discovered, in plants with and without heteromorphic sex chromosomes (e.g. DiStilio and Mulcahy, 1996; Harvey *et al.*, 1997; Mulcahy *et al.*, 1992; Polley *et al.*, 1997; Testolin *et al.*, 1995) though there are not yet enough of them to provide detailed maps of these chromosomes. As the markers used so far have been scored mostly as presence or absence of bands on gels that separate PCR products using random primers, they cannot yet tell us which X-linked loci have homologues on the Y chromosomes and which do not. Mapping data should give estimates of the fraction of X-linked loci that are located in the region that pairs with the Y, versus those that are in the differential region. In the pairing (or pseudoautosomal) region, loci should recombine in both male and female plants, whereas loci in the differential region should be transmitted unrecombined from males to their progeny. There are not yet enough markers to use this approach in any plant species, but data should soon become available.

The evolutionary model given above suggests that the region containing the sex-determining loci would initially be fully homologous between the two alternative chromosomes, except for these genes themselves. One goal of studies of plant sex chromosomes is therefore to test for this homology. A particular aspect that should be tested is whether the region involved in sex-determination in dioecious species is also found in a single chromosomal location in cosexual relatives. The alternative is that the sex-determining genes were at first on different chromosomes, and only later were brought into proximity, under selection to avoid production of recombinants. It should also become possible to discover the origins of multiple sex chromosomes, such as those of *Rumex acetosa*. The two Y chromosomes of this species may have arisen after an X-autosome fusion, as in *Drosophila miranda* (Patterson and Stone, 1952) and *R. hastatulus* (Smith, 1963), in which case one Y would be more recent than the other and might be only partially degenerated, as in *D. miranda* (see Steinemann *et al.*, 1993). Alternatively, a chromosome homologous to the *R. acetosa* X has undergone fission (and subsequently re-established a metacentric morphology). The fact that both Y1 and Y2 carry the same repetitive sequence may support the second route (Réjon *et al.*, 1994). This is not conclusive, however, as it is not yet known what processes cause such sequences to accumulate, nor how long it takes. If the process is rapid, a relatively new Y might acquire them very quickly and become indistinguishable from the older Y.

8.2 *Degeneration of Y chromosomes*

A finding that degeneration has occurred in plant, as well as in animal Y chromosomes would be of great interest, because it would be evidence that the process is very general, and because plants might be valuable material for studying the early stages of the process, which might shed light on how it happens. In most animals, the differential segment of the Y chromosome is largely genetically inert and devoid of active gene loci. The genetic deterioration of this part of the chromosome, in the homologue that is carried only in the heterozygous sex, is probably an evolutionary consequence of this lack of recombination (Charlesworth, 1991, 1996).

In animals, the process of degeneration of the Y chromosome happened long in the past. It is therefore inaccessible to study, except in species where translocations between the sex chromosomes and autosomes have occurred, a situation known in several *Drosophila* species . Among these, some involve Y-autosome translocations (e.g. Steinemann *et al.*, 1993), and the degeneration of the neo-Y arm may not, therefore, involve the same processes as in situations when a new sex chromosome system evolves. In species with X-autosome translocations (e.g. Charlesworth *et al.*, 1997), the neo-Y is not physically attached to the pre-existing Y chromosome, so its degeneration may result largely from the same kind of degenerative processes as in the initial evolution of Y chromosomes. Plant systems nevertheless provide the main source of empirical data on true newly evolving sex chromosomes. Such data should be very illuminating, as most of our current understanding of the degeneration process comes from theoretical work.

Once a region has evolved where recombination is suppressed, several processes are expected to occur that can potentially lead to the two well-known distinctive properties of Y chromosomes, loss of functional genes and accumulation of repetitive sequences. These processes may in turn lead to the cytological differentiation of the Y from other chromosomes, so that heteromorphism may be apparent. The different mechanisms that have been proposed for genetic degeneration of a proto-Y chromosome have recently been reviewed (Charlesworth, 1991, 1996). All of them depend on deleterious mutations occurring at a sufficiently high rate on the neo-Y chromosome.

One possibility is that mutations may accumulate by the process known as Muller's ratchet (Haigh, 1978; Muller, 1964). If mutation rates to deleterious alleles are high enough, mutation-free chromosomes will be present at low frequency and may fail to reproduce in any given generation, that is, they can be lost by genetic drift. In the absence of recombination, mutation-free chromosomes cannot re-appear in a population once they are lost. If this process successively happens, on a chromosome like the Y, a greater and greater mean number of mutations will accumulate over time. This process may not, however, be able to account for degeneration in *Drosophila* species because of their large population sizes (Charlesworth, 1996). Most plants, and all dioecious species, have more chromosomes than *Drosophila*. The number of genes on a proto-Y chromosome must therefore be smaller, and so the mutation rate to deleterious alleles is probably lower. The deleterious mutation rate for entire plant genomes has been estimated as about 0.5 (Charlesworth *et al.*, 1990; Johnston and Schoen, 1995). If we assume that 10% of the genes are on the X chromosome (that is, that gene numbers are roughly proportional to chromosome size), and that the proto-Y would have been the same size, the relevant mutation rate is only 0.05 per generation. Using this mutation rate to calculate the expected mean number of detrimental mutations on such a chromosome (Charlesworth, 1996), assuming a dominance coefficient of 2%, yields a value of about 2.5, probably not enough to cause Muller's ratchet to operate, unless plant population sizes were very small. If only part of the proto-Y were non-recombining, this conclusion would be even stronger.

Another possibility is hitch-hiking when favourable mutant alleles arise on the Y and rise in frequency to fixation, concomitantly fixing deleterious alleles on the same chromosome (Rice, 1987). Arguments against this mechanism have been raised (Charlesworth, 1996). A third suggestion relies on the accelerated fixation of deleterious mutations that is expected to occur on a non-recombining chromosome, because selection against deleterious alleles is equivalent to reduced effective population size (Charlesworth, 1996). Following the calculation that can be done for a *Drosophila* proto-Y, but assuming, as above, a lower number of sites under selection, even if one assumes a lower effective population size of males than for *Drosophila*, the rate of decline in fitness would probably be lower in a plant. Rapid degeneration is thus not easily reconciled with available information about deleterious mutation rates.

9. Have plant Y chromosomes degenerated?

It appears possible that these chromosomes have undergone genetic degeneration in relatively short amounts of evolutionary times, though more data are still needed, as much of the evidence for degeneration is circumstantial, and time estimates are almost entirely lacking. At present, little is known about the genetic functions of plant Y chromosomes (i.e. whether these chromosomes have genes that are homologous to genes carried on the X chromosomes, and what fraction of Y-linked loci are expressed). If degenerated loci are found on plant Y chromosomes, it will be important to test whether dosage compensation occurs and equalizes gene expression in males and females despite their different numbers of X chromosomes, as occurs in animals (Baker *et al.*, 1994; Lucchesi, 1978, 1993).

Rumex acetosa Y chromosomes are heterochromatic, but the heterochromatin is not constitutive (Clark *et al.*, 1993; Réjon *et al.*, 1994; Zuk, 1969a, 1969b, 1970). Experiments with DNAse digestion suggest transcriptional activity of the Y in this species (Clark *et al.*, 1993), though this could be due to the presence on the Y as well as the autosomes of dispersed repetitive sequences that are transcribed, such as transposable elements. The high frequency of chromosome rearrangements in this species (Wilby and Parker, 1988), and variability of its Y chromosome morphology (Wilby and Parker, 1986), are consistent with such a possibility, but it has not yet been tested.

In *S. latifolia*, the two X chromosomes differ in the time of replication, as might be expected if one of them is transcriptionally silenced (Siroky *et al.*, 1994), and they appear to be differentially methylated, possibly indicating that dosage compensation is occurring by X inactivation in females (Vyskot *et al.*, 1993). Gene expression from Y chromosomes is suggested by estimates of methylation levels (Vyskot *et al.*, 1993), which may imply that many Y-linked genes have not degenerated very greatly, if at all (though again the possibility of transposons cannot be excluded). The larger size of the Y, compared with the X chromosomes of *S. latifolia* and *S. dioica* (Costich *et al.*, 1991), and many other dioecious plants (Parker, 1990), may also suggest that plant Y chromosomes have accumulated repetitive sequences. A repetitive sequence has been found on Y chromosomes of *R. acetosa* (Réjon *et al.*, 1994). So far, however, most other species in which repetitive sequences have been detected have similar abundances on the X and autosomes (Clark *et al.*, 1993; Donnison *et al.*, 1996; Scutt *et al.*, 1997), so the evidence is not yet conclusive, and the nature and range of kinds of such sequences is currently almost totally unknown.

Two types of direct evidence suggest genetic degeneration of male-determining chromosomes in plants. One is the evidence, mentioned above, that YY progeny are inviable. Among 12 species reviewed by Westergaard (1958), species with viable and inviable YY were about equally frequent, and viability appears common in species without heteromorphic sex chromosomes (asparagus plants are viable as Y haploids; see Galli *et al.*, 1993), while YY plants are inviable in most species studied that have heteromorphic sex chromosomes (see *Table 2*). An alternative reason for inviability might be that these chromosomes carry some lethal genetic load. It is common in outbreeding organisms, such as *Drosophila*, for the viability of highly homozygous genotypes to be very low, but homozygosity for a single chromosome rarely leads to complete failure to survive, even though *Drosophila* have small chromosome numbers, so that each chromosome has many loci and could carry a high genetic load (Crow, 1993). This alternative is thus unlikely. The finding that androgenic haploid plants with only a Y chromosome are inviable in *S. latifolia* (Ye *et al.*, 1990) also suggests that the Y is deficient in some essential functions. The contrast with the viability of X-haploid plants also argues against this effect being caused by ordinary genetic load, and suggests that the Y is different from other chromosomes.

A second type of evidence comes from studies of pollen competition. When large pollen loads are used to pollinate females, the progeny often show less than 50% males (certation), even though the same cross done with a low amount of pollen produces equal numbers of male and female progeny in both *S. latifolia* and *R. acetosa* (see Correns, 1928; Smith, 1963; Wilby and Parker, 1988) as well as some other dioecious species (reviewed in Allen, 1940). This has been interpreted as being caused by Y-chromosome-bearing pollen growing more slowly than X-bearing pollen, as would be expected if plant Y chromosomes have degenerated and lost functional genes (Lloyd, 1974; Smith, 1963). At present there is no direct evidence for this interpretation, and the phenomenon is not invariably found (Carroll and Mulcahy, 1990a, 1990b). Finally, X-linked mutations are known that are not masked by the Y chromosome, as in the case of a chlorophyll-deficient mutation caused by a deletion in *R. hastatulus* (Smith, 1963), that is, males are hemizygous for this region, just like the situation for classical sex-linked loci in many animals.

10. Molecular genetics of plant Y chromosomes

It should be possible to further advance our understanding of the evolution of plant sex chromosomes and sex determination using molecular methods. One approach is the direct one of searching for sex-determining genes, though there are many difficulties. If the number of expressed genes on the Y chromosomes of the plants studied is very high, it may be extremely difficult to find the loci that are important by direct searches for them. This may well be an even greater problem in plants than animals. Plant Y chromosomes differ from those of animals in two ways that suggest that active loci may still be found on them. First, genes are actively expressed in the haploid male gametophytes during pollination (Pedersen *et al.*, 1987; Stinson *et al.*, 1987; Tanksley *et al.*, 1981), unlike the case of animals, in which there is no gene expression in male gametes (Ward and Kopf, 1993). If any such genes are on the male-determining chromosomes of plants, natural selection may therefore protect them from genetic degeneration. A large number of plant Y-linked loci may thus be protected by selection in the haploid stage (Haldane, 1933). In *R. hastatulus*, some recessive mutations on the X chromosome are not expressed in males, showing that the Y chromosome carries functional alleles (Smith, 1963), but it is not clear whether the loci are in the differential segment. Second, plant sex chromosomes evolved substantially more recently than those of most animals, that is, there has been less time for degeneration and loss of Y-chromosomal loci located in the non-pairing region. Depending on how much time is required for the genetic degeneration process to take place, which is not yet known (and may be longer in most plants than in *Drosophila*, see above), even genes that are not protected by natural selection because they are functional in pollination may still be present. This may make plant Y chromosomes particularly favourable material for testing the assumption of homology with the X, but makes it difficult to hope for progress in identifying sex determining genes by mapping Y chromosomes, though allozyme, AFLP and RAPD–PCR analysis have produced useful sex-specific markers that can be used to sex plants in crop species, including asparagus (Bracale *et al.*, 1991), hops (Polley *et al.*, 1997), yam (Chapter 10) and kiwi fruit (Harvey *et al.* 1997; Testolin *et al.*, 1995). Before this approach can be used, it should be evaluated by testing how degenerated plant Y chromosomes really are.

A further difficulty of approaches based on isolation of Y-chromosomal sequences is the probable large amount of repetitive DNA in these chromosomes of several species. Isolated Y chromosomes contain repetitive sequences common to the X and autosomes and have not so far yielded single-copy sequences that could be candidates for sex-determining genes (Scutt and Gilmartin, 1997; Scutt *et al.*, 1997; Veuskens *et al.*, 1995).

Isolation of male-specific cDNAs from developing flower buds or reproductive organs might be a way to discover sex-determining genes, and has been attempted in several species, including *S. latifolia* (Barbacar *et al.*, 1997; Matsunaga *et al.*, 1996). But the female structures of males cease development early in bud development, so sex may be determined at a very early stage (Grant *et al.*, 1994; Folke and Delph, 1997). Thus this work may mainly identify genes controlled in response to sex, rather than the controlling loci. Genes that are known to be important in floral development, including the homoeotic MADS-box genes (Coen and Meyerowitz, 1991) have also been tested, but appear not to have direct roles in sex determination (Ainsworth *et al.*, 1995; Hardenack *et al.*, 1994). This makes sense, as mutations in these genes change floral organ identities, whereas in unisexual flowers apparently normal reproductive organs merely stop developing. Identification of key sex-determination genes will therefore probably require analysis of mutants, but locating them on the Y chromosomes will remain a problem.

A more promising approach is deletion mapping of the Y chromosome. Although, as just mentioned, this cannot be expected to pinpoint the sex-determination loci very precisely, it should be possible to use this approach to define the general regions in which these genes are located, for further analysis in the future. A difficulty is the lack of chromosome banding patterns that could help identify regions affected in Y-chromosome mutants caused by deletions. But this difficulty could potentially be overcome if we possessed single-copy genetic marker loci on the Y. In *S. latifolia*, this approach has been followed, using hermaphrodite mutants to roughly locate genes suppressing female, and promoting male development, and subtractive techniques to identify sequences which may assist in such mapping (Donnison *et al.*, 1996). Another possible source of markers for such mapping is single-copy X-linked loci. If these frequently have Y-linked homologues, they could provide a source of single-copy loci on the Y.

We have recently identified an expressed locus on the sex chromosomes of *S. latifolia* (Guttman and Charlesworth, 1998). Four loci from the set described by (Matsunaga *et al.*, 1996) were tested for sex-linkage. PCR primers were designed from the GenBank sequences to amplify regions of 200–400 bp in length. PCR was performed on genomic DNA extracted from leaf tissue, and the products analysed by the method of 'multiplex cold SSCP' after restriction enzyme digestion to reduce the size of the fragments so that bands could be scored for presence or absence on gels. The bands could then be used as genetic markers that can be examined for differences between the parents, and inspected for the transmission pattern expected within families under sex linkage. Bands from an X-linked allele in the paternal parent should be present in the father and all his female offspring. The situation can be complicated if the maternal parent is also heterozygous for this allele, so that the mother shares the band with the father, the female offspring, and some, but not all, the male offspring; this is likely to be common in highly outcrossing species such as dioecious species, in which genetic diversity and frequencies of heterozygotes are high. Despite this complication, X-linked markers are distinguishable from autosomal markers that are heterozygous in one parent: such markers should not be transmitted to all offspring of one sex, but should segregate in both sexes of progeny. When bands generated from *S. latifolia* genes were examined in a family of plants, a potentially sex-linked transmission pattern was observed for one of the loci tested, MROS3 (Guttman and Charlesworth, 1998). Sequences of the parental MROS3 PCR-amplified products showed that the observed pattern was due to X-linkage, and not to heterozygosity in both parents.

MROS3 (male reproductive organ specific gene) encodes a protein with no similarity to known proteins. A male-specific homologue to the X-linked MROS3 was identified by anchored PCR, and the male-specific fragment followed the expected inheritance pattern, confirming Y chromosomal location. A 159 bp segment of this gene's sequence had high homology to the 3′ end of the MROS3 database cDNA sequence. In this region, 31 sites differed between the X and Y sequences (19.5%) plus a single insertion/deletion of three bases. There are 6.5 synonymous substitution (out of 32.75 silent sites) and 24.5 non-synonymous substitution (out of 120.25 replacement sites). The data suggest that this region has been evolving in a neutral manner, because the ratio of silent to replacement substitutions, K_a/K_s (Nei, 1987), is close to 1. From our data, the value of this ratio is 0.974 [0.989 with Jukes-Cantor correction (Jukes and Cantor, 1969). Outside the 159 bp region of high homology in Y-MROS3, there is a stretch of approximately 235 bp (corresponding to the central region in X-MROS3) with no homology between the X and Y sequence, while the portion corresponding to the 5′ end of MROS3-X consists of sequential mononucleotide repeats. This Y-linked homologue of an X-linked locus is thus clearly degenerated. As its expression is limited to anther tissue, and the gene is not expressed in pollen, selection should not be operating on this gene product during pollination, so degeneration is possible.

11. Discussion

A reasonably good understanding is now available of how separate sexes may have evolved from hermaphrodite or monoecious plants, though the relative importance of the major selective factors leading to evolution of dioecy, inbreeding avoidance and resource allocation, cannot readily be evaluated. Further understanding may come when we know more about when dioecious systems have arisen in the evolutionary history of plant families, so that we could ascertain the characteristics of the ancestral species. At present, it is clear only that they were often, but not invariably, wind-pollinated and monoecious, and that they were probably almost all self-compatible.

With the availability of molecular techniques, we may now be in a position to begin to understand more about how sex chromosomes evolve and to estimate how long this takes. This, in turn, should help us evaluate the plausibility of the proposed mechanisms for the process. If, as appears likely, plant sex chromosomes are found to be only partially genetically degenerated, they may offer opportunities to help understand the relationship between genetic degeneration and the very interesting control system that operates to ensure dosage compensation, which is now starting to be understood in animals (e.g. Bone and Kuroda, 1996; Marin *et al.*, 1996). The results of such studies may contribute to our knowledge of mutation rates to deleterious mutations, and to a growing body of understanding of what happens when evolution occurs in the absence of recombination.

References

Ainsworth, C. C., Crossley, S., Buchanan-Wollaston, V., Thangavelu, M. and Parker, J. (1995) Male and female flowers of the dioecious plant sorrel show different patterns of MADS box gene expression. *Plant Cell* **7**: 1583–1598.

Allen, C.E. (1940) Genotypic basis of sex expression in Angiosperms. *Annu. Rev. Bot.* **6**: 277–300.

Ashman, T.-L. (1992) The relative importance of inbreeding and maternal sex in determining fitness in *Sidalcea oregana* ssp. *spicata*, a gynodioecious plant. *Evolution* **46**: 1862–1874.

Baker, B.S., Gorman, M. and Marin, I. (1994) Dosage compensation in Drosophila. *Annu. Rev. Genet.* **28**: 491–522.

Barbacar, N., Hinnisdaels, S., Farbos, I., Moneger, F., Lardon, A., Delichere, C., Mouras, A. and Negrutiu, I. (1997) Isolation of early genes expressed in reproductive organs of the dioecious white campion (*Silene latifolia*) by subtraction cloning using an asexual mutant. *Plant J.* **12**: 805–817.

Barrett, S.C.H. (1992) Heterostylous genetic polymorphisms: model systems for evolutionary analysis. In: *Evolution and Function of Heterostyly* (ed. S.C.H. Barrett), Springer-Verlag, Heidelberg, pp. 1–29.

Bawa, K.S. (1980) Evolution of dioecy in flowering plants. *Annu. Rev. Ecol. Syst.* **11**: 15–39.

Belhassen, E., Dommée, B., Atlan, A., Gouyon, P.-H., Pomente, D., Assouad M.W. and Couvet, D. (1991) Complex determination of male sterility in *Thymus vulgaris* : genetic and molecular analysis. *Theor. Appl. Genet.* **82**: 137–143.

Bone, J.R. and Kuroda, M.I. (1996) Dosage compensation regulatory proteins and the evolution of sex-chromosomes in Drosophila. *Genetics* **144**: 705–713.

Bracale, M., Caporali, E., Galli, M.G. *et al.* (1991) Sex determination and differentiation in *Asparagus officinalis L. Plant Sci.* **80**: 67–77.

Bull, J.J. (1983) *Evolution of Sex Determining Mechanisms*. Benjamin/Cummings, Menlo Park, CA.

Burrows, C.J. (1960) Studies in *Pimelea*. I. The breeding system. *Trans. Roy. Soc. NZ* **88**: 29–45.

Carroll, S.B. and Mulcahy, D.L. (1990a) Progeny sex ratios in dioecious *Silene latifolia*. *Am. J. Bot.* **80**: 551–556.

Carroll, S.B. and Mulcahy, D.L. (1990b) Unexpected progeny sex ratios in dioecious *Silene latifolia*. *Am. J. Bot.* **77** (suppl.): 51.

Casper, B.B. and Charnov, E.L. (1982) Sex allocation in heterostylous plants. *J. Theoret. Biol.* **96**: 143–149.

Charlesworth, B. (1991) The evolution of sex chromosomes. *Science* **251**: 1030–1033.

Charlesworth, B. (1996) The evolution of chromosomal sex determination and dosage compensation. *Curr. Biol.* **6**: 149–162.

Charlesworth, B. and Charlesworth, D. (1978) A model for the evolution of dioecy and gynodioecy. *Am. Nat.* **112**: 975–997.

Charlesworth, B. and Laporte, V. (1998) The male-sterility polymorphism of *Silene vulgaris*. Analysis of genetic data from two populations, and comparison with *Thymus vulgaris*. *Genetics* (in press).

Charlesworth, B., Charlesworth, D. and Morgan, M.T. (1990) Genetic loads and estimates of mutation rates in very inbred plant populations. *Nature* **347**: 380–382.

Charlesworth, B., Charlesworth, D. Yu, A. and Hnilicka, J. (1997) Evidence for lack of degeneration of allozyme loci on the fourth chromosomes of *Drosophila americana*. *Genetics* **145**: 989–1002.

Charlesworth, D. (1981) A further study of the problem of the maintenance of females in gynodioecious species. *Heredity* **46**: 27–39.

Charlesworth, D. (1984) Androdioecy and the evolution of dioecy. *Biol. J. Linn. Soc.* **23**: 333–348.

Charlesworth, D. (1985) Distribution of dioecy and self-incompatibility in angiosperms. In: *Evolution - Essays in Honour of John Maynard Smith* (ed. P.J. Greenwood, and M. Slatkin), Cambridge University Press, Cambridge, pp. 237–268.

Charlesworth, D. (1989) Allocation to male and female functions in sexually polymorphic populations. *J. Theoret. Biol.* **139**: 327–342.

Charlesworth, D. and Charlesworth, B. (1978) Population genetics of partial male-sterility and the evolution of monoecy and dioecy. *Heredity* **41**: 137–153.

Clark, M.S., Parker, J.S. and Ainsworth, C.C. (1993) Repeated DNA and heterochromatin structure in Rumex acetosa. *Heredity* **70**: 527–536.

Coen, E.S. and Meyerowitz, E.M. (1991) The war of the whorls: genetic interactions controlling flower development. *Nature* **353**: 31–37.

Correns, C. (1928) Bestimmung, Vererbung und Verteilung des Geschlechtes bei den hòèheren Pflanzen. *Handb. Vererbungswissenschaft* **2**: 1–138.

Costich, D.E., Meagher, T.R. and Yurkow, E.J. (1991) A rapid means of sex identification in *Silene latifolia* by use of flow cytometry. *Plant Mol. Biol. Reporter* **9**: 359–370.

Crow, J.F. (1993) Mutation, mean fitness, and genetic load. *Oxf. Surv. Evol. Biol.* **9**: 3–42.

Damme, J.M.M. v. (1983) Gynodioecy in *Plantago lanceolata* L. II. Inheritance of three male sterility types. *Heredity* **50**: 253–273.

Darwin, C.R. (1877) *The Different Forms of Flowers on Plants of the Same Species*. John Murray, London.

DeHaan, A.A., Luyten, R.M.J.M., Bakx-Schotman, T.J.M.T. and Damme, J.M.M. v. (1997) The dynamics of gynodioecy in *Plantago lanceolata*. L. I. Frequencies of male-steriles and their cytoplasmic male sterility types. *Heredity* **79**: 453–462.

Delannay, X. (1978) La gynodioécie chez les angiosperms. *Les Naturalistes Belges* **59**: 223–237.

Delannay, X., Gouyon, P.-H. and Valdeyron, G. (1981) Mathematical study of the evolution of gynodioecy with cytoplasmic inheritance under the effect of a nuclear restorer gene. *Genetics* **99**: 169–181.

Dellaporta, S.L. and Urrea, A.C. (1994) The sex determination process in maize. *Science* **266**: 1501–1505.

Delph, L.F. (1990) Sex-ratio variation in the gynodioecious shrub *Hebe strictissima* (Scrophulariaceae). *Evolution* **44**: 134–142.

Desfeux, C. and Lejeune, B. (1996) Systematics of euromediterranean Silene (Caryophyllaceae) - evidence from a phylogenetic analysis using its sequences. *Comptes Rendus de l'Academie des Sciences Serie III – Sciences de la Vie-Life Sciences* **319**: 351–358.

Desfeux, C., Maurice, S., Henry, J.P., Lejeune, B. and Gouyon, P. H. (1996) Evolution of reproductive systems in the genus Silene. *Proc. Roy. Soc. Lond. B.* **263**: 409–414.

DiStilio, V.S. and Mulcahy, D.L. (1996) Sex linked molecular markers in *Silene alba* can be useful in detecting aberrations and amount of crossing-over between the X and Y chromosomes. *Am. J. Bot.* **83**: 274 (abstract).

Donnison, I.S., Siroky, J., Vyskot, B., Saedler, H. and Grant, S.R. (1996) Isolation of Y chromosome-specific sequences from *Silene latifolia* and mapping of male sex determining genes using representational difference analysis. *Genetics* **144**: 1893–1901.

Eckhart, V.M. (1992) The genetics of gender and the effects of gender on floral characters in gynodioecious *Phacelia linearis* (Hydrophyllaceae). *Am. J. Bot.* **79**: 792–800.

Fleming, T.H., Maurice, S., Buchmann, S.L. and Tuttle, M.D. (1994) Reproductive-biology and relative male and female fitness in a trioecious cactus, Pachycereus pringlei (Cactaceae). *Am. J. Bot.* **81**: 858–867.

Folke, S.H. and Delph, L.F. (1997) Environmental and physiological effects on pistillate flower production in *Silene noctiflora* L. (Caryophyllaceae). *Int. J. Plant. Sci.* **158**: 501- 509.

Galli, M.G., Bracale, M., Falavigna, A., Raffaldi, F., Savini, C. and Vigo, A. (1993) Different kinds of male flowers in the dioecious plant *Asparagus officinalis* L. *Sex. Plant Reprod.* **6**: 16–21.

Gould, S.J. (1989) *Wonderful Life: the Burgess Shale and the Nature of History*. W.W. Norton, New York.

Grant, S., Houben, A., Vyskot, B., Siroky, J., Pan, W.H., Macas, J. and Saedler, H. (1994) Genetics of sex determination in flowering plants. *Dev. Genet.* **15**: 214–230.

Guttman, D.S. and Charlesworth, D. (1998) An X-linked gene has a degenerate Y-linked homologue in the dioecious plant *Silene latifolia. Nature* **393**: 263–266.

Haigh, J. (1978) The accumulation of deleterious genes in a population. *Theor. Pop. Biol.* **14**: 251–267.

Haldane, J.B.S. (1933) The part played by recurrent mutation in evolution. *Am. Nature* **42**: 5–19.

Hardenack, S., Saedler, H., Ye, D. and Grant, S. (1994) Comparison of MADS box gene expression in developing male and female flowers of the dioecious plants white campion. *Plant Cell* **6**: 1775–1787.

Harvey, C.F., Gill, C.P., Fraser, L.G. and McNeilage, M.A. (1997) Sex determination in *Actinidia*. 1. Sex-linked markers and progeny sex ratio in diploid *A. chinensis*. *Sex Plant Repro.* **10**: 149–154.

Hodgkin, J. (1992) Genetic sex determination mechanisms and evolution. *Nature* **14**: 253–261.

Johnston, M.O. and Schoen, D.J. (1995) Mutation rates and dominance levels of genes affecting total fitness in two angiosperm species. *Science* **267**: 226–229.

Jukes, T. and Cantor, C. (1969) Evolution of protein molecules. In: *Mammalian Protein Metabolism* (ed. H.N. Munro). Academic Press, New York, pp. 21–132.

Kaul, M.L.H. (1987) *Male-sterility in Higher Plants*. Springer-Verlag, Berlin.

Kheyr-Pour, A. (1980) Nucleo-cytoplasmic polymorphism for male sterility in *Origanum vulgare* L. *J. Heredity* **71**: 253–260.

Koelewijn, H.P. and Damme, J. v. (1995a) Genetics of male sterility in gynodioecious *Plantago coronopus* . I. Cytoplasmic variation. *Genetics* **139**: 1749–1758.

Koelewijn, H.P. and Damme, J. v. (1995b) Genetics of male sterility in gynodioecious *Plantago coronopus*. II. Nuclear genetic variation. *Genetics* **139**: 1759–1775.
Kohn, J. (1988) Why be female? *Nature* **335**: 431–433.
Lahn, B.T. and Page, D.C. (1997) Functional coherence of the human Y chromosome. *Science* **278**: 675–680.
Lepart, J. and Dommée, B. (1992) Is *Phillyrea angustifolia* L. (Oleaceae) an androdioecious species? *Bot. J. Linn. Soc.* **108**: 375–387.
Lewis, D. (1941) Male sterility in natural populations of hermaphrodite plants. *New Phytol.* **40**: 56–63.
Lewis, D. (1942) The evolution of sex in flowering plants. *Biol. Rev.* **17**: 46–67.
Lloyd, D.G. (1974) Female-predominant sex ratios in angiosperms. *Heredity* **32**: 35–44.
Lloyd, D.G. (1975a) Breeding systems in *Cotula*. III. Dioecious populations. *New Phytol.* **74**: 109–123.
Lloyd, D.G. (1975b) The transmission of genes via pollen and ovules in gynodioecy angiosperms. *Theor. Pop. Biol.* **9**: 299–316.
Lloyd, D.G. (1979) Evolution towards dioecy in heterostylous plants. *Plant Syst. Evol.* **131**: 71–80.
Lloyd, D.G. (1984) Gender allocations in outcrossing cosexual plants. In: *Perspectives on Plant Population Ecology* (eds. R. Dirzo and J. Sarukhan), Sinauer, Sunderland, MA, pp. 277–300.
Lloyd, D.G. and Bawa, K.S. (1984) Modification of the gender of seed plants in varying conditions. *Evol. Biol.* **17**: 255–338.
Lucchesi, J.C. (1978) Gene dosage compensation and the evolution of sex chromosomes. *Science* **202**: 711–716.
Lucchesi, J.C. (1993) How widespread is dosage compensation? *Sem. Dev. Biol.* **4**: 107–116.
Marin, I., Franke, A., Bashaw, G.J. and Baker, B.S. (1996) The dosage compensation in *Drosophila* is co-opted by newly evolved x-chromosomes. *Nature* **282**: 160–163.
Matsunaga, S., Kawano, S., Takano, H., Uchida, H., Sakai A. and Kuroiwa, T. (1996) Isolation and developmental expression of male reproductive organ-specific genes in a dioecious campion, *Melandrium album* (*Silene latifolia*). *Plant J.* **10**: 679–689.
Maurice, S. and Fleming, T.H. (1995) The effect of pollen limitation on plant reproductive systems and the maintenance of sexual polymorphisms. *Oikos* **74**: 55–60.
Maurice, S., Couvet, D., Charlesworth, D. and Gouyon, P.-H. (1993) The evolution of gender in hermaphrodites of gynodioecious populations: a case in which the successful gamete method fails. *Proc. Roy Soc. Lond. B* **251**: 253–261.
Maurice, S., Belhassen, E., Couvet, D. and Gouyon, P.-H., (1994) Evolution of dioecy: can nucleo-cytoplasmic interactions select for maleness. *Heredity* **73**: 346–354.
Muenchow, G.A. and Grebus, M. (1987) The evolution of dioecy from distyly: evaluation of the loss-of-pollinators hypothesis. *Am. Nature.* **133**: 149–156.
Mulcahy, D.L., Weeden, N.F., Kesseli, R. and Carroll, S.B. (1992) DNA probes for the Y-chromosome of *Silene alba*, a dioecious angiosperm. *Sex. Plant. Reprod.* **5**: 86–88.
Muller, H. J. (1964) The relation of recombination to mutational advance. *Mutat. Res.* **1**: 2–9.
Nei, M. (1987) *Molecular Evolutionary Genetics*. Columbia University Press, New York.
Nordborg, M. (1994) A model of genetic modification in gynodioecious plants. *Proc. Roy. Soc. Lond. B* **257**: 149–154.
Pannell, J. (1997) Widespread functional androdioecy in *Mercurialis annua* L. (Euphorbiaceae). *Biol. J. Linn Soc.* **61**: 95–116.
Pannell, J.R. (1998b) The evolution of androdioecy and gynodioecy in wind-pollinated plants. *J. Evol. Biol.* (in press).
Parker, J.S. (1990) Sex-chromosome and sex differentiation in flowering plants. *Chromosomes Today* **10**: 187–198.
Patterson, J.T. and Stone, W.S. (1952) *Evolution in the Genus Drosophila*. MacMillan, New York.
Pedersen, S., Simonsen, V. and Loeschcke, V. (1987) Overlap of gametophytic and sporophytic gene expression in barley. *Theor. Appl. Genet.* **75**: 200–206.

Polley, A., Seigner, E. and Ganal, M.W. (1997) Identification of sex in hop (*Humulus lupulus*) using molecular markers. *Genome* **40**: 357–361.

Réjon, C.R., Jamilena, M., Ramos, M.G. Parker, J.S. and Rejon, M.R. (1994) Cytogenetic and molecular analysis of the multiple sex-chromosome system of *Rumex acetosa*. *Heredity* **72**: 209–215.

Renner, S.S. and Ricklefs, R.E. (1995) Dioecy and its correlates in the flowering plants. *Am. J. Bot.* **82**: 596–606.

Rice, W. R. (1987) Genetic hitch-hiking and the evolution of reduced genetic activity of the Y sex chromosome. *Genetics* **116**: 161–167.

Rieseberg, L.H., Hanson, M.A. and Philbrick, C.T. (1992) Androdioecy is derived from dioecy in Datiscaceae: evidence from restriction site mapping of PCR amplified chloroplast DNA. *Syst. Bot.* **17**: 324–336.

Sakai, A.K., Wagner, W.L., Ferguson, D.M. and Herbst, D.R. (1995) Origins of dioecy in the Hawaiian flora. *Ecology* **78**: 2517–2529.

Schultz, S. (1994) Nucleo-cytoplasmic male sterility and alternative routes to dioecy. *Evolution* **48**: 1933–1945.

Scutt, C.P. and Gilmartin, P. M. (1997) High-stringency subtraction for the identification of differentially regulated cDNA clones. *BioTechniques* **23**: 468.

Scutt, C.P., Kamisugi, Y., Sakai, F. and Gilmartin, P.M. (1997) Laser isolation of chromosomes demonstrates high sequence similarity between the X and Y sex chromosomes of dioecious *Silene latifolia*. *Genome* **40**: 705–715.

Seger, J. and Eckhart, V.M. (1996) Evolution of sexual systems and sex allocation in annual plants when growth and reproduction overlap. *Proc. Roy. Soc. Lond. B* **263**: 833–841.

Siroky, J., Janousek, B., Mouras, A. and Vyskot, B. (1994) Replication patterns of sex chromosomes in Melandrium album. *Hereditas* **120**: 175–181.

Smith, B.W. (1963) The mechanism of sex determination in *Rumex hastatulus*. Genetics **48**: 1265–1288.

Smith, B.W. (1969) Evolution of sex-determining mechanisms in *Rumex*. *Chromosomes Today* **2**: 172–182.

Steinemann, M., Steinemann, S. and Lottspeich, F. (1993) How Y-chromosomes become genetically inert. *Proc. Natl Acad. Sci. USA* **90**: 5737–5741.

Stinson, J.R., Eisenberg, A.J., Willing, R.P., Pe, M.E., Hanson, D.D. and Mascarenhas, J.P. (1987) Genes expressed in the male gametophyte of flowering plants and their isolation. *Plant Physiol.* **83**: 442–447.

Sun, M. (1987) Genetics of gynodioecy in Hawaiian *Bidens* (Asteraceae). *Heredity* **59**: 327–336.

Sun, M. and Ganders, F.R. (1986) Female frequencies in gynodioecious populations correlated with selfing rates in hermaphrodites. *Am. J. Bot.* **73**: 1645–1648.

Sun, M. and Ganders, F.R. (1988) Mixed mating systems in Hawaiian *Bidens* (Asteraceae). *Evolution* **42**: 516–527.

Tanksley, S.D., Zamir, D. and Rick, C.M. (1981) Evidence for extensive overlap of sporophytic and gametophytic gene expression in *Lycopersicon esculentum*. *Science* **213**: 453–455.

Testolin, R., Cipriani, G. and Costa, G. (1995) Sex segregation ratio and gender expression in the genus *Actinidia*. *Sex Plant Reprod.* **8**: 129–132.

Traut, W. and Willhoeft, U. (1990) A jumping sex determining factor in the fly *Megaselia scalaris*. *Chromosoma (Berl.)* **99**: 407–412.

Traveset, A. (1994) Reproductive-biology of *Phillyrea angustifolia* L. (Oleaceae) and effect of galling-insects on its reproductive output. *Bot. J. Linn Soc.* **114**: 153–166.

Veuskens, J., Marie, D., Brown, S.C., Jacobs, M. and Negrutiu, I. (1995) Flow sorting of the Y sex chromosome in the dioecious plant *Melandrium album*. *Cytometry* **21**: 363–373.

Vyskot, B., Araya, A., Veuskens, J., Negrutiu, I. and Mouras, A. (1993) DNA methylation of sex chromosomes in a dioecious plant, *Melandrium album*. *Mol. Gen. Genet.* **239**: 219–224.

Ward, C.R. and Kopf, G.S. (1993) Molecular events mediating sperm activation. *Dev. Biol.* **158**: 9–34.

Webb, C.J. (1979) Breeding systems and the evolution of dioecy in New Zealand apioid Umbelliferae. *Evolution* **33**: 662–672.

Weller, S.G. and Sakai, A.K. (1991) The genetic basis of male sterility in *Schiedea* (Caryophyllaceae), an endemic Hawaiian genus. *Heredity* **67**: 265–273.

Weller, S.G., Wagner, W.L. and Sakai, A.K. (1995) A phylogenetic analysis of *Schiedia* and *Alsinidendron* (Caryophyllaceae: Alsinoideae): implications for the evolution of breeding systems. *Syst. Bot.* **20**: 315–337.

Weller, S.G., Sakai, A.K., Rankin, A.E., Golonka, A., Kutcher, B. and Ashby, K.E. (1998) Dioecy and the evolution of pollination systems in Schiedea and Alsinidendron (Caryophyllaceae: Alsinoideae) in the Hawaiian islands. *Am. J. Bot.* **85**: 1377–1388.

Westergaard, M. (1958) The mechanism of sex determination in dioecious plants. *Adv. Genet.* **9**: 217–281.

Wilby, A.S. and Parker, J.S. (1986) Continuous variation in Y-chromosome structure of *Rumex acetosa*. *Heredity* **57**: 247–254.

Wilby, A.S. and Parker, J.S. (1988) Mendelian and non-Mendelian inheritance of newly-arisen chromosome rearrangements. *Heredity* **60**: 263–268.

Wolf, D.E., Rieseberg, L.H. and Spencer, S.C. (1997) The genetic mechanism of sex determination in the androdioecious flowering plant *Datisca glomerata* (Datiscaceae). *Heredity* **78**: 190–204.

Yampolsky, C. and Yampolsky, H. (1922) Distribution of sex forms in the phanerogamic flora. *Bibliotheca Genet.* **3**: 1–62.

Ye, D., Installé, P., Ciuperescu, C., Veuskens, J., Wu, Y., Salesses, G., Jacobs, M. and Negrutiu, I. (1990) Sex determination in the dioecious *Melandrium.* I. First lessons from androgenic haploids. *Sex. Plant Reprod.* **3**: 179–186.

Zhang, Y.H., DiStilio, V.S., Rehman, F., Avery, A., Mulcahy, D.L. and Kesseli, R. (1998) Y chromosome-specific markers and the evolution of dioecy in the genus *Silene. Génome* **41**: 141–147.

Zuk, J. (1969a) Analysis of Y chromosome heterochromatin in *Rumex thyrsiflorus*. *Chromosoma* **27**: 338–353.

Zuk, J. (1969b) Autoradiographic studies in Rumex with special reference to sex chromosomes. *Chromosomes Today* **2**: 183–188.

Zuk, J. (1970) Function of Y chromosomes in *Rumex thyrsiflorus*. *Theor. Appl. Genet.* **40**: 124–129.

3

Molecular approaches to the study of sex determination in dioecious *Silene latifolia*

Charles P. Scutt, Shona E. Robertson, Malcolm E. Willis, Yasuko Kamisugi, Yi Li, Matthew R. Shenton, Rachel H. Smith, Helen Martin and Philip M. Gilmartin

1. Background and introduction

We initially chose *Silene latifolia* (white campion) as a model for the study of plant sex determination mechanisms for several reasons. Though recent studies on maize (Chapter 2), cucumber (Chapter 13) and kiwi fruit (Chapter 11) have clearly demonstrated the accessibility of sex determination mechanisms in monoecious plants, we anticipated that a strictly dioecious species would allow a greater number of approaches to the study of sex determination. The choice of suitable dioecious species is limited, as only 4% of angiosperms are dioecious (Yamplosky and Yamplosky, 1922), and the majority of these are tree species (Bawa, 1980) and therefore impractical for molecular genetic analysis. Although a number of important crop plants are dioecious, including hops, hemp, asparagus, spinach, date palm and papaya (Ainsworth *et al.*, 1998; Dellaporta and Calderonurrea, 1993; Grant *et al.*, 1994a), we chose the wild herbacious species *S. latifolia* for our studies for the practical reasons of its suitable growth habit, relatively rapid life cycle and simplicity of cultivation.

Confirmation that we had selected an ideal plant for our studies came when we discovered the long, and somewhat hidden, history of white campion in the study of sex determination. White campion has had a varied and changing nomenclature. A close relative of the red campion *S. dioica*, white campion was most recently known as *S. alba*. Prior to its classification within the genus *Silene*, it was referred to as *Lychnis album*. However, as *Melandrium album* it enjoyed the attention of numerous researchers who have provided an immense body of classical genetic data on the mechanisms that control sex in this plant. It was only when we unravelled the historical terminology of white campion that we realized that we had stumbled on a plant with a

Sex Determination in Plants, edited by C.C. Ainsworth.

well documented classical genetic history and tremendous potential for the molecular genetic analysis of sex determination mechanisms.

The work on *S. latifolia* in our laboratory started in summer 1992 with a packet of seeds from John Chambers wild flowers. However, it was not long before we realized that there were several other groups who had initiated similar studies on the same plant at around the same time. This discovery was rather a surprise given the fact that there was nearly a 50-year gap in the literature on the analysis of sex determination in white campion; the classical work of Westergaard (Westergaard, 1940; Westergaard, 1946; Westergaard, 1958) and Warmke (Warmke, 1946) having been done during the 1940s. However, such parallel and independent investigations are by no means new for this plant. Westergaard's work on the production of tetraploids (Westergaard, 1940) was done in parallel with similar work by Warmke and Blakeslee (Warmke and Blakeslee, 1940) – neither had any knowledge of the other's work. These studies lead to the characterization of hermaphrodite mutants and the identification of X and Y sex chromosomes (Westergaard, 1958) and have provided the foundation for all our recent molecular genetic analyses. It is clear from the number of publications using this plant in the past few years that there is a considerable amount of information to be gained from the resurgence in studies on white campion; and given the complex nature of the problem, unravelling the mechanisms that determine sex in this plant will require the efforts of several groups with different perspectives and approaches.

From the seminal work of Westergaard (Westergaard, 1940) and Warmke (Warmke, 1946), a clear genetic understanding of the mechanisms of sex determination in white campion was developed. With a chromosome complement of eleven autosomal pairs and a pair of sex chromosomes, it was determined that males are the heterogametic sex with an X and a Y chromosome. Females are homogametic with a pair of X chromosomes. Chromosomes from a male plant are shown in *Figure 1*; the X and Y chromosomes are larger than the autosomes. Detailed cytogenetic analyses by Westergaard (Westergaard, 1940) revealed that the Y chromosome was the larger of the sex chromosomes, and observations of meiosis revealed that the X and Y chromosomes only pair within a short pseudo-autosomal region limited to one chromosome arm. The absence of recombination between the remainder of the X and Y chromosomes provided an explanation for the maintenance of Y chromosome-mediated sex determination. Analysis of tetraploids, generated by heat shocking flower buds 28 h after pollination (Westergaard, 1940) or by treatment of seeds with colchicine (Warmke and Blakeslee, 1940) provided the evidence that the Y chromosome acted as a dominant determinant of maleness, in a manner analogous to that of mammals (Goodfellow and Lovellbadge, 1993), rather than playing a role in a dosage-dependent sex determination mechanism as found in *Drosophila* (Sanchez *et al.*, 1994). The characterization of plant species with heterogametic females such as *Fragaria* (Westergaard, 1958), similar to the WZ sex chromosome system found in birds (Bull, 1983), and the presence of dosage-dependent sex determination systems such as *Rumex* (Ainsworth *et al.*, 1998; Westergaard, 1958), as well as sex determination mechanisms that do not involve morphologically distinguishable sex chromosomes (Ainsworth *et al.*, 1998; Westergaard, 1958) and even environmental control of sex determination (Ainsworth *et al.*, 1998; Westergaard, 1958) indicates the independent origin of sex determination mechanisms in different plant species.

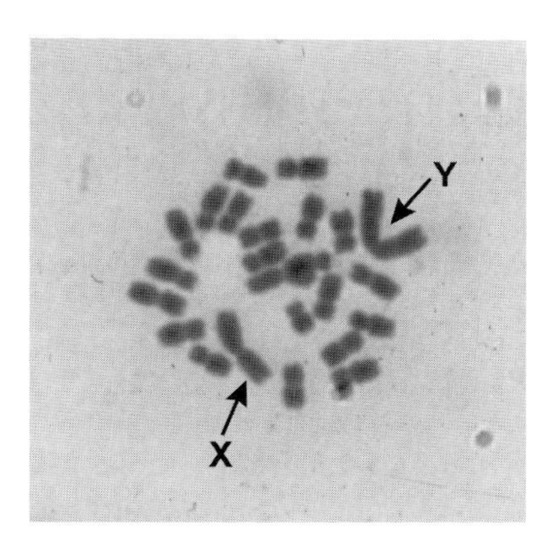

Figure 1. Silene latifolia *chromosomes. A mitotic chromosome spread from the root tip of a male seedling is shown with the X and Y sex chromosomes indicated. From Ainsworth* et al. *(1998) Sex determination in plants.* Current Topics in Developmental Biology, *vol. 38, pp. 167–223. Reprinted by permission of Academic Press, Inc.*

From the analysis of aberrant sex chromosome segregation during meiosis in triploids a number of hermaphrodite mutants were identified (Westergaard, 1940; Westergaard, 1946). Careful cytogenetic analysis of these mutants led Westergaard to propose different genetic functions for different domains of the Y chromosome. Since males have 11 autosome pairs, and an X and a Y chromosome, they carry the potential to produce both male and female reproductive structures. However, in male flowers, carpel development is arrested at an extremely early stage under the influence of the Y chromosome. In the hermaphrodite mutants, the distal arm of the Y chromosome was shortened. This argued for the presence of a locus that is involved in the suppression of female reproductive development on the Y chromosome (Westergaard, 1958). In the absence of a Y chromosome, carpels were formed, but stamens failed to develop. In combination, these studies led to the conclusion that the Y chromosome acted in a dominant manner to both suppress carpel development and promote stamen development within male flowers (Westergaard, 1958).

We (Robertson *et al.*, 1997; Scutt *et al.*, 1997b) and others (Farbos *et al.*, 1997; Grant *et al.*, 1994b) have undertaken detailed scanning electron microscopy (SEM) to investigate the structural differences between male and female flowers throughout bud development. *Figure 2* shows the key differences between male and female *S. latifolia* flower buds as revealed by SEM. Female flower buds at stage 5 (*Figure 2a*), stage 6 (*Figure 2b*) and stage 7 (*Figure 2c*), as defined previously (Farbos *et al.*, 1997; Grant *et al.*, 1994b), show the development of the five fused carpels surrounded by ten stamen primordia. By stage 9 (*Figure 2d*) the carpels have enlarged considerably but the stamens remain as primordia. *Figure 2e, f* and *g* show female flowers in between stages 9 and 10 which reveal development of the styles and absence of further stamen development. *Figure 2h* shows a female flower at late stage 10, at which point female flower development is almost complete. This developmental programme is the default pathway followed in the absence of a Y chromosome. Close examination of the stamen primordia reveals that even in the absence of a Y chromosome there is clear coordination of the development of the third floral whorl. The stamen primordia and staminal nectaries (*Figure 2i*) are clearly defined by the end of stage 9. The development of lobes in these stamen primordia is visible as is cell elongation at the base of the primordia (*Figure 2j* and *k*). Although these aspects of development are reminiscent of the formation of anthers and filaments, the arrest of development clearly occurs early in the process of stamen formation. As the female flowers mature, the stamen primordia undergo a process of degeneration that is evident from as early as stage 10 (*Figure 2h*) as shown in *Figure 2l* and *m*. This process of stamenoid degeneration has been reported previously (Grant *et al.*, 1994b).

During male flower development, the Y chromosome influences two aspects of organogenesis. *Figure 2n* shows a late stage 5 (Farbos *et al.*, 1997; Grant *et al.*, 1994b) male flower in which stamen primordia are visible surrounding a central dome. By stage 6 (*Figure 2o*) the reproductive organ primordia are clearly distinguishable. However, by stage 9 the stamens clearly consist of filaments and lobed anthers, whereas the central carpel tissue is arrested into a thin thread-like structure. The comparison between male (*Figure 2p*) and female (*Figure 2d*) flowers at stage 9 reveals the dramatic differences that arise in response to the presence of a Y chromosome. The arrested pistil in male flowers fails to undergo any significant development from stage 9 (*Figure 2q*) to stage 11 (*Figure 2r*). However, the surrounding third whorl of stamens and staminal nectaries continues to develop as a ring around the central fourth whorl (*Figure 2r*).

An additional striking feature of the biology of sex determination in white campion comes from the ability of the smut fungus, *Ustilago violacea*, to trigger stamen development in a genetically female plant (Antonovics and Alexander, 1992). A female flower infected with *U. violacea* is shown in *Figure* 3 in comparison to a female and male flower. The presence of stamens within the genetically female flower is clearly the most significant effect of the fungal infection. However, the carpels, although well developed, are smaller and mature more slowly. This is reminiscent of floral development in hermaphrodite species of *Lychnis* which are androecious and in which carpel development lags behind stamen development. The ability of the fungus to mimic the effect of one domain of the Y chromosome and stimulate stamen development indicates that the majority of genes required for male organogenesis are either autosomal or located on the X chromosome. This observation strongly suggests that the role of the Y chromosome in stamen development is regulatory rather

---

Figure 2. *Scanning electron micrographs of female and male* S. latifolia *flower development. (a) Stage 5 female flower bud with carpels removed showing petal primordia, and the ten stamen primordia surrounding the five fused carpels; bar = 100 μm. (b) Stage 6 female flower in which the five fused carpels are evident; bar = 100 μm. (c) Stage 7 female flower with sepals and petals removed. The stamen primordia are well defined and the developing carpels clearly visible. (d) Stage 9 female flower with sepals and petals removed; bar = 200 μm. (e) Female flower bud between stages 9 and 10; bar = 200 μm. (f) Female flower bud between stages 9 and 10; bar = 200 μm. (g) Female flower bud at stage 10; bar = 200 μm. (h) Female flower beyond stage 10; bar = 500 μm. (i) Stamen primordia and staminal nectaries at stage 9; bar = 50 μm. (j) Arrested stamen of a stage 9 female flower showing anther lobes; bar = 20 μm. (k) Arrested stamen of a stage 9 female flower showing anther lobes; bar = 20 μm. (l) Degeneration of stamen primordia in a stage 10 female flower; bar = 50 μm. (m) Degeneration of stamen primordia in a stage 10 female flower; bar = 20 μm. (n) Male flower at stage 5 with sepals and petals removed; bar =100 μm. (o) Male flower at stage 6 with sepals and petals removed; bar = 100 μm. (p) Male flower at stage 9 with sepals and petals removed; bar = 200 μm. (q) Arrested pistil in a stage 9 male flower; bar = 50 μm. (r) Arrested pistil in a stage 11 male flower; bar = 20 μm. Parts (f) and (r) from Robertson* et al., *(1997) Spatial expression dynamics of Men-9 delineate the third floral whorl in male and female flowers of dioecious* Silene latifolia. The Plant Journal, *vol. 12, pp. 155–168. Reprinted by permission of Blackwell Science Ltd. Part (o) from Ainsworth* et al. *(1998) Sex determination in plants.* Current Topics in Developmental Biology, *vol. 38, pp. 167–223. Reprinted by permission of Academic Press, Inc.*

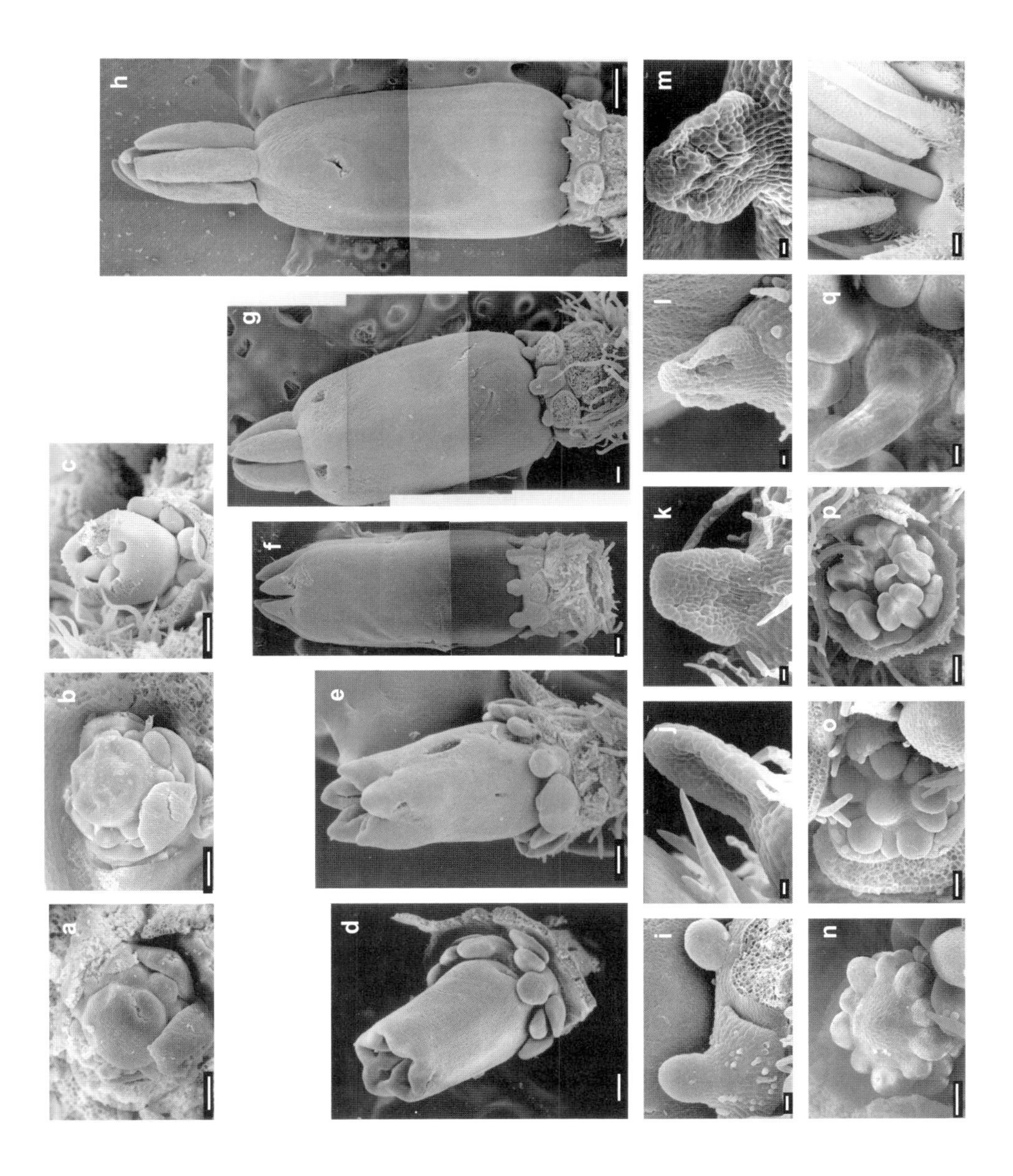
a
b
c
d
e
f
g
h
i
j
k
l
m
n
o
p
q
r

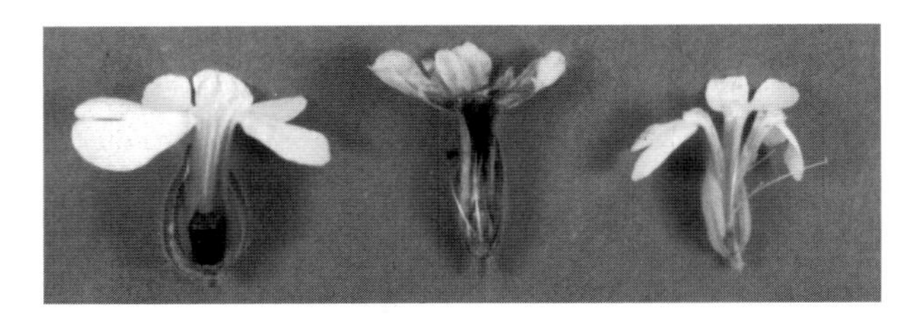

Figure 3. Silene latifolia *flowers. A female flower (left) and male flower (right), are shown for comparison with a female flower infected with* Ustilago violacea *(centre).*

than primarily encoding structural information. The stamens induced by *U. violacea* are normal in appearance, but do not produce pollen, and anther dehiscence releases fungal spores rather than pollen grains (Antonovics and Alexander, 1992). This phenomenon suggests that either the smut fungus triggers activation of all the genes required for male fertility, or that the process of infection in some way impairs the ability to produce viable pollen. The extraordinary biology of the interaction between *U. violacea* and *S. latifolia* has proved to be a valuable tool in the characterization of genes expressed during male flower development, as described below. Further characterization of the plant–fungal interaction that leads to initiation of stamen development may provide some valuable clues to the identity of the Y chromosome signal, from the identification and analysis of its fungal mimic.

One final extraordinary aspect of reproductive development in *S. latifolia* is the phenomenon of certation, initially described by Correns (Correns, 1917; Correns, 1921). Although segregation of the sex chromosome at meiosis is predicted to generate male and female progeny in equal numbers, there are clear examples of female bias within both natural and laboratory populations of white campion. Correns observed that the female bias was observed following pollination with an excess of pollen, rather than a universal event; under conditions of limiting pollen, equal numbers of male and female progeny were obtained. This observation led to the suggestion that pollen containing an X chromosome out-competes pollen containing a Y chromosome when the number of pollen grains exceeds the number of ovules available for fertilization (Correns, 1917; Correns, 1921). However, more recent analyses of sex ratios in *S. latifolia* indicate that the observed female bias may not be due to the levels of pollen and that variations between males and sex-linked mortality may play a role in this phenomenon (Carroll and Mulcahy, 1993). This explanation is in accord with the slight excess of female offspring in mammals through the hemizygous nature of the X chromosome in males and the exposure of X-linked recessive lethals.

Certation effects were avoided by Westergaard during his work on the generation of tetraploids by pollinating the female flowers at the base of the stigma to minimize the distance that the X- and Y-containing pollen tubes must grow in order to reach the ovules (Westergaard, 1940). However, given more recent work (Carroll and Mulcahy, 1993), this may have been an unnecessary and ineffective precaution. In combination, the reproductive biology of white campion provides a vast array of different and exciting avenues for molecular and genetic analysis. We have taken a broad number of approaches to investigate the mechanisms of sex determination in this plant which fall into three categories: genomic approaches, gene expression studies, and classical genetic analyses.

2. Genomic approaches to studying sex determination

One of the key advantages of *S. latifolia* as a model for the study of sex determination is the physical separation on the Y chromosome of the male-determining genes that mediate stamen development and carpel arrest, respectively. Although the majority of genes, located on either the autosomes or X chromosome are common to both males and females, male plants must posses unique Y chromosome DNA sequences. The presence of the genes controlling the developmental fate of stamen and carpel primordia on the Y chromosome, and therefore their absence from female plants, provides several possible routes for the identification and isolation of these genes. Different research groups have followed different strategies. However, as with the identification and analysis of differentially regulated genes (see below), the obligate out-crossing nature of *S. latifolia* provides plant populations with multiple allelic differences. This additional level of genomic complexity must be considered in the application of techniques to identify Y chromosome sequences as well as in their characterization.

The PCR-based strategy of random amplified polymorphic DNA (RAPD) sequence analysis has been used to identify Y chromosome-specific DNA sequences from *S. latifolia* (Mulcahy *et al.*, 1992), and these studies, which identified four male-specific polymorphic PCR products from amplifications with 60 different 10-mer arbitrary primers have recently been extended to identify additional markers (Zhang *et al.*, 1998). In these latest studies, additional RAPD markers have been used to develop sequence characterized amplified region (SCAR) markers, using longer PCR primers for the polymorphic bands, that permit reliable identification of male-specific DNA polymorphisms. The highly polymorphic nature of *S. latifolia* populations was overcome by using DNA pooled from male siblings and from pooled female siblings. This is a particularly powerful approach for identifying Y chromosome-derived sequences in the absence of any prior knowledge of Y chromosome DNA sequences. The use of RAPDs, and identification of amplified fragment length polymorphism (AFLP) markers (see Future prospects, p. 67), will prove to be a valuable and powerful approach to the physical mapping of mutations within genes located on the Y chromosome. However, given that the majority of sequences identified by either RAPD or AFLP analysis are short sequences, and that despite its large size the Y chromosome probably has a relatively small number of functional genes, the likelihood of identifying key Y chromosome sex determination genes by these approaches is small.

A further approach for the identification of Y chromosome-specific DNA sequences is genomic subtraction (Straus and Ausubel, 1990), although given the size of the Y chromosome and the limited number of Y chromosome genes involved in sex determination, the probability of identifying such key regulatory genes by this approach again seems unlikely. However, a related approach, representational difference analysis (RDA) has been used (Donnison *et al.*, 1996; Chapter 4) to identify Y chromosome-specific DNA sequences that have proved useful in mapping Y chromosome deletion mutants.

A more direct approach for the identification of Y chromosome-derived sequences is chromosome flow sorting which has been used for gender determination (Dolezel and Gohde, 1995) as well as for the physical isolation of Y chromosome DNA sequences (Veuskens *et al.*, 1995). Another approach is chromosome micro-dissection which we

(Scutt *et al.*, 1997a) and others (Buzek *et al.*, 1997) have used to directly isolate sex chromosome-derived DNA sequences. These two techniques have the advantage that the Y chromosomes are physically separated from the X chromosomes and autosomes and do not rely on hybridization to provide Y chromosome-specific DNA sequences; a potential disadvantage is that the physically isolated Y chromosomes are large and share considerable sequence identity to both X chromosomes and autosomes (see below). However, both these approaches are likely to be useful techniques for identifying Y chromosome-specific genes, either through the use of total Y chromosome DNA as a probe to screen cDNA and genomic libraries, or through the provision of additional markers for the physical mapping of Y chromosome genes.

In order to identify Y chromosome-specific sequences we have, in collaboration with Professor Fukumi Sakai (formerly at the National Institute of Agrobiological Resources, Tsukuba, Japan and currently at the Wood Research Institute, Kyoto, Japan), used argon ion laser beam chromosome micro-dissection (Scutt *et al.*, 1997a). This approach involved the preparation of male and female mitotic chromosome preparations from root tips of *S. latifolia* seedlings on polyester membranes. These chromosomes were then visualized using an inverted microscope of an ACAS 470 Cell Workstation (Meridian Instruments) designed for laser ablation studies. Metaphase preparations with well separated chromosomes were selected and the autosomes ablated by argon ion laser irradiation. The fully automated system permitted accurate control of the laser position through movement of the automated microscope stage via a mouse. This system enabled the accurate ablation of autosomes and X chromosomes from male chromosome spreads and autosomes from female chromosome spreads. The remaining sex chromosomes were recovered by increasing the laser power and cutting an octagonal disc of polyester membrane from around the remaining chromosome(s). This 0.5 mm disc was recovered with forceps and pools of twenty isolated chromosomes used for degenerate oligonucleotide primer-PCR (DOP-PCR) to amplify chromosome-specific DNA sequences from isolated X or Y sex chromosomes (Scutt *et al.*, 1997a). Following the physical isolation of both X and Y sex chromosomes, we used these chromosomes as templates to generate chromosome-specific amplification products. We have used these sex chromosome-derived DNA probes in several ways with the objectives of identifying expressed genes located on the Y chromosome, and investigating the physical relationship between the X and Y chromosomes.

Following the amplification of X chromosome and Y chromosome sequences by DOP-PCR, we initially used the Y chromosome probes to screen 500 000 phage plaques from an early male flower bud cDNA library. This cDNA library was prepared from premeiotic male flower buds and we reasoned that this stage of flower bud would be the most likely source of cDNA sequences representing Y chromosome genes involved in the control of sex determination. From this library screen we identified seven positive hybridization signals. Further analysis of the encoded cDNA clones revealed that six were similar and one was unique. The unique sequence showed no homology to any database accessions. However, the more abundant sequence showed similarity to chorismate synthase, an enzyme involved in aromatic amino acid biosynthesis, from *Corydalis sempervirens* (EMBL accession number X63595). Surprisingly the database search did not identify similarity to chorismate synthase from other species. Subsequent database searches indicated that the *C. sempervirens* chorismate synthase accession initially identified does not show similarity to other chorismate synthase sequences. Therefore, the function of both cDNA sequences isolated by screening with Y chromosome-derived probes remains unknown. Genomic Southern analysis with these two sequences also revealed hybridization to both male and female genomic DNA. We interpret this observation as indicative either that these two sequences are encoded by the homologous pseudo-autosomal region of the Y chromosome, and are therefore present on both X and Y sex chromosomes, or that these genes are located on the Y chromosome but also represented by closely related sequences on the X chromosome or autosomes. A further precedent for this last possibility is discussed below in relation to the genomic organization of the male-specific cDNA *Men-9.*

With the objective of isolating Y chromosome-derived sequences with homology to male-specific cDNA, we undertook a parallel series of experiments. Biotinylated male flower cDNA was annealed to the PCR-derived Y chromosome DNA fragments, and following hybridization the hybrids were selected using streptavidin-coated magnetic beads. The selected Y chromosome-derived genomic DNA fragments were subsequently reamplified and used as a probe on blots of both genomic DNA and total RNA. Surprisingly, we observed that the resulting subset of Y chromosome sequences hybridized to both male and female genomic DNA repeat sequences with similar patterns, and that hybridization to total RNA from male and female flowers revealed strong hybridization to both samples. This indicated not only that the Y chromosome probe contained highly repetitive sequences common to both male and female genomes, but that these sequences were expressed in both male and female flowers. This observation suggested that the Y chromosome probe could not readily be used to identify unique sequences since it contained repetitive expressed sequences present in both male and female genomes.

Hybridization of the Y chromosome-derived probe and the X chromosome-derived probe to male and female genomic DNA digested with a range of restriction enzymes revealed that both probes identified repetitive sequences with male and female genomic DNA. Hybridization of either the X chromosome probe or the Y chromosome probe to male and female genomic DNA yielded the same hybridization patterns, irrespective of sex, indicating that the sex chromosome-derived probes contain sequences represented in both male and female genomes. Since the Y chromosome is unique to male plants, this observation indicates that repetitive sequences located on the Y chromosome are also present on either the X chromosome or the autosomes, or both. It was not surprising that the X chromosome probe hybridized to both male and female DNA with similar patterns, since the X chromosome is present in both sexes. Interestingly, the hybridization signals obtained with the X and Y chromosome-derived probes differ from each other. This observation suggests that chromosome dissection, and subsequent PCR amplification of genomic DNA, resulted in amplification of different target sequences from the two chromosomes.

The isolation and characterization of repeat sequences from each of the sex chromosome-derived probes enabled us to directly address the genomic organization of these sequences using hybridization to genomic Southern blots as well as fluorescent *in situ* hybridization. These studies confirmed that, for the sequences analysed, there were no differences in abundance between male and female individuals, and that both X and Y chromosome-derived repeat sequences are present not only on each of the sex chromosomes, but also on the autosomes. Although these studies did not lead to the identification of any unique sex chromosome-specific sequences, they did reveal a likely common origin of the X and Y chromosomes and highlight the extent to which these sex chromosomes share DNA sequence homology, not only with each other, but also with the autosomes. An interesting final observation on this finding is that despite the extensive similarity between the X and Y chromosomes throughout their length, as highlighted by fluorescent *in situ* hybridization with Y chromosome-derived probes, the X and Y chromosomes fail to form chiasmata during meiosis other than over the short pseudo-autosomal region (Westergaard, 1940). Given this cytogenetic evidence, it is particularly surprising that the chromosome micro-dissection studies and genomic *in situ* hybridization (GISH) studies with total male genomic DNA as a probe in the presence of an excess of female genomic DNA as competitor did not reveal any differential hybridization to the Y chromosome (Scutt *et al.*, 1997a).

3. Gene expression studies during dioecious flower development

The availability of separate male and female flowers from white campion provide an ideal starting point for the identification of cDNA sequences that are expressed specifically in one sex only. Clearly, the final objective of such studies is the isolation of key sex determination genes located on the Y chromosome. However, such approaches will also identify cDNA sequences derived from genes which are expressed as a consequence of sex rather than which control it. Various approaches to cDNA subtraction have been employed by several different research groups (Hinnisdaels *et al.*, 1997; Matsunaga *et al.*, 1996; Scutt *et al.*, 1997b), who have all focused on the identification and characterization of male-specific cDNA sequences. It is interesting that very little attention has been focused on the female-specific genes, particularly given the fact that carpel arrest in male flowers must involve the suppression of key early carpel development genes.

Our approach to the identification of early male-specific genes has primarily involved the construction and screening of subtracted libraries. In order to maximize the efficiency of these cDNA subtractions we developed the technique of high stringency subtraction in which tissue-specific genes with constitutive homologues are not removed along with constitutive sequences during the subtraction process (Scutt and Gilmartin, 1997) in order to minimize contamination of the subtracted library with homologous sequences common to both males and females. Given the number of multigene families encoding related, yet functionally distinct proteins, we considered that this was a necessary precaution to avoid subtracting key sequences that could possibly be related to constitutively expressed sequences. Libraries were constructed using cDNA prepared from male and female flower buds up to stage 9 (*Figure 2*) and subtraction undertaken to yield a male-enriched cDNA library. This subtracted library was subsequently screened differentially using first strand cDNA from male and female flower buds. Ten cDNA clones from this library have been characterized in detail. We have prefixed these clones with the acronym *Men* to indicate the *m*ale-*en*hanced nature of their expression dynamics.

The first eight of these sequences, *Men-1* to *Men-8* (Scutt *et al.*, 1997b) and *Men-10* (Scutt and Gilmartin, 1998) are expressed exclusively in male but not female flowers. *Men-9* (Robertson *et al.*, 1997) is also expressed at a lower level in female flowers and will be discussed in more detail below. Following the isolation of these male-enhanced sequences, we undertook a detailed analysis of their expression dynamics by Northern analysis of RNA from male and female flowers as well as from female flowers infected with the smut fungus, *U. violacea.* In addition, we also investigated the temporal expression dynamic of all ten clones at different stages of male flower development and further characterized the spatial expression dynamics of selected clones by *in situ* hybridization.

Of the ten characterized sequences, all are expressed in the earliest stages of male flower buds that we have examined by Northern analysis. The lower size limit for such analyses are buds between those shown in *Figure 2o* and *2p* due to the small size of the buds and the quantity of RNA required for Northern analysis. However, these studies demonstrate that the ten *Men* clones are expressed at the earliest stages of male flower bud development. Northern analysis of *Men* gene expression dynamics throughout male flower bud development shows that these genes fall into two categories with respect to temporal expression. Expression of four of the *Men* genes, *Men-1*, *Men-4*, *Men-6* (Scutt *et al.*, 1997b) and *Men-10* (Scutt and Gilmartin, 1998) is maintained throughout male flower development. However, expression of the remaining six *Men* genes peaks in 1–2 mm flower buds and dramatically decreases as the male flowers mature, as assayed by Northern blot analysis of total RNA. However, it is important to consider that these assays are based upon equal loadings of RNA and that the relative contribution of stamen RNA to the total flower RNA is reduced as the sepals and petals develop. The apparent dramatic decrease in expression of some *Men* genes as monitored by Northern analysis may exaggerate the relative expression differences between young and mature flower buds. However, these analyses have confirmed that we have identified a group of male-enhanced cDNA sequences that are expressed at the early stages of male flower bud development that provide valuable molecular markers for analysis of the differentiation of male flowers.

The major difference between male and female flower buds is the differential presence of stamens and carpels in male and female flowers, respectively. We therefore anticipated that, as male-enhanced sequences, these cDNA clones would represent genes that are expressed in the third floral whorl during stamen development. However, one clone, *Men-6*, is not only expressed in developing stamens, but is also expressed in male, but not female petals (Scutt *et al.*, 1997b). This sequence therefore represents the first example of a male-specific gene expressed in a non-reproductive tissue of a dioecious plant.

One further aspect of the biology of white campion that we have exploited in our analyses of male-specific gene expression is the induction of stamen development in female flowers infected with *U. violacea* (*Figure* 3). Analysis of *Men* gene expression in anthers induced by *U. violacea* rather than by the Y chromosome has permitted a further distinction to be made between the different *Men* genes. *Men-1, Men-4, Men-6* (Scutt *et al.*, 1997b) and *Men-9* genes (Robertson *et al.*, 1997) are expressed in stamens from male flowers and from female flowers infected with *U. violacea*. The remaining six *Men* genes are not expressed in smut fungus-induced stamens (Scutt *et al.*, 1997b; Scutt and Gilmartin, 1998). This observation is represented in *Figure 4* where total RNA from female, smut-infected female, and male flowers has been probed with *Men-5* (*Figure 4a*) and *Men-4* (*Figure 4b*). This differential expression provides a molecular distinction between stamens induced in male flowers by the Y chromosome, and those induced in female flowers in response to the as yet unidentified signal stimulus provided by *U. violacea.*

There are two possible explanations for the differences in expression dynamics of the different *Men* genes in the two forms of stamens. The first possibility is that *Men* genes that are not expressed in *U. violacea*-induced stamens act downstream of the point of influence of the fungus and represent the key Y chromosome sex determination genes

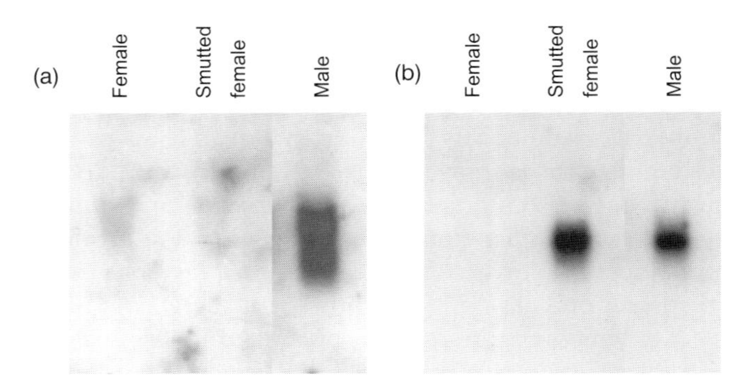

Figure 4. *Northern blot analysis of* Men *gene expression. Total RNA (10μg) from female flower buds, smut fungus-infected female flower buds, and male flower buds was probed with radiolabelled* Men-5 *(a) and* Men-4 *(b) cDNA probes.*

involved in the establishment of sex determination. The second possibility is that those sequences not expressed in *U. violacea*-induced stamens are usually expressed in cells and tissues that are beyond the scope of influence of the smut fungus. The latter possibility could reflect structural differences in the Y chromosome- and smut fungus-induced stamens. Smut-infected female anthers have little or no tapetum and no pollen is formed (Antonovics and Alexander, 1992). Those *Men* genes expressed exclusively in Y chromosome-induced anthers may therefore represent genes expressed at the later stages of sex determination. However, it is interesting to note that there appears to be an inverse correlation between temporal expression and expression in smut-induced anthers. Of the four *Men* genes that are induced by *U. violacea*, three (*Men-1*, *Men-4* and *Men-6*) (Scutt *et al.*, 1997b) are the same three genes that are expressed through all stages of male flower development including mature flowers. Of the *Men* sequences not induced by the smut fungus, *Men-2*, *Men-3*, *Men-5*, *Men-7*, *Men-8* (Scutt *et al.*, 1997b) and *Men-10* (Scutt and Gilmartin, 1998) are expressed primarily at the earliest stages of male flower development with expression reduced dramatically in buds over 2 mm.

Clearly the *Men* genes that respond to *U. violacea* in genetically female flowers cannot be Y chromosome-encoded genes. However, in order to investigate the possibility that any of the other *Men* sequence genes are encoded by Y chromosome genes, we undertook genomic Southern analysis of male and female genomic DNA. None of the genes *Men-1* to *Men-10* (Robertson *et al.*, 1997; Scutt *et al.*, 1997b; Scutt and Gilmartin, 1998) showed consistently different hybridization profiles with male and female genomic DNA, though differences due to allelic polymorphisms were present. These observations led us to conclude that none of these sequences is encoded on the Y chromosome. However, a recent report (Guttman and Charlesworth, 1998) indicates that the *Men-9* gene is located on the X chromosome and that a degenerate copy is present on the Y chromosome (see below).

Sequence analysis of the various *Men* clones has identified similarities between some of these sequences and known genes, while others represent novel sequences. *Men-1* is particularly interesting in that the longest open reading frame within the mRNA predicts a polypeptide of only 36 amino acid residues. *Men-1* is expressed throughout male flower development and is also expressed in *U. violacea*-induced stamens. The same gene was also isolated independently by other workers and termed

MROS1 (Matsunaga *et al.*, 1997). However, this sequence was reported to be expressed in mature pollen grains. Since this gene is also expressed in anthers induced by *U. violacea* that do not produce pollen, the full cell type-specific expression dynamics of this gene remain to be resolved. However, the extremely short open reading frame encoded by *Men-1* raises the possibility that the predicted 36 amino acid peptide could play a regulatory role in male flower development, or that the RNA itself could be a riboregulator (Scutt *et al.*, 1997b).

Men-2, *Men-3*, *Men-4* (Scutt *et al.*, 1997b) and *Men-10* (Scutt and Gilmartin, 1998) all show similarities to glycine-rich proteins and may represent male-specific cell wall proteins. *Men-5*, *Men-6* (Scutt *et al.*, 1997b) and *Men-9* (Robertson *et al.*, 1997) show no homology to known proteins. *Men-7* is similar to known lipid-transfer proteins, and *Men-8* is related to a previously characterized tapetum-specific gene from *Brassica* (Scutt *et al.*, 1997b). In agreement with the observation that anthers produced in response to *U. violacea* within female flowers do not produce a tapetum (Antonovics and Alexander, 1992), the tapetum-specific gene *Men-8* is only expressed in anthers of male flowers.

The *Men-9* gene was independently isolated by three different groups (Hinnisdaels *et al.*, 1997; Matsunaga *et al.*, 1996; Robertson *et al.*, 1997) and therefore has three different designations: *Men-9*, *MROS3* and *CCLS-4*. Although none of our analyses suggested a sex chromosome location for this gene, the expression of this gene in *U. violacea*-infected anthers and the expression in the staminal nectaries of female flowers (Robertson *et al.*, 1997) as described below, indicate that the expressed copy of this gene cannot reside on the Y chromosome. It is interesting to note that genomic Southern analysis of eight males and eight females from a completely outbred population (*Figure 5*) show virtually no restriction fragment length polymorphisms for *Men-9*. This is surprising given the identification of two different alleles of *Men-9* from this same population (Robertson *et al.*, 1997). The uniformity of the *Men-9* hybridization signal between individuals is particularly striking in comparison to the highly polymorphic nature of the *Men-10* gene (Scutt and Gilmartin, 1998).

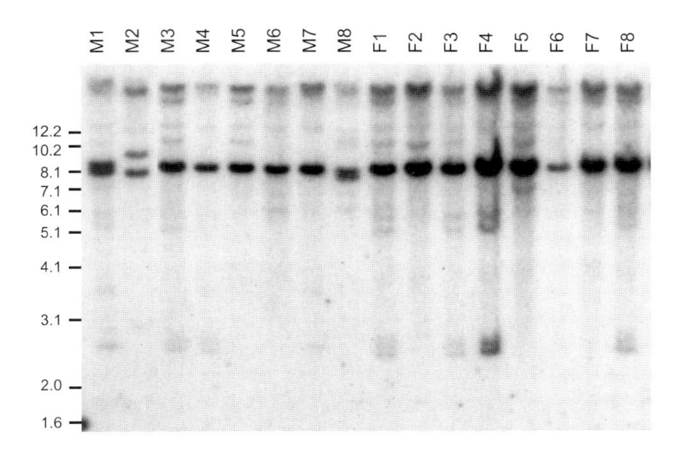

Figure 5. *Genomic Southern analysis of* Men-9. Hind*III-digested genomic DNA from eight male (M) and eight female (F) plants probed with radiolabelled* Men-9 *cDNA. Size markers in kb are indicated.*

Our analyses of *Men-9* have been the most detailed of all the *Men* genes because of its unusual expression profile revealed by Northern analysis. *Men-9* was originally isolated as a male-specific clone by cDNA subtraction (Scutt and Gilmartin, 1997). However, when used as a probe on total RNA from male, female and *U. violacea*-infected female flower buds we observed strong expression in male flower buds, induction of expression in female flower buds following infection with *U. violacea*, as well as weak expression in female flower buds. Our initial thoughts were that *Men-9* was expressed in stamens, but would also be expressed in a non-reproductive tissue common to both male and female plants. In order to investigate this possibility, we undertook detailed *in situ* hybridization analyses of *Men-9* expression (Robertson *et al.*, 1997). These data are summarized in *Figure 6*. Expression in female flowers is limited to the third floral whorl, which in the absence of a Y chromosome, is limited to development of the staminal nectaries and the arrested stamen primordia. *Figure 6a* shows expression in a single cell layer of the third whorl of a 7 mm female flower bud at stage 10 (Farbos *et al.*, 1997; Grant *et al.*, 1994b). Expression of *Men-9* delimits the third whorl and ceases at the junctions between the

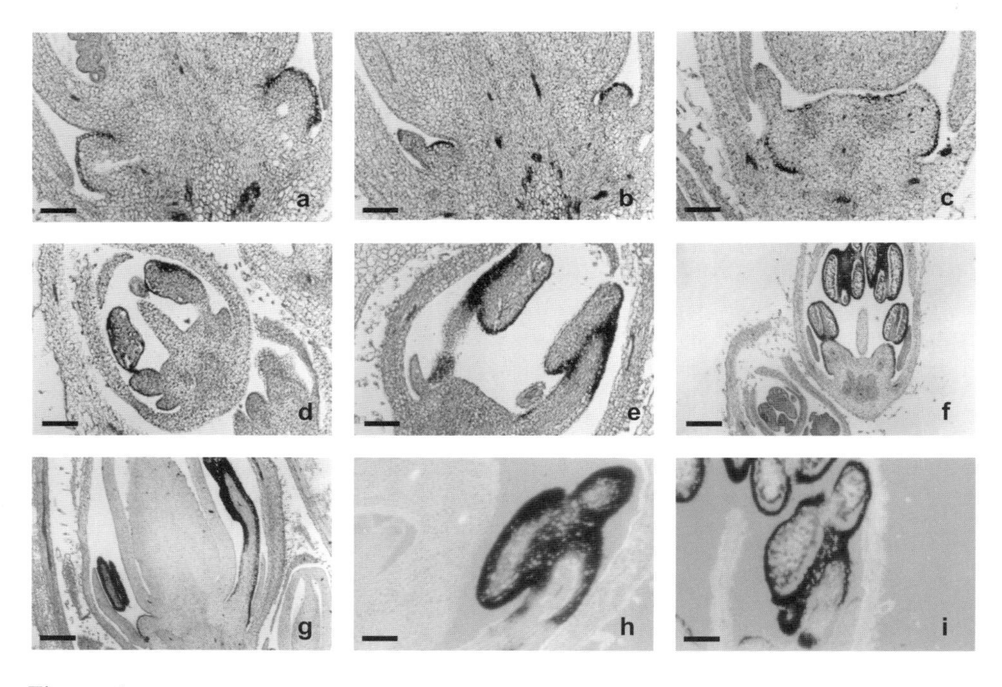

Figure 6. In situ *hybridization analysis of* Men-9. *The* Men-9 *cDNA probe was hybridized to sections of female, smut fungus-infected female, and male flower buds. (a) Female flower bud at stage 10; bar = 300 μm. (b) Female flower bud at stage 10, sectioned through arrested stamen; bar = 350 μm. (c) Female flower bud at stage 10, sectioned through the edge of the flower bud; bar = 300 μm. (d) Male flower bud at stage 8; bar = 100 μm. (e) Male flower bud at stage 9; bar = 120 μm. (f) Male flower bud at stage 10; bar = 450 μm. (g) Smut fungus-infected female flower with carpel development at stage 10; bar = 350 μm. (h) Anther of smut fungus-infected female flower; bar = 300 μm. (i) Anther of male flower; bar = 300 μm. Parts (d) and (f) from Robertson* et al., *(1997). Spatial expression dynamics of Men-9 delineate the third floral whorl in male and female flowers of dioecious* Silene latifolia. The Plant Journal, *vol. 12, pp. 155–168. Reprinted by permission of Blackwell Science Ltd.*

second and the fourth whorls. *Figure 6b* shows a section through one of the ten arrested stamens. Expression at the tip of the arrested stamen is visible as well as in the single upper cell layer of the staminal nectaries. *Figure 6c* shows a section through the edge of the flower bud in which expression over the surface of the third whorl and in the tip of the arrested stamen can be seen. Expression of *Men-9* is limited to the doughnut-shaped ring of tissue that surrounds the developed carpels.

In male flowers, the higher levels of *Men-9* transcripts are derived from expression in discrete zones within the developing anthers. *Figure 6d* shows a male flower bud at stage 8 (Farbos *et al.*, 1997; Grant *et al.*, 1994b). By stage 9 (Farbos *et al.*, 1997; Grant *et al.*, 1994b), expression has extended to the epidermis of the developing anthers and filaments (*Figure 6e*). By stage 11, expression in the endothecium and epidermis of the developing anthers has increased (*Figure 6f*). Expression is also clearly visible in the epidermis of the staminal nectaries of the third whorl up to the junction of the third whorl and the second and fourth whorls (*Figure 6f*). This latter aspect of the expression profile corresponds to the expression profile in female flowers. From our Northern analyses it was clear that *Men-9* was expressed in female flowers infected with *U. violacea. Figure 6g* shows a smut fungus-infected mature female flower with both carpels and stamens. The expression profile in these smut-induced stamens is very similar to that observed in mature male stamens, namely within the anthers and in the epidermis of the filaments. Expression in the epidermis of the staminal nectary dome is also clearly visible. For comparison, expression in the fungus-induced anthers in a female flower (*Figure 6h*) and the Y chromosome-induced anthers of a male flower (*Figure 6i*) are shown. Developing pollen within the male flower anther are clearly visible, but absent in the female flower anther.

Men-9 therefore has two domains of expression. The first is Y chromosome-independent and delimits the extent of the third floral whorl. The second domain is Y chromosome- or *U. violacea*-dependent within the endothecium and epidermis of the developing anthers. In order to further investigate the regulatory elements that control expression of *Men-9* we have isolated the corresponding genomic clone, constructed several promoter–reporter gene expression constructs, and initiated studies on the DNA–protein interactions that occur between *Men-9* promoter elements and nuclear proteins from both male and female flowers (Y. Li and P.M. Gilmartin, unpublished data). Although *Men-9* is not likely to be directly involved in the initiation of sex determination, the unusual expression dynamics of this gene provide us with a novel perspective on the patterns of male-specific gene expression in the presence and absence of the Y chromosome. It is quite remarkable that of the several thousand possible male-specific genes from *S. latifolia*, *Men-9* has been independently isolated and characterized by three groups. We have isolated two different alleles of *Men-9* (Robertson *et al.*, 1997) and a corresponding genomic clone (Y. Li and P.M. Gilmartin, unpublished data). In comparison to the analyses undertaken by other groups on allelic sequences, those on *CCLS-4* (Hinnisdaels *et al.*, 1997) show identical expression patterns to those reported by us (Robertson *et al.*, 1997). However, the expression dynamics of *MROS3* (Matsunaga *et al.*, 1996), differ in that no expression was observed in female flowers. Despite the intensive attention focused on *Men-9*, the function of this gene remains unknown and awaits development of a transformation system for *S. latifolia* which will permit homologous antisense and sense over-expression studies as well as detailed analyses of the available promoter–reporter gene constructs (see below).

A further approach that we have taken to the characterization of gene expression in male flowers has focused on the arrested pistil. In male flowers the pistil is arrested in response to a Y chromosome-derived signal (Westergaard, 1940) as shown in *Figures 2q, 2r* and *6d*. However, this Y chromosome function is not mimicked by *U. violacea* infection in female flowers (*Figure 6g*). The molecular basis for the Y chromosome-mediated arrest of carpel development remains to be determined, but the hermaphrodite mutants lacking a domain of the Y chromosome (Farbos *et al.*, 1997; Grant *et al.*, 1995; Westergaard, 1940) would indicate a Y chromosome-mediated genetic basis for carpel arrest. However, evidence from studies with 5-aza-cytidine (Janousek *et al.*, 1996) have shown that hypomethylation of DNA in males leads to the development of androhermaphrodites. The development of carpels in the presence of a Y chromosome following treatment of the plants with 5-aza-cytidine suggests an epigenetic control of carpel arrest.

In order to investigate the gene expression profile of the arrested carpels, we made a cDNA library from 2000 dissected arrested pistils (Shenton, 1996). This library was used to analyse the abundance of the nominally constitutive β-ATPase gene. Our studies on the relative abundance of this gene in the arrested pistil library compared to cDNA libraries from whole male and whole female flower buds revealed that it was more highly represented in the arrested pistil library. However, *in situ* hybridization studies showed that expression in the arrested pistil was no higher than in other floral tissues. Further studies on the expression of cDNA sequences isolated from the arrested pistil led us to conclude that the higher relative transcript abundance of constitutive clones within the arrested pistil arises through a reduced transcript complexity of this tissue (Shenton, 1996).

One further approach that we have taken to identify male-specific genes involved in the regulation of male flower development is the isolation and characterization of gene families encoding known developmental regulators, and analysis of their expression dynamics to identify sex-specific family members. Two classes of sequences that we have focused on are the glycogen synthase, or shaggy kinases previously shown to be differentially regulated during petunia flower development (Decroocq-Ferrant *et al.*, 1995), and the family of *Myb* transcription factor clones, some of which are expressed in a flower-specific manner (Jackson *et al.*, 1991). For both these approaches we used degenerate oligonucleotide primers designed to conserve domains within the particular gene families for PCR. The PCR products were subsequently used to identify longer clones either by cDNA library screening or by PCR amplification of cDNA ends. From these studies three different *Myb* homologues were isolated from male flower bud cDNA libraries. Preliminary analysis of these sequences indicates that two are flower-specific, but expressed in both male and female flowers (Robertson, 1997). From our work on the isolation of homologues of the glycogen synthase/shaggy kinases, we identified four cDNA sequences (Smith, 1997). One of these sequences is expressed at too low a level to determine the expression dynamics; the second appears to be expressed at equally low levels in both male and female flower buds and dissected floral organs, as assayed by Northern analysis. The third sequence shows highest expression in stigmas, but is also expressed at a lower level in all male and female floral organs, as assayed by RNAse protection analysis. The fourth gene is expressed in all floral organs, with highest expression in the receptacles of female flowers as determined by RNAse protection assays (Smith, 1997). In summary, our analyses of multigene families encoding defined regulatory proteins have so far failed to identify gene family members that are up-regulated during male flower development. However, on-going studies on other multigene families may lead to the identification of differentially regulated gene family members that could prove useful in our analyses of sex-specific differential gene regulation.

4. Genetic approaches

The strong classical genetic background to studies of sex determination in *S. latifolia* has been fundamental to the more recent molecular genetic studies and it is particularly appropriate to integrate a combination of classical and molecular approaches in any attempts to understand the molecular basis of sex determination in this plant. The availability of mutants provided the first insights into the control of sex determination in white campion (Warmke, 1946; Warmke and Blakeslee, 1940; Westergaard, 1940; Westergaard, 1946), and more recent studies that have recreated previously described mutant phenotypes (Farbos *et al.*, 1997; Grant *et al.*, 1995) have demonstrated the potential for integrating modern methods with classical approaches. However, to date, all mutants of *S. latifolia* that have been described are chromosome deletion mutants. Although these plants have contributed significantly to the study of sex determination, and have considerable future potential, the limitation of deletion mutants is the imprecise nature of the gene disruption. In order to identify further mutant phenotypes, we have used chemical mutagenesis of *S. latifolia* seeds. Of 40 000 seeds treated with ethane methane sulphonate (EMS), 20 000 were sown and 16 000 germinated. Given the hemizygous nature of the X and the Y sex chromosomes in male plants, mutations in genes on either of the sex chromosomes could be identified in the M1 population. This screen will also identify dominant mutations. The primary objective of this mutagenesis screen is to identify hermaphrodite and male sterile mutants that arise through defects in Y chromosome sex determination genes. However, there is a body of literature describing a number of sex-linked mutations including *angustifolia, aurea, variegated* and *abnormal* (Shull, 1914; Westergaard, 1940; Winge, 1927; Winge, 1931). These mutants, along with some of the original hermaphrodite mutants were obtained in the progeny of inter-species crosses between *S. latifolia* and *S. dioica*. Unfortunately, these mutants are no longer available but we anticipate identification of new candidates for these previously defined mutants within our population.

5. Future prospects

From the recent resurgence of interest in the mechanisms of sex determination in white campion it is apparent that the combined effort of the different research groups has contributed a significant amount of new molecular information on the structure, organization and expression of genes involved at various stages in the control of white campion flower development. However, there is still a significant way to go before we can begin to understand the molecular basis of sex determination in this plant.

One of the most critical stumbling blocks at present is the development of a transformation system for white campion. We and others have exerted a considerable amount of effort to address this problem. Although we have made significant progress towards developing such a system by optimizing the tissue culture and regeneration procedures for white campion (*Figure 7*), our attempts to transform this plant using *Agrobacterium tumefaciens*-mediated gene transfer have all failed, since white campion is not particularly susceptible to *A. tumefaciens* infection. In contrast, we have successfully transformed *S. latifolia* root cultures with a cauliflower mosaic virus 35S promoter fused to the β-glucuronidase reporter gene using *A. rhizogenes* (H. Martin and P.M. Gilmartin, unpublished data). However, we have been unable to regenerate the resulting transgenic roots into whole plants. Our current attempts to transform white

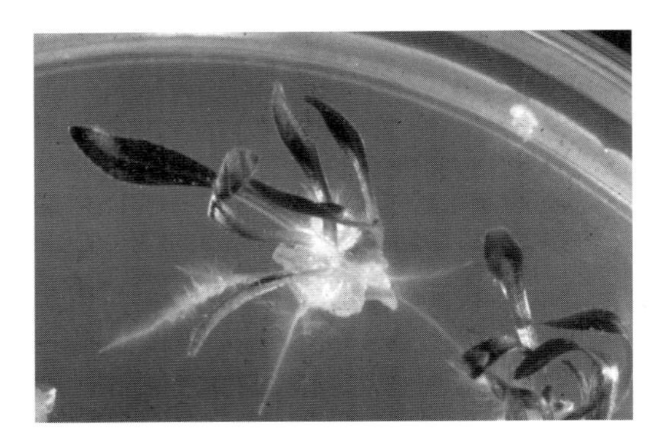

Figure 7. *Regeneration of* Silene latifolia *from tissue culture. Regenerating shoot from petiole explant with developing roots.*

campion are focused on particle bombardment using 35S-luciferase reporter gene constructs as a visible selectable marker. Development of a rapid and routine gene delivery system for *S. latifolia* is essential for progression of a number of lines of investigation.

We have extended our analyses on the expression of several different *Men* genes to the isolation and characterization of the genomic clones and corresponding promoter regions (Y. Li and P.M. Gilmartin, unpublished data). The availability of a transformation system for white campion will permit the transgenic expression of promoter–reporter gene fusions that will permit the identification of key regulatory elements required for male-specific gene expression. These studies, in combination with *in vitro* DNA–protein interaction analyses, will permit the identification and characterization of regulatory genes through which male-specific genes are regulated. A further application of transgenic technology will be the introduction of transposons that will facilitate the production of hermaphrodite and male sterile mutants through insertional inactivation giving direct access to the key Y chromosome sex determination genes. The phenotypic characterization of these mutants will be facilitated by comparison to available chromosome deletion and EMS-induced mutants. However, the availability of a DNA probe for the mutagenic transposon will facilitate direct isolation of the inactivated Y chromosome gene. The functional significance of any genes isolated in this way could be determined by reintroduction of the Y chromosome gene into female plants where the effect on flower development could be directly assayed.

An alternative approach to the identification of Y chromosome genes involved in the control of sex determination will rely on the availability of new mutants which, in combination with AFLP and microsatellite analysis, will facilitate the development of an integrated physical and genetic map of the Y chromosome. Although the amount of effort required to gain a map of sufficient resolution will be extensive, an obvious benefit from such studies will be an alternative route to the isolation and characterization of key Y chromosome genes. However, the relative ease in generating and maintaining large populations of plants, coupled to dramatic recent advances in technologies developed for human disease gene mapping, could facilitate the map-based cloning of mutant sex determination genes from *S. latifolia.*

Other recent advances in micro-array technology will also prove to be powerful tools for the analysis of sex-specific gene expression profiles. The ability to directly monitor the expression dynamics of the entire genome during flower development as well as to make direct comparisons between male and female flower gene expression profiles will greatly facilitate the global analysis of sex-specific gene expression patterns, as well as provide additional approaches for defining the chromosomal origin of expressed male sequences. As such technologies become more routinely accessible, their application to the study of sex determination in white campion will revolutionize the scale of analyses that can be undertaken. One such approach that is currently available is fluorescent differential display (FDD). We have previously attempted to use traditional differential display (Liang and Pardee, 1992) without success to identify male-specific cDNA sequences from *S. latifolia* flower buds (R.H. Smith and P.M. Gilmartin, unpublished data). After much effort we concluded that although identification of differential bands was routine, cloning and characterization of true differentials was far from routine. However FDD is a highly sensitive, routine and reliable approach for large-scale comparisons of mRNA expression profiles. In collaboration with Masaki Furuya (Hitachi Advanced Research Laboratory, Hatoyama, Japan) we have recently applied FDD technology to the identification of male-specific sequences from *S. latifolia* (C.P. Scutt, M. Furuya and P.M. Gilmartin, unpublished data). The ability to compare vast numbers of transcripts simultaneously, coupled to the relatively straight-forward isolation of true differential sequences using a few micrograms of total RNA makes this technique a powerful approach for analysis of global gene expression profiles from very small amounts of tissue. We used this technique to compare transcript profiles from male and female flower buds covering stages up to and including stage 5 of flower bud development (Farbos *et al.*, 1997; Grant *et al.*, 1994b). From comparative analysis of approximately 28 000 random PCR amplification products derived from male and female cDNA, we identified 135 differential bands, confirmed 56 as true differentials and subsequently demonstrated that these represent 25 unique transcripts. Analysis of the expression dynamics and genome organization of these 25 male-specific genes has identified two potential candidates for Y chromosome-encoded genes. Despite the success of these experiments, the true potential of such technologies remains to be fully exploited.

Silene latifolia has been studied, on and off, for nearly a century as a model for investigating sex determination in plants, and the potential for applying modern molecular genetic approaches to the problem has caused a dramatic resurgence in attention to the unique biology of this species. It is the critical mass and the synergism generated by several parallel lines of investigation, together with individual and distinctive scientific approaches of different groups that will provide us with the necessary insight to understand the molecular basis of sex determination in dioecious white campion.

Acknowledgments

We are extremely grateful to the following funding bodies for financial support for this work. CPS was supported by a postdoctoral fellowship from BBSRC and the UK Israel Research Fund; YK was supported by a postdoctoral fellowship from the HFSPO; YL was supported by a postdoctoral fellowship from the Leverhulme Trust; MRS, SER and MEW were supported by PhD studentships from BBSRC; RHS was supported by a University of Leeds PhD studentship award; HM was supported by funds from BBSRC and the University of Leeds. Additional support for our work on *Silene latifolia* came from the Royal Society and the Hitachi Advanced Research Laboratory.

References

Ainsworth, C., Parker, J. and Buchanan-Wollaston, V. (1998) Sex determination in plants. *Curr. Top. Dev. Biol.* **38**: 167–230.

Antonovics, J. and Alexander, H.M. (1992) Epidemiology of anther-smut infection of *Silene alba* (*S. latifolia*) caused by *Ustilago violacea* – patterns of spore deposition in experimental populations. *Proc. R. Soc. London, Ser. B.* **250**: 157–163.

Bawa, K.S. (1980) Evolution of dioecy in flowering plants. *Ann. Rev. Ecol. Syst.* **11**: 15–39.

Bull, J.J. (1983) Evolution of sex determining mechanisms. Benjamin/Cummings Pub. Co, London.

Buzek, J., Koutnikova, H., Houben, A., Riha, K., Janousek, B., Siroky, J., Grant, S. and Vyskot, B. (1997) Isolation and characterization of X chromosome-derived DNA sequences from a dioecious plant *Melandrium album. Chromosome Res.* **5**: 57–65.

Carroll, S.B. and Mulcahy, D.L. (1993) Progeny sex-ratios in dioecious *Silene latifolia* (Caryophyllaceae). *Am. J. Bot.* **80**: 551–556.

Correns, C. (1917) Ein Fall experimenteller Verschiebung des Geschlechtsverhaltnisses. *Sitzungsber. d. konig. Preuss. Akad. d. Wiss.* **51**: 685–717.

Correns, C. (1921) Zweite Fortsetzung der Versuche zur experimentellen Verschiebung des Geschlechtsverhaltnisses. *Sitzungsber. d. konig. Preuss. Akad. d. Wiss. Phys.-Mat. Klasse* **18**: 330–354.

Decroocq-Ferrant, V., Van Went, J., Bianchi, M.W., de Vries, S.C. and Kreis, M. (1995) *Petunia hybrida* homologues of *Shaggy/zeste-white 3* expressed in female and male reproductive organs. *Plant J.* **7**: 879–911.

Dellaporta, S.L. and Calderonurrea, A. (1993) Sex determination in flowering plants. *Plant Cell* **5**: 1241–1251.

Dolezel, J. and Gohde, W. (1995) Sex determination in dioecious plants melandrium-album and m-rubrum using high-resolution flow-cytometry. *Cytometry* **19**: 103–106.

Donnison, I.S., Siroky, J., Vyskot, B., Saedler, H. and Grant, S.R. (1996) Isolation of Y chromosome-specific sequences from *Silene latifolia* and mapping of male sex-determining genes using representational difference analysis. *Genetics* **144**: 1893–1901.

Farbos, I., Oliveira, M., Negrutiu, I. and Mouras, A. (1997) Sex organ determination and differentiation in the dioecious plant *Melandrium album* (*Silene latifolia*): a cytological and histological analysis. *Sex. Plant Reprod.* **10**: 155–167.

Goodfellow, P.N. and Lovellbadge, R. (1993) Sry and sex determination in mammals. *Ann. Rev. Genet.* **27**: 71–92.

Grant, S., Houben, A., Vyskot, B., Siroky, J., Pan, W.H., Macas, J. and Saedler, H. (1994a) Genetics of sex determination in flowering plants. *Develop. Genet.* **15**: 214–230.

Grant, S., Hunkirchen, B. and Saedler, H. (1994b) Developmental differences between male and female flowers in the dioecious plant *Silene latifolia. Plant J.* **6**: 471–480.

Grant, S.R., Hardenack, S., Ye, D., Houben, A. and Saedler, H. (1995) Differences in gene-expression between the sexes of the dioecious plant, *Silene latifolia. Develop. Biol.* **170**: 748.

Guttman, D.S. and Charlesworth, D. (1998) An X-linked gene with a degenerate Y-linked homologue in a dioecious plant. *Nature* **393**: 263–266.

Hinnisdaels, S., Lardon, A., Barbacar, N. and Negrutiu, I. (1997) A floral third whorl-specific marker gene in the dioecious species white campion is differentially expressed in mutants defective in stamen development. *Plant Mol. Biol.* **35**: 1009–1014.

Jackson, D., Culianez-Macia, F., Prescott, A.G., Roberts, K. and Martin, C. (1991) Expression patterns of *Myb* genes from *Antirrhinum* flowers. *Plant Cell* **3**: 115–125.

Janousek, B., Siroky, J. and Vyskot, B. (1996) Epigenetic control of sexual phenotype in a dioecious plant, *Melandrium album. Mol. Gen. Genet.* **250**: 483–490.

Liang, P. and Pardee, A.B. (1992) Differential display of eukaryotic messenger RNA by means of the polymerase chain reaction. *Science* **257**: 967–971.

Matsunaga, S., Kawano, S., Takano, H., Uchida, H., Sakai, A. and Kuroiwa, T. (1996) Isolation and developmental expression of male reproductive organ-specific genes in a dioecious campion, *Melandrium album* (*Silene latifolia*). *Plant J.* **10**: 679–689.

Matsunaga, S., Kawano, S. and Kuroiwa, T. (1997) MROS1, a male stamen-specific gene in the dioecious campion *Silene latifolia* is expressed in mature pollen. *Plant Cell Physiol.* **38**: 499–502.

Mulcahy, D.L., Weeden, N.F., Kesseli, R. and Carroll, S.B. (1992) DNA probes for the Y chromosome of *Silene latifolia*, a dioecious angiosperm. *Sex. Plant Reprod.* **5**: 86–88.

Robertson, S.E. (1997) Gene expression during flower development in *Silene latifolia*. PhD thesis, University of Leeds.

Robertson, S.E., Li, Y., Scutt, C.P., Willis, M.E. and Gilmartin, P.M. (1997) Spatial expression dynamics of *Men-9* delineate the third floral whorl in male and female flowers of dioecious *Silene latifolia*. *Plant J.* **12**: 155–168.

Sanchez, L., Granadino, B. and Torres, M. (1994) Sex determination in *Drosophila melanogaster* – x-linked genes involved in the initial step of sex-lethal activation. *Develop. Genet.* **15**: 251–264.

Scutt, C.P. and Gilmartin, P.M. (1997) High-stringency subtraction for the identification of differentially regulated cDNA clones. *Biotechniques* **23**: 468–471.

Scutt, C.P. and Gilmartin, P.M. (1998) The *Men-10* cDNA encodes a novel form of proline-rich protein expressed in the tapetum of dioecious *Silene latifolia*. *Sex. Plant Reprod.* **11**: 236–248.

Scutt, C.P., Kamisugi, Y., Sakai, F. and Gilmartin, P.M. (1997a) Laser isolation of plant sex chromosomes: studies on the DNA composition of the X and Y sex chromosomes of *Silene latifolia. Genome* **40**: 705–715.

Scutt, C.P., Li, Y., Robertson, S.E., Willis, M.E. and Gilmartin, P.M. (1997b) Sex determination in dioecious *Silene latifolia* – effects of the Y chromosome and the parasitic smut fungus (*Ustilago violacea*) on gene expression during flower development. *Plant Physiol.* **114**: 969–979.

Shenton, M.R. (1996) Gene expression in the developmentally arrested pistil of male *Silene latifolia* flowers. PhD thesis, University of Leeds.

Shull, G.H. (1914) Sex-limited inheritance in *Lychnis dioica*. *Zeitschr. fuer Ind. Abst. und Vererb.-lehre* **12**: 265–272.

Smith, R.H. (1997) Isolation of genes expressed during male and female flower development in dioecious *Silene latifolia*. PhD thesis, University of Leeds.

Straus, D. and Ausubel, F.M. (1990) Genomic subtraction for cloning DNA corresponding to deletion mutations. *Proc. Natl Acad. Sci. USA* **87**: 1889–1893.

Veuskens, J., Marie, D., Brown, S.C., Jacobs, M. and Negrutiu, I. (1995) Flow sorting of the *Y* sex-chromosome in the dioecious plant *Melandrium album*. *Cytometry* **21**: 363–373.

Warmke, H.E. (1946) Sex determination and sex balance in *Melandrium*. *Am. J. Bot.* **33**: 648–660.

Warmke, E.H. and Blakeslee, A.F. (1940) The establishment of a 4n dioecious race in *Melandrium*. *Am. J. Bot.* **27**: 751–762.

Westergaard, M. (1940) Studies on cytology and sex determination in polyploid forms of *Melandrium album*. *Arch. Dan. Bot.* **5**: 1–131.

Westergaard, M. (1946) Aberrant Y chromosomes and sex expression in *Melandrium album*. *Hereditas* **32**: 419–443.

Westergaard, M. (1958) The mechanism of sex determination in dioecious flowering plants. *Adv. Genet.* **9**: 217–281.

Winge, O. (1927) On a Y-linked gene in *Melandrium*. *Hereditas* **9**: 274–282.

Winge, O. (1931) X- and Y-linked inheritance in *Melandrium*. *Hereditas* **15**: 127–165.

Yamplosky, E. and Yamplosky, H. (1922) Distribution of sex forms in the phanerogamic flora. *Bibliog. Genet.* **3**: 1–62.

Zhang, Y.H., DiStilio, V.S., Rehman, F., Avery, A., Mulcahy, D. and Kesseli, R. (1998) Y chromosome specific markers and the evolution of dioecy in the genus *Silene*. *Genome* **41**: 141–147.

4

Male sex-specific DNA in *Silene latifolia* and other dioecious plant species

Iain S. Donnison and Sarah G. Grant

1. Introduction

Silene latifolia is a model species for the study of sex determination in plants. Not only is *S. latifolia* dioecious, but sex is determined by loci located on morphologically distinct chromosomes. Male plants have 22 autosomes and a pair of sex chromosomes, X and Y. Female plants have 22 autosomes and two X-chromosomes. Although most higher plant species are hermaphroditic, approximately 11% have single sex flowers and 4% are dioecious (Dellaporta and Calderon-Urrea, 1993). However, dimorphic sex chromosomes have only been shown to be sex-determining in a small number of species. Single sex flowers are considered a modern evolutionary development. For example, most dioecious species still possess rudimentary stamens in female flowers and a rudimentary gynoecium in male flowers. Furthermore, only one plant family (*Cannabidaceae*), and very few genera, are entirely dioecious and those species which are dioecious, are scattered amongst many families and genera, and have many close relatives which are hermaphroditic. The molecular mechanism by which sex chromosomes determine sex in a dioecious plant species such as *S. latifolia* is not yet understood although a number of laboratories are making progress in this area. In this chapter we describe the isolation and characterization of male sex-specific sequences from *S. latifolia* and their similarity to comparable sequences from other plant species.

2. Genetics of sex

S. latifolia is one of a small number of plant species in which the possession of polymorphic sex chromosomes is unequivocal (Parker, 1990). The genetics of sex in *S. latifolia* were well characterized in the 1940s and 1950s by Westergaard (1946, 1948, 1958) and have been discussed more recently (Dellaporta and Calderon-Urrea, 1993; Grant *et al.*, 1994). The studies of Westergaard were based on a collection of sterile and hermaphroditic mutant plants, none of which are now available, which led to the proposal of a model

Sex Determination in Plants, edited by C.C. Ainsworth.

in which the Y chromosome comprises three distinct regions and two sex-determining functions. Region one has homology to the X chromosome and is responsible for the limited degree of pairing between X and Y chromosomes during meiosis. Region 2 contains genes necessary for anther maturation which do not occur on the X chromosome or autosomes. Region 3 contains genes necessary for the suppression of female flower tissues.

3. Isolation of male-specific DNA fragments

On the assumption that differences in the DNA sequences of the X and Y chromosomes are responsible for sex determination, we previously used representational difference analysis (RDA), a genomic subtraction protocol, (Lisitsyn *et al.*, 1993), to isolate male sex-specific restriction fragments from *S. latifolia* (Donnison *et al.*, 1996). RDA uses PCR to amplify polymorphic DNA sequences that exist between two genomes. The technique is ideal for the study of a complex genome, such as that of *S. latifolia*, as complexity is reduced by the analysis of only a subset of fragments, generated by restriction endonuclease digestion, which are small enough to be amplified by PCR. In RDA experiments using *S. latifolia*, eight male-specific fragments were cloned from genomic DNA digested with *Bgl*II and one from genomic DNA digested with *Bam*HI. The male-specific restriction fragments identified by RDA, however, were also homologous to other sequences shared between male and female plants. The total number of these bands varied from few (<10) in clones Bgl10 and Bgl16, to many (>10) in clones Bam37, Bgl7, Bgl35. Thus, a part at least of each sequence must be repeated on a homologous region of the X chromosome or on the autosomes. In two other studies with *S. latifolia*, when Y chromosomes were either microdissected or isolated by selective chromosome laser ablation and Y chromosome DNA cloned, all isolated sequences were found to be repetitive and hybridized to DNA from both sexes (Grant *et al.*, 1994; Scutt *et al.*, 1997). Male sex-specific fragments isolated by a variety of techniques from other dioecious plants species including *Rumex acetosa* (Ruis Rejón *et al.*, 1994), *Cannabis sativa* (Sakamoto *et al.*, 1995), *Humulus lupulus* (Polley *et al.*, 1997) and *Atriplex garettii* (Ruas *et al.*, 1998) have also been found to hybridize to both male and female genomic DNA in Southern blots. In another study, seven non-homologous repeat sequences were isolated from *R. acetosa* and estimated to represent 18% of the genome (Clark *et al.*, 1992). When these clones were hybridized separately to chromosomes of *R. acetosa*, each clone was found to be dispersed along all chromosomes, X, Y and autosomes, with no bias (Clark *et al.*, 1992). More recently Guttman and Charlesworth (1998) have identified in *S. latifolia*, a Y chromosome locus (*MROS3*) which has degenerated compared to a homologous sequence on the X chromosome. In the degenerate Y chromosome locus there is an additional region of approximately 235 base pairs (bp) which is unique to the Y chromosome version of *MROS3*. However, primers specific to this additional region can be used to amplify, by PCR, identical bands from male and female plants indicating that this fragment also occurs on the X chromosomes or autosomes (Guttman and Charlesworth, 1998).

4. Mapping of the Y chromosome in *Silene latifolia*

Five of the RDA-derived male-specific restriction fragments isolated from *S. latifolia* have been used to map male sex-determining genes in a collection of 18 hermaphroditic and 25 asexual mutants (Donnison *et al.*, 1996). Hermaphroditic and sterile mutant plants were derived from a programme in which X-ray mutagenized pollen was used to pollinate female plants of the same genotype (Grant *et al.*, 1994). When the clones were used to probe genomic DNA isolated from hermaphroditic and sterile mutant plants, a number of mutants were identified in which up to three of the five male-specific sequences were found to be absent when the five clones were used in hybridization. In other words, it is predicted that these plants no longer contained a fragment of the Y chromosome that corresponded to the cloned DNA fragment. Bgl10 was absent from nine of the 19 hermaphroditic mutants and present in all sterile mutants, whereas Bgl7 and Bgl16 were absent from seven of the 19 hermaphroditic mutants and four of the 25 sterile mutants. The pattern of deletion in the latter two clones was identical in both sets of mutants; that is any mutant, which lacked the male-specific band of Bgl7, always lacked the male-specific band of Bgl16 (and *vice versa*). Neither the Bam37 nor the Bgl35 male-specific bands were absent from any of the mutants tested.

That two of the male-specific sequences, Bgl7 and Bgl16, can be deleted in both hermaphroditic and sterile mutants, suggests that the gynoecium suppresser gene(s) (deleted in hermaphrodites) and a stamen promoter gene (deleted in steriles) are linked on one arm of the Y chromosome. When used as a probe, Bgl10 also shared the same deletion pattern as a putative cDNA clone, isolated independently from a male-specific cDNA subtraction experiment (Ye, D., Koutnikova, H., Saedler, H. and Grant, S., unpublished). Thus, it would appear that there might be islands of difference on the Y chromosome with at least four of the clones so far isolated focused on these islands. The karyotype data for the hermaphroditic and sterile mutant plants (Donnison *et al.*, 1996; Grant *et al.*, 1994) suggest that deletions which span the Bgl7, Bgl10, Bgl16 and some unpublished cDNA markers, are more often large enough to be visible at the light microscope level. For example, three hermaphroditic mutants featured a large visible deletion of the Y chromosome and lacked these four markers whilst in another hermaphroditic mutant no visible deletion was observed yet the four markers were all absent. A mutant which had a small deletion and lacks the Bgl7-Bgl10-Bgl16-cDNA marker island would make a good candidate for future work.

5. Homology of male-specific DNA fragments with other DNAs and proteins

A number of the male-specific sequences isolated by RDA from *S. latifolia* are homologous to DNA sequences from other organisms. For example, Bam37, Bgl7 and Bgl16 are similar to reverse-transcriptase sequences from retrotransposons; Bgl10, Bgl38 and Bgl43 are similar to microsatellite sequences which contain simple sequence repeats that occur in a number of sequences in the nucleotide databases (*Table 1*). The other *S. latifolia* male-specific fragments either show no significant homology to other sequences or do so only over relatively short regions.

5.1 *Bgl16*

The sex-specific fragment Bgl16 is 57% identical to a male-associated DNA sequence MADC1 from another dioecious species, *C. sativa* (Sakamoto *et al.*, 1995). Bgl16 is one of the least repetitive of the *S. latifolia* male-specific clones and has therefore been selected for further study. Radiolabelled Bgl16 DNA has recently been used to probe

Table 1. *Comparison of male-specific* S. latifolia *sequences isolated by RDA (Donnison* et al., *1996) to sequences in the nucleotide or protein databases*

Clone and accession no.	Clone length (bp)	Homology of cloned sequence to sequence published in DNA and protein databases (accession no., species, description)	% of homology	Length of homology
Bam37 (X99864)	449	X99865, *S. latifolia*, Bgl7	62	370-bp
		O22148, *A. thaliana*, putative reverse transcriptase	48	143-aa
Bgl7 (X99865)	469	X99864, *S. latifolia*, Bam37	62	370-bp
		O22148, *A. thaliana*, putative reverse transcriptase	50	167-aa
Bgl10 (X99866)	330 <1031>	M80586, *Cylophaga xylanolytica*, 16s rRNA internal sequence	61	133-bp
		AC004617, human, Y chromosome sequence	59	130-bp
		AA824445, human, simple sequence repeat in cDNA	83	30-bp
Bgl16 (X99867)	292 <1306>	D50414, *C. sativa*, male-specific sequence MADC1	57	253-bp
		U90128, maize, ACCase gene intron containing LINE-like retrotransposon colonist2	51	369-bp
		X91204, *A. thaliana*, LINE-like retrotransposon RT4	62	238-bp
		L47186, *A. thaliana*, LINE-like retrotransposon Ta15	51	255-bp
Bgl21 (X99871)	693	AB015476, *A. thaliana*, P1 clone MNC6	57	101-bp
Bgl35 (X99868)	759	X64886, red clover mottle virus, complete RNA sequence	62	98-bp
Bgl38 (AJ010062)	320	AC005183, human, simple repeat region	63	213-bp
		L10713, pig, trinucleotide repeat	64	108-bp
Bgl43 (X99869)	378	Z27493, pig, microsatellite repeat region	54	142-bp
		AC001652, *Drosophila melanogaster* repeat	58	204-bp
		AC004414, human, repeat sequence	62	174-bp
Bgl49 (X99870)	212	Z70689, human, cosmid DNA sequence U19D8	58	130-bp
		AC005155, human, PAC clone DJ0877J02	54	177-bp

-bp, indicates identity to nucleic acid sequence; -aa, indicates similarity to amino acid sequence;
< > indicates length of extended sequence following isolation of genomic clone.
Similarities identified by BLASTN, BLASTX, FASTA and BESTFIT programmes from the GCG package (Genetics Computer Group package, University of Wisconcin, Madison).

a male *S. latifolia* genomic library in lambda FixII. A number of lambda clones containing the Bgl16 fragment polymorphism have been identified and the region either side of the original BglII site sequenced directly using internal Bgl16 primers. In comparison to the nucleotide databases a number of other sequences have been identified which share sequence motifs with the extended *S. latifolia* Bgl16 and *C. sativa* MADC1 sequences. Many of these other sequences have been identified as retrotransposons and include Colonist2 from maize (Lutz and Gengenbach, 1997) and RT4 (Knoop *et al.*, 1996) and Ta*15* from *Arabidopsis thaliana* (Wright *et al.*, 1996). Retrotransposons replicate by reverse transcription of a mRNA intermediate and re-integration of a DNA copy back into the host genome, and are normally the most abundant mobile elements in higher plants and other eukaryotes (Xiong and Eickbush, 1990; Wright *et al.*, 1996). By such replication, retrotransposons can significantly contribute to the proportion of repetitive DNA in the genome and might, therefore, be predicted to contribute to the increase in size which has occurred in the *S. latifolia* and *C. sativa* Y chromosomes relative to the X chromosomes and autosomes. Retrotransposons which are flanked by long-terminal-repeat (LTR) sequences are classified as LTR retrotransposons and are the predominant source of repetitive DNA in maize, probably making up more than 50% of the nuclear genome (SanMiguel *et al.*, 1996). The sequences of Bgl16 and MADC1, however, appear characteristic for non-LTR retrotransposons which resemble the long interspersed nuclear element (LINE)-1 in human DNA (Fanning and Singer, 1987) and similar sequences which have been identified in plants (Knoop *et al.*, 1996; Kubis *et al.*, 1998; Schwarz-Sommer *et al.*, 1987; Wright *et al.*, 1996).

The DNA homology between Bgl16, MADC1, Colonist2, Ta15, RT4 and other putative *A. thaliana* retrotransposon sequences (EMBL accession numbers: AC003673, AL022223 and AB009052) is usually 50–60%, however, in some regions the figure is significantly higher (*Figure 1* and data not shown). Moreover, for much of the comparison, all similarities appear within the same frame and indicate a higher order conservation. On translation, the extended sequence for Bgl16 is significantly similar to the protein sequence of reverse transcriptase from a number of LINE-like retrotransposons including Cin4 of maize (Schwarz-Sommer *et al.*, 1987) and LRE2 of human (Holmes *et al.*, 1994) (*Figure 2*). However, even though most, and if not all of the non-LTR reverse-transcriptase conserved amino acid sequence domains occur in Bgl16, the similarity only occurs by shifting the reading frame in two positions over the core 285 amino acids. There are also two deletions: one of 8 amino acids between the first frame shift in the core sequence and the second of 90 amino-acids which occurs immediately downstream of the core sequence and coincides with a third frame shift (data not shown). The inference is, therefore, that the original sequence has degenerated somewhat or that errors occurred during transposition.

PCR experiments using primers (5Bgl16: GATCTTGTGACACCTCCA and 3Bgl16: CCTGCAGTGTCATACCAC) designed to amplify the internal Bgl16 sequence reveal that this fragment can be amplified from both male and female plants of *S. latifolia* and *S. dioica*. When the Bgl16 PCR fragments have been sequenced, they have been found to be identical between the two sexes and between the two *Silene* species.

The Bgl16 and MADC1 sequences are both relatively low in abundance yet both appear to occur on the Y chromosome of two plant species that have developed dioecy independently. Significant numbers of LINE-like retrotransposons have been identified in *A. thaliana* (Knoop *et al.*, 1996; Wright *et al.*, 1996), some of which show homology to Bgl16 and MADC1, and although they share common motifs they are rarely similar enough to cross hybridize to each other in Southern hybridizations. The inference, therefore, is that many similar retrotransposons may also exist in *S. latifolia* and *C. sativa*, however, in most cases the *S. latifolia* and *C. sativa* sequences are sufficiently diverged so that they do not readily cross hybridize in genomic Southern hybridizations. The 730 bp MADC1 DNA fragment was digested with *Xba*I and *Xho*I to generate three smaller fragments and probed to *C. sativa* genomic DNA (Sakamoto *et al.*, 1995). The

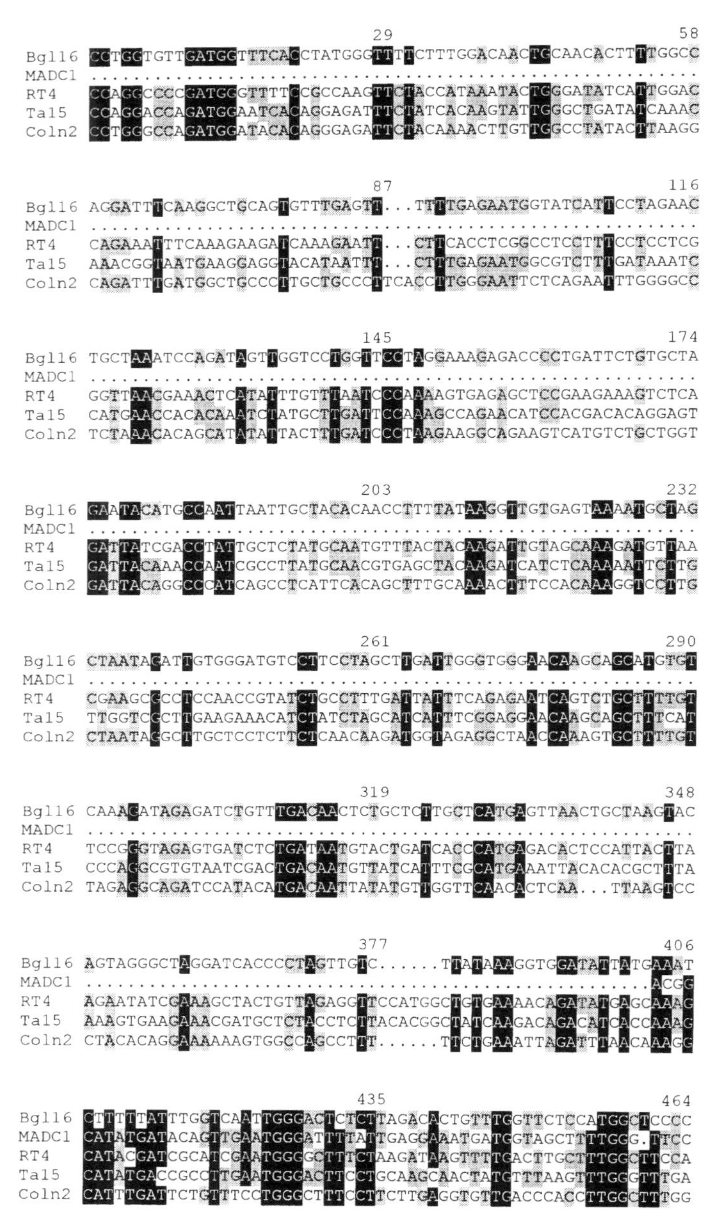

Figure 1. *Nucleic acid similarity between* S. latifolia *male-specific fragment Bgl16,* C. sativa *male-specific fragment MADC1 and LINE-like retrotransposons RT4 (*A. thaliana*), Ta15 (*A. thaliana*) and Colonist2 (*Zea mays*). Comparison covers the predicted reverse transcriptase amino acid sequence domains I–VII (Xiong and Eickbush, 1990). An additional domain preceding domain I, and conserved among LINE-like retrotransposons, is also included (Wright* et al.*, 1996). Black boxes indicate nucleotides conserved in at least four of the five sequences, grey boxes indicate nucleotides conserved in three of the five sequences; dots indicate the limits of available sequence information or gaps introduced to improve the sequence alignment. EMBL accession numbers are: Bgl16 (X99867), MADC1 (D50414), RT4 (X91204), Ta15 (L47186) and Colonist2 (U90128). Multiple sequence alignment drawn using GeneDoc (Nicholas and Nicholas, 1997).*

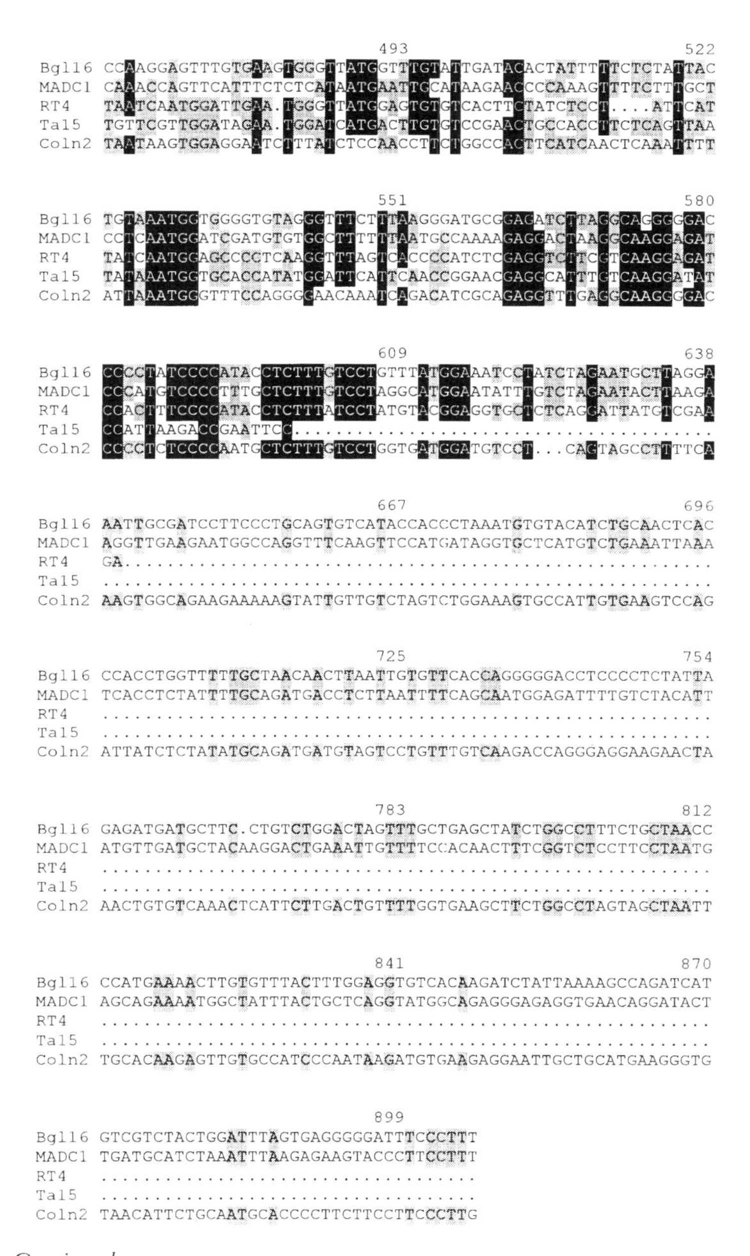

Figure 1. Continued

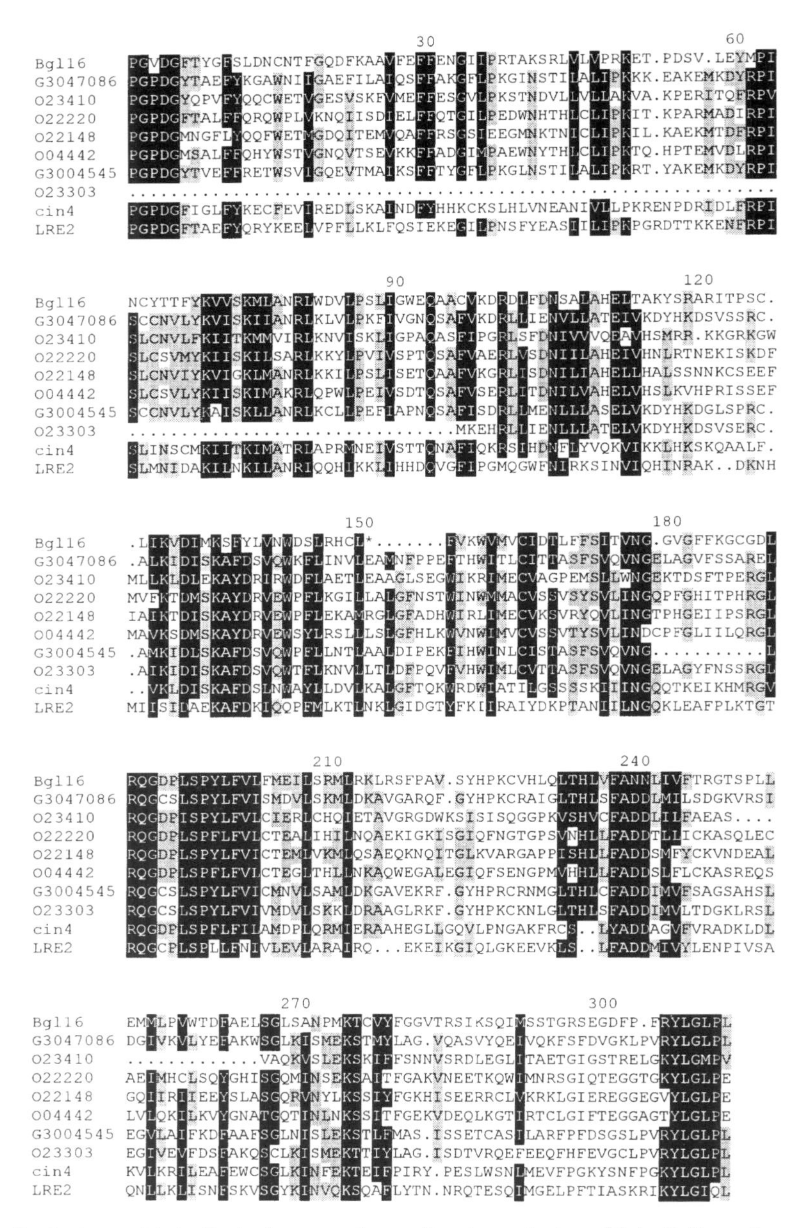

Figure 2. *Amino-acid similarity between the predicted translation of* S. latifolia *male-specific fragment Bgl16 (X99867), cin4* (Z. mays) *(Y00086), LRE2 (*Homo sapiens*) (Q12881) and seven reverse-transcriptase-like protein sequences of* A. thaliana *(labelled by EMBL accession numbers). Comparison covers the reverse transcriptase amino acid sequence domains I–VII (Xiong and Eickbush, 1990). An additional domain preceding domain I, and conserved among LINE-like retrotransposons, is also included (Wright* et al., *1996). Black boxes indicate nucleotides conserved in at least eight of the 10 sequences; grey boxes indicate nucleotides conserved in six or seven of the 10 sequences; dots indicate the limits of available sequence information or gaps introduced to improve the sequence alignment. Two frameshifts were made to the Bgl16 sequence to optimize the translation (at residues 164 and 268). Multiple sequence alignment drawn using GeneDoc (Nicholas and Nicholas, 1997).*

first 220 bp fragment, which equates to a partial fragment of the Bgl16-like reverse transcriptase (domains III-IV, recognized as regions with sequence similarity across all known reverse transcriptases; Xiong and Eickbush, 1990), hybridizes more strongly than the other fragments and in a non-sex-specific pattern. Hybridization with the third fragment of 164 bp, which equates to a region downstream of the reverse transcriptase core sequence (domain VII), resulted in fewer bands, some of which were male sex-specific (Sakamoto *et al.*, 1995). Furthermore, when MADC1 was hybridized to DNA in common between male and female plants, the bands were more intense in males suggesting that this or similar retrotransposons have indeed accumulated on the *C. sativa* Y chromosome.

5.2 *Bam37 and Bgl7*

Bam37 and Bgl7 sequences are significantly more abundant than Bgl16 in the *S. latifolia* genome and also show homology to reverse transcriptases of non-LTR retrotransposons. However, the homology to reverse transcriptases sequenced from other species is much weaker than for Bgl16 because the Bam37 and Bgl7 fragments are located ~260–410 amino acids downstream of domain VII (Xiong and Eickbush, 1990) of the reverse transcriptase core sequence, and similarities between reverse transcriptases from different elements become smaller in this region. Moreover, the similarity to other reverse transcriptases is only apparent from the deduced amino acid sequence (*Figure 3*). As for Bgl16, the similarity of Bam37 and Bgl7 to other reverse transcriptases occurs only by shifting the reading frame in two positions. However, since Bam37 is known to be highly repetitive in the *S. latifolia* genome and the similarity to other retrotransposons downstream of domain VII is much weaker, it is more likely (than for Bgl16) that a Bgl7/Bam37-like transposable element is still mobile. In addition, there does not appear to be a close relationship between Bgl16 and either Bam37 or Bgl7, as sets of primer pairs taken from the 3′ end of Bgl16 and the 5′ end of Bam37 or Bgl7, do not result in a PCR fragment of the predicted size as would be generated if the fragments were contiguous. However, given that Bgl7 and Bgl16 share an identical pattern of deletion in the hermaphroditic and sterile mutant plants (Donnison *et al.*, 1996), the two sequences are predicted to map to the same region of the Y chromosome.

PCR experiments using primers (5Bam37: GGTCAACCTCCTTCGGATCG and 3Bam37: TTGGAGGTGGCTGGTGCTTC) designed to amplify the internal Bam37 sequence reveal that this fragment can be amplified from male plants of *S. latifolia* and *S. dioica*. When the Bam37 PCR fragments have been sequenced, they have been found to be identical between the two *Silene* species.

5.3 *Bgl10, Bgl38 and Bgl43*

Bgl38 contains two simple sequence repeats (microsatellites) CAAAN and CCA identified by RepeatMasker (Smit, AFA and Green, P RepeatMasker at http://ftp.genome.washington.edu/RM/RepeatMasker.html) as repeats CAAAA and TGG known to exist in mammalian genomes. Bgl10 and Bgl43 also contain a number of CAAAA or CAA and closely related degenerate repeats.

In Bgl38, there are eight copies of repeat CAAAN including a perfect stretch of five repeats, and 20 copies of CCA, including two perfect stretches of four repeats (*Figure*

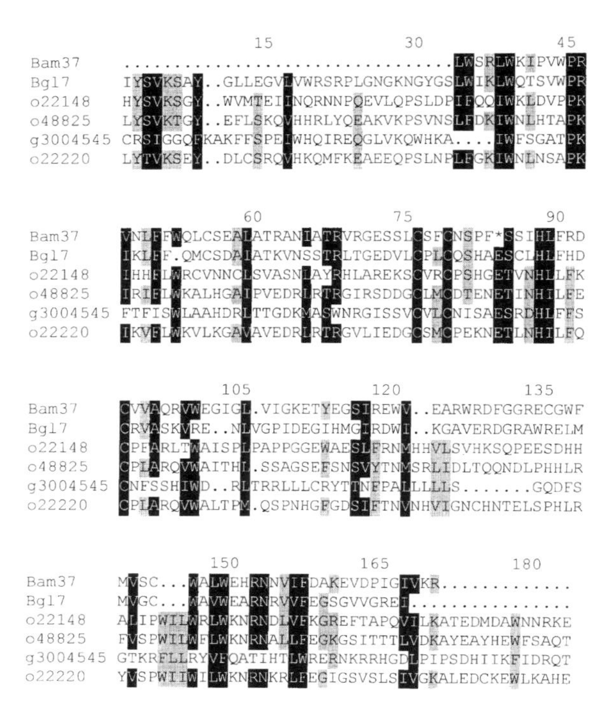

Figure 3. Amino acid similarity between the predicted translation of S. latifolia *male-specific fragments Bam37 and Bgl7, and four reverse-transcriptase-like protein sequences of* A. thaliana *(labelled by EMBL accession numbers). Comparison covers the reverse transcriptase amino acid sequence ~260–410 amino acids downstream of domain VII (Xiong and Eickbush, 1990). Black boxes indicate nucleotides conserved in at least five of the six sequences; grey boxes indicate nucleotides conserved in four of the six sequences; and dots indicate the limits of available sequence information or gaps introduced to improve the sequence alignment. Two frameshifts were made to the Bam37 and Bgl7 sequences to optimize the translation (at residues 17 and 106, and 34 and 135, respectively). Multiple sequence alignment drawn using GeneDoc (Nicholas and Nicholas, 1997).*

4). The CCA repeats are primarily responsible for the DNA similarity over 100 bp between Bgl38 and human and pig sequences (*Table 1*). In Bgl43 there are 21 perfect and 29 degenerate copies of CAC but only two of which are perfect stretches, each consisting of two repeats (*Figure 5*). In Bgl10 there are 16 perfect and 15 degenerate copies of CAC with one perfect stretch of three repeats. The Bgl38 CAAAN repeat block identifies highly similar blocks in human (AF011889: 89% over 38 bp), *Xenopus laevis* (X01175: 81% over 48 bp), *Cebus albifrons* (primate) (M81409: 86% over 36 bp), *Drosophila melanogaster* (AL021106: 85% over 34 bp) and mouse (X53336: 88% over 34 bp). In large genomic fragments of human and other primate DNA (Af011889, M81409), the CAAAN repeat occurs within regions containing LINE-1 retrotransposons. The same simple sequence repeat also occurs on the sex determining Y chromosome of humans and mice.

Bgl10, like Bgl16, was found to be relatively non-repetitive in the *S. latifolia* genome (~ five copies) and was chosen as a probe to identify a longer genomic fragment from a *S. latifolia* genomic DNA library. A number of apparently identical clones have been

1 agatctcatt ccccgtcatc tatctcactt cttactcatt tatctctcc**C**

51 **AAAA**gaaatc aatctctttc atttatctct cc**CAAAGCAA** **AACAAATCAA**

101 **AACAAAC**caa --**CAAAA**cat ataccctccc tcccttaat caa**CCACCA**a

151 cacaccc**CCA** catcaccgtc gttctccttc gcaacccgcc g**CCACCA**ctg

201 ca**CCA**acaac aacagcgg**CC** **A**g**CCA**cac**CC** **A**taccgt**CCA** **CCACCACCA**t

251 cttcgtcgct gac**CCA**aaa**C** **C**accgtcact gac**CCACCAC** **CACCA**ccgaa

301 actcccctct tcgt**CCA**gat ct

Figure 4. *Sequence of* S. latifolia *male-specific fragment Bgl38 (AJ010062). CAAAN and CCA repeats are in capitals and bold, and neighbouring repeats which diverge by one base from either sequence are underlined.*

isolated which contain the original identified *Bgl*II restriction polymorphism specific to male plants. Approximately 350 bp have been sequenced upstream and downstream of the Bgl10 fragment; one of these additional sequences contains a degenerate stretch of four CAAAA repeats, which primarily shows homology to a number of similar repeats within cDNAs from human and other species.

Microsatellite sequences such as (CAC)n and (CAA)n have been used previously for DNA finger-printing in humans (Schäfer *et al.*, 1988), other animals (Buitkamp *et al.*, 1991; Coppieters *et al.*, 1995) and plants (Smith and Devey, 1994). Trinucleotide repeats have also been identified in transcribed sequences in *Drosophila* (Haynes *et al.*, 1987), human (Sutherland and Richards, 1995) and more recently plants (Gortnor *et al.*, 1996). Such sequences can often be of highly variable length and when their sequence occurs within a gene they have been implicated in a number of genetically determined human diseases (Fu *et al.*, 1991; Sutherland and Richards, 1995).

1 agatctat**CA** **C**tcgttgacg aacttacgat atgtcataaa gcgctccgcg

51 taccaaacat gcggcccaat cat**CAC**tagg tggttggcag aaggtgcagc

101 aatgaggtat ctacgcgtcc c**CAC**tcctaa c**CAC**gt**cCAC** catagccggc

151 catgtcagc**C** **AC**ccc**CAC**ct tatcccccgt ccgaggtgag cccatcaacg

201 ac**CACCAC**ga tggccgtaac **CAC**cagtggc aacagcaa**CA** **CCAC**ggtag**C**

251 **A**ccaacaaca gttaacgaca a**CAC**aaa**CAC** ga**CAC**cctct ttctctcccc

301 tttcccgc**CA** **C**ctacggc**CA** **C**cct**CAC**ggc **CAC**ccatgac aacccctcaa

351 ct**CAC**ctcta ttccctcaac ctagatct

Figure 5. *Sequence of* S. latifolia *male-specific fragment Bgl43 (X99869). CAC repeats are in capitals and bold, and neighbouring repeats which diverge by one base from CAC are underlined.*

6. Conclusions

Several recent studies by ourselves (Donnison *et al.*, 1996; Grant *et al.*, 1994) and others (Polley *et al.*, 1997; Ruas *et al.*, 1998; Sakamoto *et al.*, 1995), have identified sex-specific DNA fragments considered to be on the Y or sex-determining chromosomes or loci of dioecious plant species. The dramatic increases in recent DNA database submissions has meant that we are now better able to assess the characteristics of male sex-specific DNA sequences. Perhaps, and not surprisingly, many of these sequences are highly repetitive as shown by Southern analysis, and show similarity to repeat sequences from many different organisms. There are also comparatively less abundant sequences which are homologous to non-LTR retrotransposons from non-dioecious plant species, whilst another group of sequences shows no significant homology to any database submissions.

In *S. latifolia* and *C. sativa,* the evolution of dioecy has been accompanied by the development of sex chromosomes which are most likely of autosomal descent (Westergaard, 1958). However, during evolution, the sex chromosomes have become greatly enlarged compared to any autosome ancestor. An increase in the proportion of repetitive DNA is the most likely explanation for this. Since the evolution of dioecy in these plants is a relatively modern evolutionary phenomenon, it is probably unlikely that sex-specific repeat sequences have yet evolved to the extent to which they exist in animal systems (Devlin *et al.*, 1998; Eicher *et al.*, 1989). Moreover, the X and Y chromosomes, of several animal species contain disproportionately high numbers of retrotransposons compared to autosomes (Charlesworth *et al.*, 1994; Eicher *et al.*, 1989; Steinemann and Steinemann, 1992). A genetic runaway scenario is predicted, and referred to as Muller's ratchet (Charlesworth, 1978), once an absence of chromosome pairing, as in sex chromosomes, starts to occur. Again, an expansion of repetitive DNA would be predicted to contribute to the mechanism of the proposed ratchet. In *S. latifolia* we have identified three different non-LTR retrotransposon fragments which are male sex-specific and another three male-specific fragments which contain simple sequence repeats of the CAAAA and CAC types found in many other species. One of the *S. latifolia* retrotransposon fragments (Bgl16) is also homologous to a male-specific fragment from *C. sativa*. A male-specific sequence from another dioecious plant species, *Atriplex garretti*, contains a number of motifs which are repeated and of a more complex nature than that found in *S. latifolia* sequences (Ruas *et al.*, 1998). Northern analysis using RNA extraction from a number of *S. latifolia* tissues did not indicate the presence of Bam37, Bgl7 or Bgl16 transcripts. Moreover, protein sequence homology to retrotransposons of other organisms was only preserved over the length of the *S. latifolia* fragments by several frame shifts. It is thus unlikely that any of these sequences are still capable of transposition. However, the repetitive nature of Bam37 might suggest that either a similar sequence is still capable of transposition or was capable until recently.

The *S. latifolia* Y chromosome is predominantly euchromatic, unlike the more heterochromatic Y chromosomes of mammals and *R. acetosa* (Grant *et al.*, 1994; Parker, 1990; Vyskot *et al.*, 1993). Heterochromatin, such as indicated by C banding, is associated with repetitive DNA in plants (Clark *et al.*, 1993; Hutchinson and Lonsdale, 1982; Jones and Flavell, 1982). Retrotransposons and other transposable elements have been proposed to be the major driving force behind the degeneration of genetic information on the Y chromosome and the conversion of a euchromatic chromosome to a heterochromatic one (Steinemann and Steinemann, 1992). However, on the basis of the development of a neo-Y/X2 chromosome system in *Drosophila miranda,* this conversion probably occurs only slowly (Steinemann and Steinemann, 1992). This is intriguing, given that we know that the *S. latifolia* Y chromosome has expanded recently in evolutionary terms, that it contains a number of repetitive sequences including retrotransposons (Donnison *et al*., 1996), and that one Y chromosome locus has recently been shown to have degenerated compared to a homologue on the X chromosome (Guttman and Charlesworth, 1998). Therefore, the indication may be that the sex chromosomes in *S. latifolia* are on an evolutionary path which will lead to these chromosomes becoming heterochromatic, as has occurred in other sex chromosome systems. In *Beta vulgaris*, the only plant species in which LINE-like retrotransposons have been localized by *in situ* hybridization, non-LTR retrotransposons are clustered on all chromosomes unlike LTR-Ty1-copia retrotransposons which are more evenly distributed (Schmidt *et al*., 1995). Future *in situ* localization studies in a dioecious species with sex chromosomes such as *S. latifolia* with Bam37, Bgl7 and Bgl16 probes should provide an interesting comparison to *B. vulgaris*. Moreover, given the recent advent of dioecy in *S. latifolia*, it should make an excellent model system in which to study any putative role that retrotransposons and possibly other repeats have in the establishment of sex chromosomes enabling comparisons to animal systems to be made.

References

Buitkamp, J., Ammer, H. and Geldermann, H. (1991) DNA fingerprinting in domestic animals. *Electrophoresis* **12**: 169–174.

Charlesworth, B. (1978) Model for evolution of Y chromosomes and dosage compensation. *Proc. Natl Acad. Sci. USA* **75**: 5618–5622.

Charlesworth, B., Sniegowski, P. and Stephan, W. (1994) The evolutionary dynamics of repetitive DNA in eucaryotes. *Nature* **371**: 215–220.

Clark, M.S., Parker, J.S. and Ainsworth, C.C. (1992) Repeated DNA and heterochromatin structure in *Rumex acetosa*. *Heredity* **70**: 527–536.

Coppieters, W., Van de Weghe, A., Depicker, A., Coppieters, J., Peelman, L., Van Zeveren, A. and Bouquet, Y. (1995) Polymorphic CAC/T repetitive sequences in the pig genome 1. *Anim. Genet.* **26**: 327–330.

Dellaporta, S.L. and Calderon-Urrea, A. (1993) Sex determination in flowering plants. *Plant Cell* **5**: 1241–1251.

Devlin, R.H., Stone, G.W. and Smailus, D.E. (1998) Extensive direct-tandem organization of a long repeat DNA sequence on the Y chromosome of Chinook salmon (*Oncorhynchus tshawytscha*). *J. Mol. Evol.* **46**: 277–287.

Donnison, I.S., Siroky, J., Vyskot, B., Saedler, H. and Grant S.R. (1996) Isolation of Y chromosome specific sequences from *Silene latifolia* for mapping male determining genes. *Genetics* **144**: 1893–1901.

Eicher, E.M., Hutchinson, K.W., Phillips, S.J., Tucker, P.K. and Lee, B.K. (1989) A repeated segment on the mouse Y chromosome is composed of retroviral-related, Y-enriched and Y-specific sequences. *Genetics* **122**: 181–192.

Fanning, T.G. and Singer, M.F. (1987) LINE-1 – a mammalian transposable element. *Biochim. Biophys. Acta.* **910**: 203–212.

Fu, Y.H., Kuhl, D.P.A., Pizzuti, A. ***et al.*** (1991) Variation of the CGG repeat at the fragile-x site results in genetic instability – resolution of the sherman paradox. *Cell* **67**: 1047–1058.

Gortner, G., Pfenninger, M., Kahl, G. and Weising, K. (1996) Northern blot analysis of simple repetitive sequence transcription in plants. *Electrophoresis* **17**: 1183–1189.

Grant, S., Houben, A., Vyskot, B., Siroky, J., Pan, W.-H., Macas, J. and Saedler, H. (1994) Genetics of sex determination in flowering plants. *Dev. Genet.* **15**: 214–230.

Guttman, D.S. and Charlesworth, D. (1998) An X-linked gene with a degenerate Y-linked homologue in a dioecious plant. *Nature* **393**: 263–266.

Haynes, S.R., Rebbert, M.L., Mozer, B.A., Forquignon, F. and Dawid, I.B. (1987) Pen repeat sequences are GGN clusters and encode a glycine-rich domain in a *Drosophila* cDNA homologous to the rat helix destabilizing protein. *Proc. Natl Acad. Sci. USA* **84**: 1819–1823.

Holmes, S.E., Dombroski, B.A., Krebs, C.M., Boehm, C.D. and Kazazian, H.H. (1994) A new retrotransposable human L1 element from the LRE2 locus on chromosome 1q produces a chimeric insertion. *Nat. Genet.* **7**: 143–148.

Hutchinson, H. and Lonsdale, D.M. (1982) The chromosomal distribution of cloned highly repetitious sequences from hexaploid wheat. *Heredity* **48**: 371–376.

Jones, J.D.G and Flavell, R.B. (1982) The mapping of highly-repeated DNA families and their relationship to C-bands in chromosomes of *Secale cereale*. *Chromosoma* **86**: 595–612.

Knoop, V., Unseld, M., Marienfeld, J., Brandt, P., Sünkel, S., Ullrich, H. and Brennicke, A. (1996) cpia-, gypsy- and LINE-like retrotransoson fragments in the mitochondrial genome of *Arabidopsis thaliana*. *Genetics* **142**: 579–585.

Kubis, S.E., Heslop-Harrison, J.S., Desel, C. and Schmidt, T. (1998) The genomic organization of non-LTR retrotransposons (LINEs) from three *Beta* species and five other angiosperms. *Plant Mol. Biol.* **36**: 821–831.

Lisitsyn, N., Lisitsyn, N. and Wigler, M. (1993) Cloning the difference between two complex genomes. *Science* **259**: 946–951.

Lutz, S.M. and Gengenbach, B.G. (1997) EMBL accession number U90128.

Nicholas, K.B. and Nicholas, H.B. Jr (1997) GeneDoc: analysis and visualization of genetic variation, http://www.cris.com/~Ketchup/genedoc.shtml

Parker, J.S. (1990) Sex-chromosomes and sexual differentiation in flowering plants. *Chromosomes Today* **10**: 187–198.

Polley, A., Seigner, E. and Ganal, M.W. (1997) Identification of sex in hop (*Humulus lupulus*) using molecular markers. *Genome* **40**: 357–361.

Ruis Rejón, C., Jamilena, M., Garrido Ramos, M., Parker, J.S. and Ruis Rejón, M. (1994) Cytogenetic and molecular analysis of the multiple sex chromosome system of *Rumex acetosa*. *Heredity* **72**: 209–215.

Ruas, C.F., Fairbanks, D.J., Evans, R.P., Stutz, H.C., Andersen, W.R. and Ruas, P.M. (1998) Male-specific DNA in the dioecious species *Atriplex garrettii* (*Chenopodiaceae*). *Am. J. Bot.* **85**: 162–167.

Sakamoto, K., Shimomura, K., Komeda, Y., Kamada, H. and Satoh, S. (1995) A male-associated DNA sequence in a dioecious plant, *Canabis sativa* L. *Plant Cell Physiol.* **36**: 1549–1554.

SanMiguel, P., Tikhonov, A., Jin, Y-K. *et al.* (1996) Nested retrotransposons in the intergenic regions of the maize genome. *Science* **274**: 765–768.

Schäfer, R., Zischler, H. and Epplen, J.T. (1988) (CAC)5, a very informative oligonucleotide probe for DNA fingerprinting. *Nucleic Acids Res.* **16**: 5196.

Schmidt, T., Kubis, S. and Heslop-Harrison, J.S. (1995) Analysis and chromosomal localisation of retrotransposons in sugar beet (*Beta vulgaris* L.): LINEs and Ty1-copia-like elements as major components of the genome. *Chromosome Res.* **3**: 335–345.

Scutt, C.P., Kamisugi, Y., Sakai, F. and Gilmartin, P.M. (1997) Laser isolation of plant sex chromosomes: studies on the DNA composition of the X and Y sex chromosomes of *Silene latifolia*. *Genome* **40**: 705–715.

Schwarz-Sommer, Z., Leclercq, L., Göbel, E. and Saedler, H. (1987) Cin4, an insert altering the structure of the *A1* gene in *Zea mays*, exhibits properties of nonviral retrotransposons. *EMBO J.* **6**: 3873–3880.

Smith, D.N. and Devey, M.E. (1994) Occurrence and inheritance of microsatellites in *Pinus radiata*. *Genome* **37**: 977–983.

Steinemann, M. and Steinmann, S. (1992) Degenerating Y chromosome of *Drosophila miranda*: a trap for retrotransposons. *Proc. Natl Acad. Sci. USA* **89**: 7591–7595.
Sutherland, G.R. and Richards, R.I. (1995) Simple tandem DNA repeats and human genetic disease. *Proc. Natl Acad. Sci. USA* **92**: 3636–3641.
Vyskot, B., Araya, A., Veuskens, J., Negrutiu, I. and Mouras, A. (1993) DNA methylation of sex chromosomes in a dioecious plant, *Melandrium album*. *Mol. Gen. Genet.* **239**: 219–224.
Westergaard, M. (1946) Aberrant Y chromosomes and sex expression in *Melandriumn album*. *Hereditas* **32**: 419–443.
Westergaard, M. (1948) The relation between chromosome constitution and sex in the offspring of triploid *Melandrium*. *Hereditas* **34**: 257–279.
Westergaard, M. (1958) The mechanism of sex determination in dioecious flowering plants. *Adv. Genet.* **9**: 217–281.
Wright, D.A., Ke, N., Smalle, J., Hauge, B.M., Goodman, H.M. and Voytas, D.F. (1996) Multiple non-LTR retrotransposons in the genome of *Arabidopsis thaliana*. *Genetics* **142**: 569–578.
Xiong, Y. and Eickbush, T.H. (1990) Origin and evolution of retroelements based upon their reverse transcriptase sequences. *EMBO J.* **9**: 3353–3362.

5

The Y chromosome of white campion: sexual dimorphism and beyond

André Lardon, Abdelmalik Aghmir, Sevdalin Georgiev, Françoise Monéger and Ioan Negrutiu

1. Introduction

White campion (*Silene latifolia*) is a dioecious plant with heteromorphic X and Y sex chromosomes. The Y chromosome contains key genes controlling the sexual dimorphism. In male plants, a filamentous structure replaces the pistil, while in female plants the stamens degenerate early in flower development. The white campion experimental system (latest review by Ainsworth *et al.*, 1998) represents a suitable model in studying the organization and the evolution of an XY system in plants. The Y chromosome has been divided into four domains (Westergaard, 1958), harbouring distinct features: (Y^{I}) female suppression, containing the gynoecium suppression function (GSF); (Y^{II}) male promoting function(s), containing the stamen promoting function (SPF); (Y^{III}) male fertility functions; (Y^{IV}) pseudo-autosomal region (*Figure 1*). Despite long-lasting interest in studies on sexual development in this species (Correns, 1928; Gedes and Thomson, 1889; Ruddat *et al.*, 1991; van Nigtevecht, 1966; Westergaard, 1958; Winge, 1931), three main problems await proper answers or clarification: (1) the origin of dioecy and of the sex chromosomes among *Silene* species, (2) the morphological and molecular bases of sexual dimorphism and (3) the structural features of Y and X chromosomes.

1.1 *The phylogenetic debate*

The phylogenetic debate concerns both the monophyletic origin of the Silenoideae subfamily and the origin of the dioecious condition within the subfamily (Baker, 1958; Desfeux *et al.*, 1996). The main gynoecium traits (*Table 1*), the observed reproductive systems (hermaphroditic, gynodioecious and dioecious species), cytogenetic investigations among 99 species (Degraeve, 1980) or molecular phylogenies based on internal transcribed spacer (ITS) region of rDNA (Desfeux and Lejeune, 1996), have

Sex Determination in Plants, edited by C.C. Ainsworth.

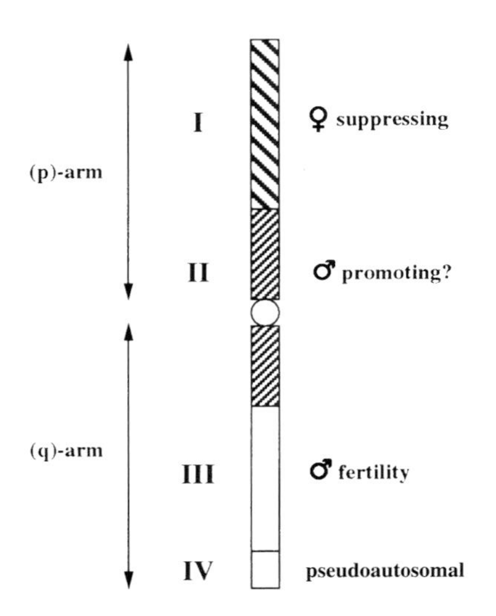

Figure 1. *Functional and structural domains (I–IV) of the Y chromosome according to Westergaard (1958). Domains I and III were established based on Y deletion mutants with hermaphroditic and male sterile phenotypes, respectively. Domain II was positioned by default and in conformity with the female flower phenotype in XX plants. The pseudo autosomal region corresponds to the X–Y pairing domain during meiosis in wild-type male flowers.*

classified white campion successively as a *Lychnis*, *Melandrium* or *Silene* species. At present, the prevailing view is based on the molecular phylogeny, which groups *Silene-Lychnis-Melandrium* in a single monophyletic genus, *Silene*. Furthermore, the available evidence suggests that dioecy has evolved at least twice within the subfamily (sections Otites and Elisanthe, respectively; Chater and Walters, 1964; Desfeux *et al*., 1996). With the recent isolation of repetitive (Donnison *et al*., 1996; Scutt *et al*., 1997) and coding sequences identified on the X and Y chromosomes of white campion (Guttman and Charlesworth, 1998; Monéger *et al*., 1998), we foresee rapid progress on these challenging evolutionary issues within the *Silene* species.

1.2 *Morphological bases of sexual dimorphism*

Morphological differences between the sexes were largely ignored until recently, and the situation has generated confusion in both comparative studies between unisexual species

Table 1. *Controversial generic boundaries within the complex* Silene–Lychnis–Melandrium *based on main morphological gynoecium traits and the successive phylogenetic classification of white campion versus molecular systematics using ITS of rDNA sequences (according to data from Desfeux and Lejeune, 1996)*

	Number of styles		Septa in the ovary		Capsule teeth		ITS sequences
	3	5	Present	Absent	Entire (5)	Split (10)	
Genus	*Silene*	*Lychnis*	*Silene* or *Lychnis*	*Melandrium*	*Lychnis*	*Silene*	Monophyletic grouping
Nomenclature for white campion	*Lychnis alba*		*Melandrium album*		*Silene latifolia*		*Silene latifolia*

(Dellaporta and Calderon-Urrea, 1993) and in defining the type of functions underlying male or female organ arrest in flowers of opposite sex (Grant *et al.*, 1994; Mittwoch, 1967). By using combined histological, scanning electron microscopy and genetic analyses, we were able to dissect precisely the morphological bases of sexual dimorphism in white campion (*Figure 2*; also see Farbos *et al.*, 1997). We show that the arrest of the male and female programmes are independent events in both space and time.

The female flower. In the absence of SPF, located on the Y chromosome, the male developmental arrest in the female flower results from the lack of parietal initials and the subsequent degeneration of sporogenous cell initials during anther differentiation. Thus, the female plant (XX constitution) contains the genetic information necessary to initiate anther development up to the early sporogenous stage. Subsequently, anther development is arrested because the genes required to proceed beyond that stage are not expressed or are missing. Thus, the earliest male differentiation function on the Y chromosome is most likely involved in initiating the parietal differentiation.

The male flower. In the presence of GSF, located on the Y chromosome, the female developmental arrest in the male flower results from a block in carpel initiation. The male plant (XY constitution) contains all the genetic information to produce a functional hermaphroditic flower, while the GSF causes a sudden arrest of cell proliferation in whorl 4 of male flowers at the time of partitioning between whorls 3 and 4, at

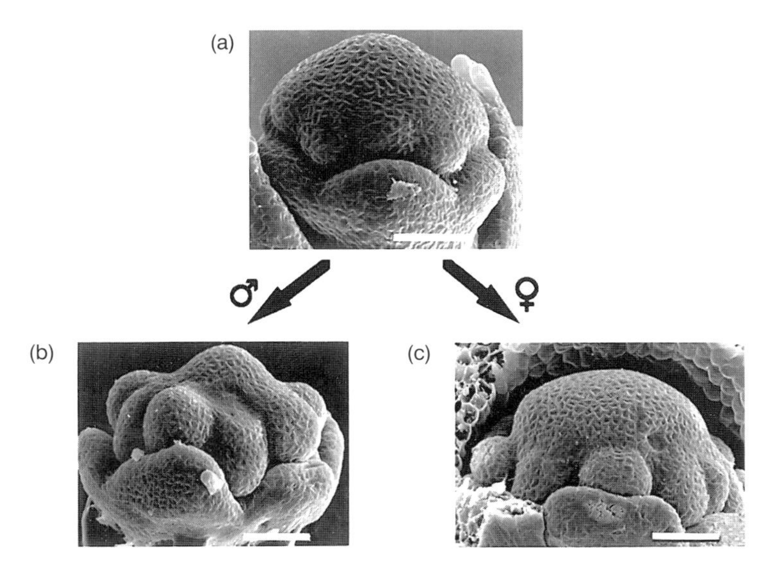

Figure 2. *Morphological traits as evidenced by scanning electron microscopy that underlie the timing of sexual dimorphism during very early stages of flower formation. (a) A bipotential flower meristem at a stage where stamen primordia are being initiated. No differences exist at this stage between male and female flower buds. (b) A male flower bud is produced by a sudden arrest of cell proliferation in whorl 4 once whorl 3 has been spatially specified. Five stamen primordia are clearly visible, 3 of which are anti-sepalous and 2 are anti-petalous. (c) A female flower bud at the same developmental stage as in (b) in which active cell proliferation takes place in whorl 4. Stamen primordia (5 are visible here) develop normally at this stage, but will become arrested during the next developmental stage. Bars = 50 μm.*

the flower meristem centre. As a result, a filamentous structure in the centre of the male flower is located in the position where five carpels normally appear in the female flower. We argue that the GSF operates at or just downstream of the ABC pathway in the flower meristem (Lardon *et al.*, in press).

1.3 *The structural features of the Y (and X) chromosome*

By the 1950s, numerous studies based on genetic and classical cytogenetic approaches had generated a loose division of the Y chromosome into four functional/structural domains which is still valid now (*Figure 1*). The great majority of the sexual mutants produced at that time were derived from polyploid lines (Westergaard, 1958), which obviously complicated the interpretation of the results. Better mapping of the Y chromosome also suffered from another general limitation of previously reported mutants, namely that they were not analysed simultaneously with respect to genetic, cytogenetic and morphological characteristics. More recently, genetic, cytogenetic and molecular data suggest that the Y and X chromosomes have all the main features (including dosage compensation by X inactivation) of the heteromorphic sex chromosomes known in animal systems (Buzek *et al.*, 1997; Vyskot *et al.*, 1993 and ref. therein; see also Chapter 6). Preliminary molecular evidence tends to suggest that the X and Y chromosomes in white campion have a high degree of commonality in DNA sequence, at least for some major classes of satellite DNA motifs which have been cloned and tested by FISH (Scutt *et al.*, 1997). Furthermore, physical markers hybridizing to the Y chromosome have been reported (Donnison *et al.*, 1996; Farbos *et al.*, 1998), but further refinements are needed in order to significantly improve the mapping of the Y chromosome. Again, the analysis of genes located on the X and Y chromosomes represents both a key to defined regions of the sex chromosomes and a critical tool in evolutionary studies tracing the origin of the sex chromosomes.

2. Generating the experimental tools: screening for Y-deleted sexual mutants

Here we summarize our efforts to generate a collection of Y deletion mutants in a diploid genetic background by γ-irradiation of pollen. By combining irradiation of pollen (containing X or Y gametes) with screening in the M1 generation, the Y chromosome becomes the main target in this type of mutagenesis experiment due to its permanent haploid condition. In total, we generated more than 50 mutants exhibiting altered sexual phenotypes and analysed in detail 15 hermaphroditic mutants, two asexual mutants and 15 male sterile mutants (*Figure 3*). Most of these exhibited defined deletions on the Y chromosome. Subsequent deletion mapping allowed the limits of the position of *GSF* and *SPF* loci to be determined and provided tools for molecular studies.

2.1 *Mutations affecting the GSF*

Hermaphroditic mutants (*bsx* mutants) were screened phenotypically and were estimated to be present in the M1 plant population at a frequency of 1%. The subsequent genetic analyses (*Table 2*) have enabled us to distinguish two loci with GSF properties:

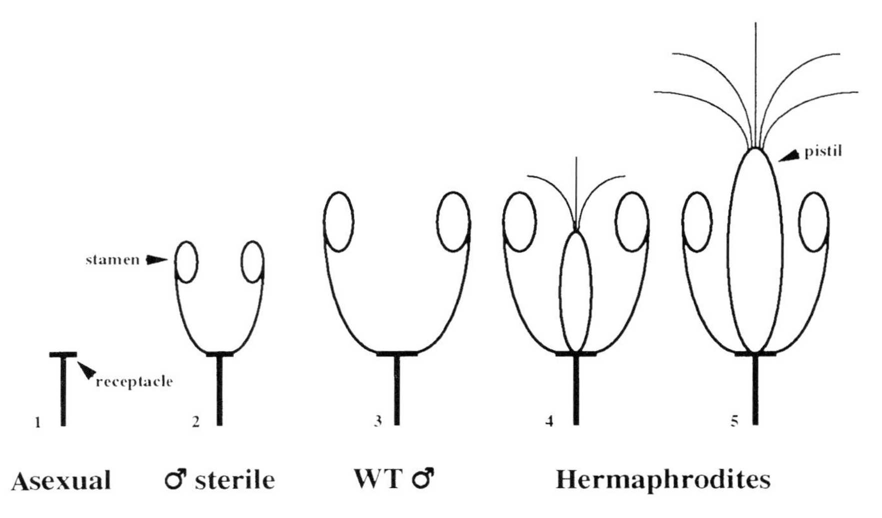

Figure 3. *Schematic representation of sexual mutants produced by pollen irradiation and showing alterations in male and/or female organs. Three main types of mutants were identified: asexual (1), male sterile (2) and hermaphroditic (4 and 5). The latter exhibit variable carpel number (also see text). Such mutations were shown to result from deletions on the Y chromosome, the generated phenotypes being in agreement with the structural and functional division of the Y chromosome as shown in Figure 1.*

a Y-linked locus (*GSF-Y*) and an autosomal locus (*GSF-A*). No *bsx* mutants in an XX background were identified. Y-linked and potentially Y-linked mutants represented by far the largest group (13 out of 15), as expected due to the experimental design. The dominant autosomal mutation, *GSF-A*, was obtained at a frequency at least 10-fold lower than that at the Y-linked locus (Lardon *et al.*, 1998).

The *bsx* mutants carried deletions ranging from small (approx. 5%) to as large as half the length of the Y chromosome (p)-arm. In addition, alterations in the *GSF* loci have generated variable states of carpel number restoration, both between and within the mutants (*Table 2*). Since the size of the Y deletions did not correlate with the penetrance of GSF mutations, we believe that *GSF-Y* represents a defined, single locus on the (p)-arm, with no redundant loci being present elsewhere along the distal half of the arm. Furthermore, mutation of GSF generated patterns typical for methylation-controlled genes, with incomplete penetrance and the presence of variegated phenotypes (Eden and Cedar 1994; Ronemus *et al.,* 1996). This result extends and confirms previous data on 5-azacytidine treatments in wild-type seedlings which generated mosaic hermaphrodites (Janousek *et al.*, 1996). We also showed that carpel restoration is a progressive process across successive sexual generations. The result was a narrower range of variation in carpel number, perhaps due to the reduction or elimination of the remnant effect of GSF. However, the restoration of carpel number was incomplete after one or two meiotic generations. This could be due to the fact that DNA methylation patterns are inherited in plants and show typical progressivity and specificity. An alternative explanation is based on the reported redundancy of *GSF* loci corresponding to the existence of Y and autosomal loci acting in concert. The presence of two partially redundant *GSF* loci makes para-mutation a possible

Table 2. *GSF-Y and GSF-A mutations in white campion: a summary of genetic, cytogenetic and levels of carpel number restoration in M1 as compared with M2 generation*

GSF locus	Genetic transmission (number of individual mutants)	Y deletions (% of (p)-arm deletion)	Level of penetrance (carpel number)	
			M1	M2
GSF-Y	Transmitted (8)	5–51	3.6	4.5
	Not transmitted (5)	5–25	3.6	–
GSF-A	Transmitted (1)	–	0.5	Same as M1

mechanism of GSF action (Matzke *et al.*, 1996). In such a context, the Y-linked locus could act as the main para-mutagenic locus.

An additional interpretation of the discrepancy between carpel number in female wild-type flowers (five carpels) and average carpel number in hermaphroditic mutants might involve an evolutionary component (also see Jürgens *et al.*, 1996). Depending whether the ancestral condition was five or three carpels (*Table 1*), the evolution of dioecy might have been associated with a modification of carpel number from three to five in the wild-type female plant. If this were the case, the variation in carpel number might reflect this particular situation as well (*Tables 1 and 2*). This hypothesis is summarized below:

GSF mutation
↓
Variant (a) ☿/**5C** → ♂/0C + ♀/5C ——➤ ☿/varC, **<5**

considers that the ancestral condition was five carpels. Our experiments indicate that back-mutations in *GSF-Y* and *GSF-A* loci exhibit incomplete penetrance and variegation; or

GSF mutation
↓
Variant (b) ☿/**3C** → ♂/0C + ♀/**5C** ——➤ ☿/varC

considers that the ancestral condition was three carpels, the back-mutations revealing part of these evolutionary changes.

2.2 *Mutations affecting the SPF*

No SPF mutants have been reported so far, which makes the position of the *SPF* locus on the Y chromosome purely speculative. We have generated mutants showing phenotypes that display the developmental defects characterizing the sexual dimorphism in this species (Farbos *et al.*, 1998), namely replacement of pistil by a filamentous structure and vestigial anthers (*Figure 4*). The mutants were named *asexua* (*asx*), and we demonstrated that they are affected at the *SPF* locus. In the *asx* mutants, anthers initiate the formation of sporogenous cells but lack any parietal cell initials and layers. Sporogenous differentiation is abruptly arrested, followed by differentiation of the

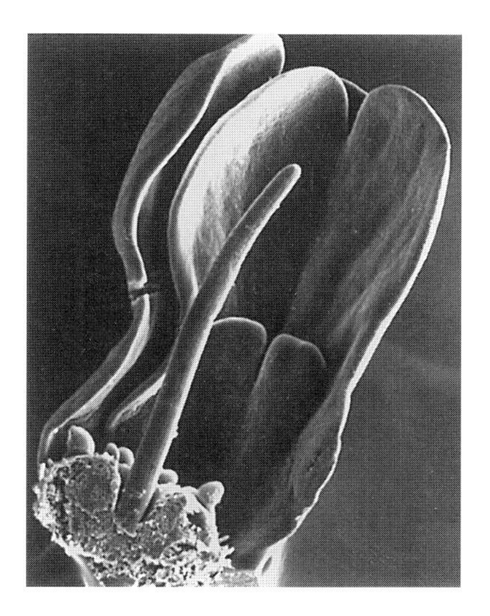

Figure 4. *A maturing flower of an asexual mutant with vestigial stamens visible as small bulges and an elongating filamentous structure replacing the pistil. The sepals, three petals and four stamens have been removed, prior to sectioning the bud, to show the cavity of the flower receptacle surrounding the filament. The mutant displays the two developmental blocks responsible for sexual dimorphism in wild-type male and female flowers.*

sporogenous cells into parenchymatous cells. Subsequently, the anthers degenerate slowly. The *asx* mutants represent a novel class of mutations in plants in which the first stages of sporogenous/parietal tissue specialization in the anther are affected. By comparing stamen development in wild-type female and *asx* mutant flowers, we demonstrated that they share the same block in anther development, which results in the production of vestigial anthers. Mutants in SPF are shown to result from interstitial deletions on the Y chromosome. These deletions cover the central domain on the (p)-arm of the Y chromosome (see Section 2.3). It seems that this is the earliest function in the male developmental programme which is located on the Y chromosome and that it is likely to be responsible for male dimorphism in white campion.

2.3 *Preliminary mapping of GSF and SPF on the Y chromosome*

GSF and *SPF* loci were tentatively located on the (p)-arm of the Y chromosome. The two classes of mutants, *bsx* and *asx*, exhibit mutually exclusive phenotypes, which prompted us to propose the following configuration between the two loci as a working hypothesis (*Figure 5*). The model takes into account the fact that the two largest deletions we obtained, *bsx1* and *asx1* – covering 51% and 40% of the (p)-arm, respectively – could cover together almost entirely the (p)-arm. By using FISH analysis with the X-43 subtelomeric repeat sequence (Buzek *et al.*, 1997) and probe 5E4 hybridizing to the central region of the (p)-arm (Farbos *et al.*, 1998), we could assign the *SPF* locus to a location proximal to the 'pY51' position and the *GSF* locus towards the tip of the (p)-arm. If the model is correct, the Y^{II} domain as shown in *Figure 1* is located entirely on the (p)-arm of the Y chromosome.

3. GSF – an evolutionary innovation, SPF – an evolutionary accident?

Our results indicate that the GSF is a system containing more than one gene/locus and whose primary component is located on the Y chromosome. The autosomal *GSF-*

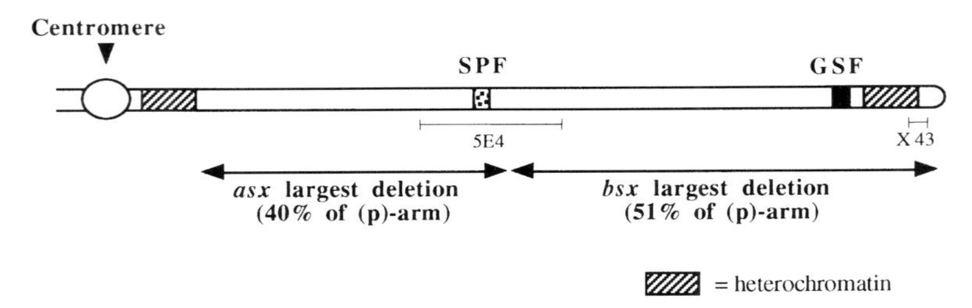

Figure 5. *Proposed model for the organization of the (p)-arm of the Y chromosome based on deletion mutants with hermaphroditic (locus GSF) and asexual (locus SPF) phenotypes and data from experiments with two molecular markers (X-43 and 5E4). The two blocks of heterochromatin were observed by Giemsa staining.*

A locus may correspond to a partially redundant component. This is supported by the fact that the Y deletion series of *bsx* mutants and the autosomal *bsx-A* mutant independently allow carpel development when mutated, albeit at significantly different levels. Schematically, the effect of the two mutations on carpel number is summarized below (also see *Table 3*):

GSF mutation
↓
GSF-Y = high levels of carpel number restoration: 0C ⟶ 2–5C
GSF-A = low levels of carpel number restoration: 0C ⟶ 0–3C

Interestingly, the *GSF-A* locus seems only to be active in the presence of the Y (see below), since no *GSF-A* operates in XX female flowers. Such a situation supports the 'incomplete dominance' hypothesis, according to which full dioecy was established by more than one mutation in one sexual pathway (Charlesworth, 1996). Indeed, independent mutations at two GSF loci give dominant restoration of carpel development.

The data suggest that the *SPF* locus, a key factor controlling the sporogenous/parietal specialization in pre-meiotic anthers, is most likely to be a gene normally operating within the anther differentiation pathway and which is missing or is not expressed in females (XX constitution). The evolutionary history of SPF is so far unclear. The alternative to the possibility above is that the Y locus for *SPF* represents a suppressor of a suppressor of parietal cell formation within the anther.

The existence of independent pathways for male and female developmental arrest is reflected by the fact that sex reversal generates either hermaphroditic or asexual mutants (*Table 3*). The GSF is a negative regulator of carpel formation. The SPF is an activating function of stamen differentiation. This tandem of dominant functions with opposite growth effects is inherited as a unit and generates a Y chromosome–dependent pattern of sex expression. Thus, according to prevailing models on evolution of separate sexes and incipient sex chromosomes in plants, the GSF represents the dominant female sterile (f^S) mutation and the SPF represents the wild-type male fertile (m^F) function (Charlesworth, 1991, 1996). The corresponding Y/X genetic formula should therefore be (f^Sm^F/f^fm^s). In reality, and based on the hermaphroditic and asexual phenotypes described here, and their respective karyotypes, we consider that the genetic formula is in fact (f^Sm^F/--). This reflects the fact that the Y-located f^S and m^F

Table 3. *Sex transformation events in white campion based on natural mutations, interactions with a plant pathogen and γ-ray induced mutations*

	Stamen promoting function (SPF)	Gynoecium suppressing function (GSF)
In WT♂	Present	Present
In WT♀	Absent	GSF-A, inactive
Loss-of-function	SPF^-	GSF^-
In WT♂	Asexual (*asx*)	Hermaphroditic (*bsx*)
In WT♀	na	na
Gain-of-function	SPF^+	GSF^+
In WT♂	na	na
In WT♀	Hermaphroditic[a]	Low carpel number to asexual[b]

[a] Not obtained experimentally by mutagenesis, but naturally occurring through infection of female plants by the fungus *Mycrobotryum violacea* (Ruddat *et al.*, 1991).
[b] Mutation obtained in XX background.
na, not applicable.

loci seem to have no functional counterparts on the X chromosome. Since normal carpels develop in XX plants, we conclude that *GSF* loci are (novel) functions recruited on top of the gynoecium differentiation pathway. The Y, X and autosomal formula of sexual dimorphism control in white campion could therefore be

$$(f^{SY\leftrightarrow A}\, m^F/\text{--}) \text{ and } (f^{SA\leftrightarrow Y}\, f^{SA\leftrightarrow Y})$$

where $^{Y\leftrightarrow A}$ and $^{A\leftrightarrow Y}$ mean that under wild-type conditions the *GSF-A* locus needs direct interaction with *GSF-Y* for normal activity.

4. Conclusions

Based on the results discussed above, our conclusions and working hypotheses are the following:

(i) *GSF* and *SPF* loci are located on the differential, (p)-arm of the Y chromosome. This configuration constitutes the basis of the chromosomal component controlling the described sexual dimorphism and its stability.
(ii) *GSF* and *SPF* loci might be the only reproductive functions present within most of the (p)-arm. The differential arm may constitute the case of a plant chromosomal region with very low gene density specifically associated with the control of sexual development in white campion via constraints imposed by the heteromorphic sex chromosome system.
(iii) We speculate that the (q)-arm of the Y chromosome contains regions of extensive homology with the X chromosome. If this were the case, the fact that meiotic pairing was restricted to the pseudoautosomal region must be due to an active and specific mechanism preventing pairing outside that terminal region of X and Y chromosomes.

References

Ainsworth, C., Parker, J. and Buchanan-Wollaston, V. (1998) Sex determination in plants. *Curr. Topics Develop. Biol.* **38**: 167–223.

Baker, H.G. (1958) Hybridization between dioecious and hermaphrodite species in the Caryophyllaceae. *Evolution* **12**: 423–427.

Buzek, J., Koutnikova, H., Houben, A., Riha, K., Janousek, B., Siroky, J., Grant, S. and Vyskot, B. (1997) Isolation and characterization of X chromosome-derived DNA sequences from a dioecious plant *Melandrium album*. *Chromosome Res.* **5**: 57–65.

Charlesworth, B. (1991) The evolution of sex chromosomes. *Science* **251**: 1030–1032.

Charlesworth, B. (1996) The evolution of sex determination and dosage compensation. *Curr. Biol.* **6**: 149–162.

Chater, A.O. and Walters, M. (1964) *Silene* L. In: *Flora Europaea*, Vol. 1 (ed. T.G. Tutin). Cambridge University Press, Cambridge, pp. 158–181.

Correns, C. (1928) Bestimmung, vererbung und verteilung des geschlechtes bei höheren planzen. In: *Hanbd. Vererbungswissensch* (ed. C. Borntraeger). Borntraeger Verlag, Berlin, pp. 1–138.

Degaeve, N. (1980) Etude de diverses particularités caryotypiques des genres *Silene, Lychnis* et *Melandrium*. *Bol. Soc. Brot. Sér. 2* **53**: 595–643.

Dellaporta, S.L. and Calderon-Urrea, A. (1993) Sex determination in flowering plants. *Plant Cell* **5**: 1241–1251.

Desfeux, C. and Lejeune, B. (1996) Systematics of Eumediterranean *Silene* (*Caryophyllaceae*): evidence from a phylogenetic analysis using ITS sequences. *C.R. Acad. Sci. Paris* **319**: 351–358.

Desfeux, C., Maurice, S., Henry, J.-P., Lejeune, P. and Goujon, P-H. (1996) Evolution of reproductive systems in the genus Silene. *Proc. Roy. Soc. Lond. B.* **263**: 409–414.

Donnison, I.S., Siroky, J., Vyskot, B., Saedler, H. and Grant, S.R. (1996) Isolation of Y chromosome-specific sequences from *Silene latifolia* and mapping of male sex-determining genes using representational difference analysis. *Genetics* **144:** 1893–1901.

Eden, S. and Cedar, H. (1994) Role of DNA methylation in the regulation of transcription. *Curr. Opinions Genet. and Develop.* **4**: 255–259.

Farbos, I., Oliveira, M., Negrutiu, I. and Mouras, A. (1997) Sex organ determination and differentiation in the dioecious plant *Melandrium album*: a cytological and hystological analysis. *Sex. Plant Reprod.* **10**: 155–167.

Farbos, I., Veuskens, J., Oliveira, M., Vyskot, B., Hinnisdaels, S., Aghmir, A., Mouras, A. and Negrutiu, I. (1998) Sexual dimorphism in white campion: deletion on the Y chromosome results in a floral asexual phenotype. *Genetics* (in press).

Gedes, P. and Thomson, A. (1889) *The Evolution of Sex*. W. Scott, London.

Grant, S., Hunkirchen, B. and Saedler, H. (1994) Developmental differences between male and female flowers in the dioecious plant *Silene latifolia. Plant J.* **6**: 471–480.

Guttman, D.S. and Charlesworth, D. (1998) An X-linked gene with a degenerated Y-linked homologue in a dioecious plant. *Nature* **393**: 263–265.

Janousek, B., Siroky, J. and Vyskot, B. (1996) Epigenetic control of sexual phenotype in a dioecious plant, *Melandrium album*. *Mol. Gen. Genet.* **250**: 483–490.

Jürgens, A., Witt, T. and Gottsberger, G. (1996) Reproduction and pollination in Central European population of *Silene* and *Saponaria* species. *Bot. Acta* **109**: 316–324.

Lardon, A., Georgiev, S., Aghmir, A., Le Merrer, G. and Negrutiu, I., (1998) Sexual dimorphism in white campion: compex control of carpel number is revealed by Y chromosome deletions. *Genetics* (in press).

Lardon, A., Delichère, C., Monéger, F. and Negrutiu, I. (1998) Sex determination or sexual dimorphism? On facts and terminology. In: *Sexual Plant Reproduction and Biotechnological Applications*: *Recent Advances by Molecular Biology, Biochemistry and Morphology* (ed. M. Cresti). PRO EDIT, Heidelberg, pp. 45–51.

Matzke, M.A., Matzke, A.J.M. and Eggleston, W.B. (1996) Paramutation and transgene silencing: a common response to invasive DNA? *Trends Plant Sci.* **1:** 382–388.

Mittwoch, U. (1967) *Sex Chromosomes*. Academic Press, London.

Monéger, F., Barbacar, N., Delichère, C., Georgiev, S., Lardon, A. and Negrutiu, I. (1998) Functional analysis of the Y chromosome in *Silene latifolia. Flowering Newsletters* **25**: 20–25.

Ronemus, M.J., Galbiati, M., Ticknor, C., Chen, J.C. and Dellaporta, S.L. (1996) Demethylation-induced developmental pleiotropy in *Arabidopsis*. *Science* **273**: 654–657.

Ruddat, M., Kokontis, J., Birch, L., Garber, E.D., Chiang, K-S., Campanella, J. and Dai, H. (1991) Interactions of *Mycrobotryum violacea* (*Ustilago violacea*) with its host plant *Silene alba. Plant Sci.* **80:** 157–165.

Scutt, C.P., Kamisugi, Y., Sakai, F. and Gilmartin, P.M. (1997) Laser isolation of plant sex chromosomes: studies on the DNA composition of the X and Y sex chromosomes of *Silene latifolia. Genome* **40:** 705–715.

van Nigtevecht, G. (1966) Genetic studies in dioecious *Melandrium*. I. Sex-linked and sex influenced inheritance in *M. album* and *M. dioicum. Genetica* **37**: 281–306.

Vyskot, B., Araya, A., Veuskens, J., Negrutiu, I. and Mouras, A. (1993) DNA methylation of sex chromosomes in *Melandrium album. Mol. Gen. Genet.* **239**: 219–224.

Westergaard, M. (1958) The mechanism of sex determination in dioecious flowering plants. *Adv.Genet.* **9**: 217–281.

Winge, Ö. (1931) X- and Y-linked inheritance in *Melandrium*. *Hereditas* **15**: 127–165.

The role of DNA methylation in plant reproductive development

Boris Vyskot

1. Introduction

Although plants and animals share many structural and functional features as both represent groups of higher eukaryotic organisms, their patterns of body formation and development are rather different. The crucial stage of animal sex differentiation occurs during gastrulation when primordial germ cells are formed and later migrate to the gonads. In higher animals, usually as a result of the presence of specific sex chromosomes, either male or female gonads are established which immediately begin to influence the resulting gender phenotype. This germline represents the only cells from which genomes are transmitted to the sexual progeny. However, no germinal line is set aside during the embryogenesis in plants. Some cell lines, the meristems, maintain their ability to divide (similar to stem cells in animals) and give rise to both vegetative parts of plants (roots, stem, branches, leaves) and later complex reproductive organs (flowers). The majority of plant species are hermaphrodites, forming one type of bisexual flower; the rest are mostly monoecious (possessing male and female flowers on the same individual) or dioecious (male and female flowers are formed on different individuals). Flowers originate from apical or axillary shoot meristem buds, when subject to appropriate internal and environmental factors. This implies that any change in the genome of meristematic cells that occurs during the life of the plant can be transmitted to the sexual progeny. Moreover, plant cells possess an extremely high regeneration capacity, totipotency, whereby any cell of the plant body, both somatic and generative, can in principle give rise to a new individual. This indicates that the changes occurring during plant development, which lead to a high degree of differentiation, must be reversible. On the other hand, plants – probably due to their inability of locomotive movement – have evolved a high tolerance to both genetic and epigenetic changes. These include even such gross alterations as haploidy, polyploidy, aneuploidy and chromosome aberrations which are generally not tolerated by higher animals.

Sex Determination in Plants, edited by C.C. Ainsworth.

A number of genes controlling plant development have been identified recently which include genes affecting the body organization of the embryo, plant architecture, and flowering processes. The regulation of spacio-temporal gene expression in plants and animals seems to involve similar mechanisms. However, a difference between these two groups of organisms is that the homeotic genes, encoding transcriptional factors which control development, are different; plants have mainly dispersed MADS-box genes while animals have clustered homeobox genes. Development in plants and animals is also under epigenetic control mediated by chromatin modifications of corresponding gene regions (including DNA methylation and nucleosomal histone acetylation). These chromatin modifications represent both the cell memory (transmission of information on gene expression through mitosis and development) and the imprinting mechanism (transmission of regular or stochastic epigenetic information to the sexual progeny).

2. DNA methylation and development

DNA methylation is one of the epigenetic mechanisms which play a long-term role in the control of gene expression (for review, see Adams, 1996). Flowering plants often possess large nuclear genomes which are rich in repetitive DNA sequences and up to 30% of cytosine residues are methylated. However, only a small proportion of these methylated cytosine residues, those involving upstream gene regions, affect the expression of corresponding genes. The majority of other cytosines which are methylated [as 5-methylcytosine (5-mC)], are accumulated in non-coding repetitive DNA sequences. These sequences are cytologically observable as constitutive heterochromatin. The vast majority of results demonstrating the biological role of DNA methylation has involved artificial experimental systems where inactivation of ectopic transgenes and partial inhibition of DNA methyltransferases has been achieved either by binding drugs or by antisense RNA strategies.

DNA methylation is also involved in genomic imprinting, a regular process which occurs widely in mammals, but which is restricted to extraembryonic (endosperm) tissue in plants (for a review, see Matzke and Matzke, 1993). While in vertebrate genomes the methylation of cytosines occurs mostly in CG doublets, plant DNA is methylated in CG doublets, CNG triplets and even in non-palindromic DNA sequences (Gruenbaum *et al.*, 1981; Meyer *et al.*, 1994). Cytosine methylation is heritable during cell divisions by means of maintenance DNA methyltransferases which work only on hemimethylated templates. This may represent a mechanism by which information from an epigenetic source can be transmitted during the development of an individual.

2.1 *Methylation patterns during mammalian ontogenesis*

DNA methylation patterns are subject to gross changes during mammalian development. For example, all tissue-specific genes which have been analysed in sperm are methylated (Yisraeli and Szyf, 1984), while the oocyte DNA is relatively undermethylated. Monk *et al.* (1987) have also shown that the level of methylation in the embryo after three cell divisions of the zygote was intermediate to the values observed in sperm and oocytes. However, the genome is largely undermethylated in the blastula. This indicates that extensive demethylation of the genome takes place in the early embryo between the eight-cell and blastula stages, which may represent a resetting of the genome to initiate the processes of embryonic development (for review, see Razin and Cedar, 1993). The final methylation patterns of gene sequences in different tissues are set up during late embryonic development and may even be tuned in adult life. Both in males and females, the germline is established from primordial cells that contain genomes which are strikingly unmethylated (Kafri *et al.*, 1992).

Several endogenous genes have been shown to be subject to parental imprinting in the mouse and human; either only paternal or only maternal alleles are active in the progeny. In these cases, the methylation state of alleles is often determined by their parent-of-origin. If the parent-specific scheme of the monoallelic gene expression fails, serious human genetic disorders can occur (for review, see Reik, 1996). In experiments on transgenic mice it was found that the maternally inherited transgene was methylated and silenced, while the paternally derived constructs were not modified (Monk, 1990). Other recent data indicate that one of most important roles of DNA methylation is a defence mechanism against transposons: if their promoters are methylated, the transposons are not mobile and, due to frequent 5-mC to thymine transition mutations, they could even be destroyed (Yoder *et al.*, 1997). Another example of the role of DNA methylation in mammals is the inactive (lyonized) X chromosome which contains hypermethylated CpG islands and which displays some other features of chromatin inertness: histone H4 and H3 hypoacetylation and late replication (for review, see Jamieson *et al.*, 1996).

At least two kinds of evidence have been presented that an erasure of DNA methylation patterns during very early vertebrate development leads to embryonic lethality. 5-Azacytidine applied to early chick embryos was found to perturb methylation patterns, to induce polypeptides which are not normally present in morula stage, and finally, to cause developmental arrest (Zagris and Podimatas, 1994). Murine embryonic cells possessing a homozygous mutation in the gene encoding DNA methyltransferase by gene targeting were introduced into the germline of mice and caused lethal embryos at mid-gestation (Li *et al.*, 1992).

2.2 *Role of DNA methylation in plant growth and development*

Many data showing the role of methylation in plant development have been obtained with DNA methylation inhibitors or transgenic plants that express antisense constructs of the cytosine-DNA methyltransferase gene. The phenotypic consequences on the plants of these two experimental approaches were similar in different plant species (e.g. dwarfing, early flowering, abnormal flowers, reduced fertility) and in some cases these changes as well as a global hypomethylation were inherited by the sexual progeny (Finnegan *et al.*, 1996; Janousek *et al.*, 1996; Sano *et al.*, 1990; Vyskot *et al.*, 1995). Despite suffering significant developmental defects, undermethylated plants are, contrary to the methylation in mammals, viable. This difference may indicate that plants and mammals use DNA methylation in different ways (Richards, 1997). The plasticity of plant developmental programmes allows their survival, despite a significant modification of gene expression. While imprinting of parental genomes (usually realized via differential methylation patterns) in mouse is strictly required for regular development (Jaenisch, 1997), imprinting in plants appears to affect mainly the endosperm and not the embryo (Kermicle and Alleman, 1990). However, many data show that 5-mC plays a similar role in plant development as in mammals.

Drozdenyuk *et al.* (1976) followed the 5-mC content in wheat seeds during germination and found that the level of cytosine methylation decreased significantly between dry seeds and day 3 of germination (25.1% to 21.0%, respectively). Changes in DNA methylation may be associated with the regulation of gene activity in the differentiating plant cells at various stages of ontogenesis. Similar results were demonstrated by Follmann *et al.* (1990) who found a rapid and substantial drop in the 5-mC content during the first round of DNA replication in germinating wheat seeds (from 23.7% to 15.2%). Such undermethylation is compatible with the onset of global gene expression and cell division in the germinating embryos but it contrasts with the increase in DNA methyltransferase activity observed at the same time (Theiss and Follmann, 1980). The level of 5-mC also showed an increase during somatic embryogenesis in carrot (Munksgaard *et al.*, 1995). Messeguer *et al.* (1991) compared the 5-mC content in tomato seeds (average 5-mC content 27%) with immature tissues (20%), mature leaves (25%) and pollen (22%). These data suggest that after fertilization and during development and seed maturation, embryonic DNA could be subjected to *de novo* methylation. Germination and subsequent cell division and growth then result in a drop in methylation levels. As is the case in mammals and some fungi, DNA methylation in plants also represents a defence mechanism against intrusive DNA sequences; both transgenes and mobile genetic elements are often methylated (for comprehensive reviews, see Meyer, 1995).

A global hypomethylation (by about 20% compared to the control) was induced in tobacco plants after a treatment of seeds with a hypomethylating drug, 5-azacytidine (5-azaC). These plants generally suffered from dwarfism and a low pollen fertility and their flowers were malformed, strikingly resembling homeotic transformations, since two adjacent whorls were usually affected (Vyskot *et al.*, 1995). Some flowers did not form a proper calyx and corolla and others displayed petaloid stamens or pistils (*Figure 1*). In this case, it is clear that several genes involved in inflorescence and flower formation were affected and even the dwarfism could be caused by a temporarily changed activation of genes responsible for the transition of the vegetative meristem to the inflorescence in this monopodial plant species. Since the other parts of these hypomethylated plants were similar to the controls, it seems that specific genes playing a role in the process of flowering are controlled by DNA methylation and thus

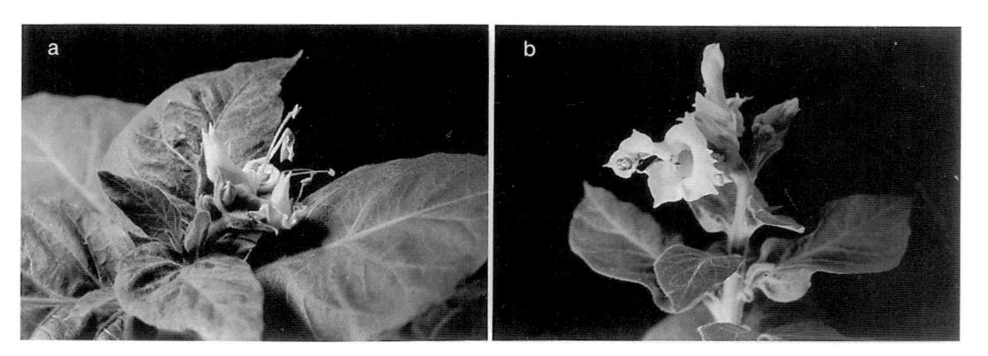

Figure 1. *Homeotic-like floral epimutations as a result of cytosine hypomethylation induced by 5-azaC treatment on tobacco seeds: (a) abortion of outer whorls, corolla absent, (b) petaloid reproductive organs.*

are highly vulnerable to methylation disturbances. In other experiments, 5-azaC was applied on tobacco calli to assess their regeneration potency through protoplast culture and to monitor the stability of hypomethylation in a model, highly repetitive DNA sequence. Regenerated plants maintained the hypomethylation status induced in the calli and their phenotypes were similar to the plants obtained from 5-azaC-treated seeds: reduced plant height and floral malformations including petaloid stamens accompanied by a low pollen fertility (Koukalova *et al.*, 1994).

When studying a single-gene dwarf mutant of maize, Sano *et al.* (1989) found that the amount of 5-mC was about 8% lower than in the wild type. To check whether DNA methylation was correlated with the dwarf mutation, they applied 5-azaC to maize seeds and induced both hypomethylation and dwarfism. A possible relationship between cold-induced flowering (vernalization) and DNA hypomethylation was described in two crucifer species, *Arabidopsis thaliana* and *Thlaspi arvense*. Their late-flowering ecotypes and mutants blossom early after cold treatment or application of 5-azaC. Both cold-treated and 5-azaC-treated *A. thaliana* plants had reduced levels of 5-mC in the genome compared to non-vernalized plants (Burn *et al.*, 1993). Recessive DNA hypomethylation mutant plants of *A. thaliana*, isolated after mutagenization, showed a reduction in 5-mC content of over 70% (Vongs *et al.*, 1993). Although these mutations directly influenced neither DNA methyltransferase activity nor *S*-adenosylmethionine level, the homozygous hypomethylated mutants in the self-pollinated progeny exhibited a variety of severe phenotypic changes (Kakutani *et al.*, 1995). These included reduction or increase in apical dominance, late flowering, reduced fertility and abnormal flowers (Kakutani *et al.*, 1996). The role of DNA methylation in floral development has been unambiguously demonstrated in hypermethylated epimutants of the *SUPERMAN* gene in *A. thaliana* in which deletions lead to an increase in the number of stamens and carpels. Jacobsen and Meyerowitz (1997) isolated a novel group of mutants in which reduced levels of corresponding mRNA were found. Genomic sequencing revealed no nucleotide differences from the wild-type, but there was found to be extensive cytosine methylation of these mutant alleles. Moreover, a revertant of this epimutant showed a large decrease in specific hypermethylation. Jacobsen and Meyerowitz (1997) also studied the DNA methylation status of a transgenic *A. thaliana* line possessing an antisense cytosine methyltransferase which displayed the mutant *superman* phenotype. Although this plant was largely hypomethylated, they found the *SUPERMAN* gene to be hypermethylated as was the case in the epimutants. These data clearly show that DNA methylation plays a role in floral development and that site-specific hypermethylation can still occur even in a globally hypomethylated genome.

Based on DNA sequence homology between prokaryote and mouse cytosine methyltransferases, a small multigene family of cytosine methyltransferases has been identified in *A. thaliana* (Finnegan and Dennis, 1993). An antisense construct of one of these methyltransferase cDNAs was transferred into *A. thaliana* which led to a significant reduction of 5-mC content (up to 10% compared to the control). These plants displayed various phenotypic and developmental abnormalities, including altered plant size and shape, decreased fertility, and homeotic floral changes which were associated with ectopic expression of some floral homeotic genes in leaves. Moreover, removal of the antisense gene in segregants did not completely restore methylation level, indicating that methylation patterns are meiotically transmitted and that remethylation is a slow process (Finnegan *et al.*, 1996). Recent data show that reduced levels of DNA methylation, due to either mutation or presence of the methyltransferase antisense gene, cause early flowering (vernalization) and that the promotion of flowering is directly proportional to the decrease in methylation (Finnegan *et al.*, 1998). Similar experiments with antisense cytosine methyltransferase have also been presented by Ronemus *et al.* (1996). The transgenic *Arabidopsis* plants were substantially hypomethylated (in both CG and CCG sequences) which induced various modifications in vegetative and reproductive development: altered heterochrony, changes in meristem identity and organ number, and female sterility.

Angiosperm plants produce seeds by double fertilization: one of the two male gametes fertilizes the egg cell thus forming a diploid zygote (later embryo), while the other male gamete unites with the central cell of the embryo sac. In the majority of plant species this central cell is diploid, so the fusion product yielding the endosperm (extra-embryonic nurse tissue) is triploid. The ratio of two maternal genomes to one paternal genome, as well as the monoallelic expression of some genes dependent on the parent-of-origin in the endosperm, are often necessary for viable embryo and subsequent seed development (Haig and Westoby, 1991). The liliaceous plant *Gagea lutea* forms the embryo sac with a triploid chalazal and a haploid micropylar polar nucleus as parts of the central cell, which becomes pentaploid after fusion with the male gamete. The *G. lutea* endosperm nuclei are strikingly heterogeneous in structure: some parts possess a decondensed structure (euchromatin), while others form sticky, highly condensed bodies (facultative heterochromatin). This heterochromatin comes from the three inactivated maternal genomes of the chalazal polar nucleus in the embryo sac. In the young endosperm, the heterochromatin is present as smaller elongated bodies. As the seed matures, the endosperm nuclei become even more condensed and the heterochromatin forms large masses near the periphery of the nucleus (Buzek *et al.*, 1998a). Immunolabelling with an antibody raised against 5-mC showed that the distribution of 5-mC-rich regions in the nuclei was not homogenous. The signal in the euchromatin was generally very low, while some parts of the heterochromatin (but not all) displayed a heavy cytosine methylation. This indicates that the heavily methylated DNA regions are included in the heterochromatin, but DNA methylation cannot be responsible for all the heterochromatin. Experiments involving antisera raised against H4 histone acetylated at N-terminal lysine positions 5, 8 or 12 showed that nuclear regions, characterized by a highly intensive DAPI staining (heterochromatin), generally showed a weak immunolabelling (low histone acetylation), while the euchromatic parts possessed a high degree of H4 acetylation (Buzek *et al.*, 1998b). Taken together, these results show that the process of the facultative heterochromatin formation in plants is tightly connected with H4 histone deacetylation rather than DNA hypermethylation.

3. Experimental studies on dioecious *Melandrium album*

Melandrium album Garcke (syn. *Silene latifolia* Poiret, white campion, Caryophyllaceae family) is a classical model species displaying strict dioecy. *M. album* possesses a pair of easily distinguishable heteromorphic sex chromosomes: males are heterogametic ($2n = 22 + XY$) and females are homogametic ($2n = 22 + XX$). The Y chromosome harbours both male-determining and female-suppressing genes, while the role of the X chromosome has not been elucidated yet (reviewed in Ainsworth *et al.*, 1998; Grant *et al.*, 1994a; Westergaard, 1958). Contrary to some other dioecious plant species, no chemical treatment on white campion (e.g. with plant hormones) has been found to lead to a sex reversal.

3.1 *Epigenetic sex reversal*

Based on our previous data on tobacco, which indicated that genes involved in flower development are highly vulnerable to DNA methylation changes (Vyskot *et al.*, 1995), similar experiments were performed on *M. album*. We presumed that the absence of female sex organ formation in male plants is connected with an inactivation and hypermethylation of corresponding genes via suppressor gene products encoded by genes on the Y chromosome. We tested whether a hypomethylating drug (5-azaC at 10 μM) applied during the early seed germination could eliminate the female sex organ suppression in male plants. While there was no visible phenotypic effect of the drug on female plant development, 5-azaC induced a sex-reversal (androhermaphroditism) in 21% of male plants which then produced both male and bisexual flowers (*Figure 2*). The sexual character of flowers was rather dependent on their position within the inflorescence; bisexual flowers were more frequent in the later floral branches. Karyological analysis of these androhermaphroditic plants confirmed their standard male karyotype ($2n = 22 + XY$) without any observable aberrations. RFLP, studied using 5-mC-sensitive enzymes, and thin layer chromatographic analyses revealed that their genomes had a substantially reduced level of 5-mC (about 10% decrease in CG doublets when compared to control males).

Since these androhermaphrodites were fertile, a series of crosses and self-fertilization experiments were performed in order to follow the sexual transmission of the epimutation (Janousek *et al.*, 1996). The androhermaphroditism was inherited by sexual progeny when plants were self-pollinated or used as pollen parents, but it was never transmitted through the female line (*Figure 3*). It is worthwhile mentioning that the same results were obtained when pollen grains from both bisexual and male flowers were used in pollinations. This implies that the sexual inflorescence mosaicism could not be attributed to the mosaic methylation character of plant meristems induced by the 5-azaC treatment on seeds, but was a consequence of incomplete

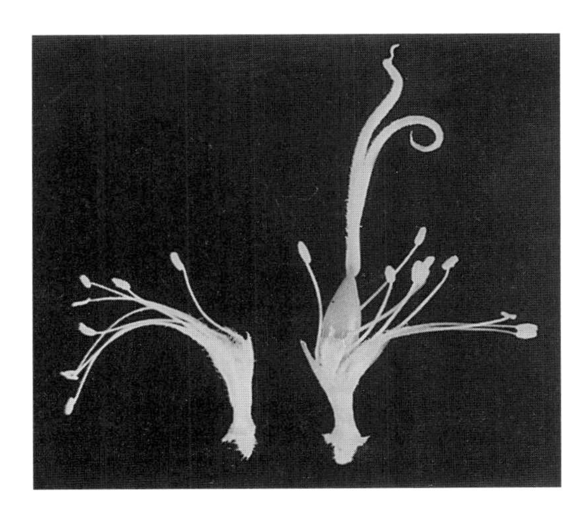

Figure 2. *Two types of flowers formed by 5-azaC induced* M. album *androhermaphrodites: male, staminate flower (left) and bisexual flower (right). The bisexual flowers possessed one to five carpels and nearly standard female fertility (seed setting).*

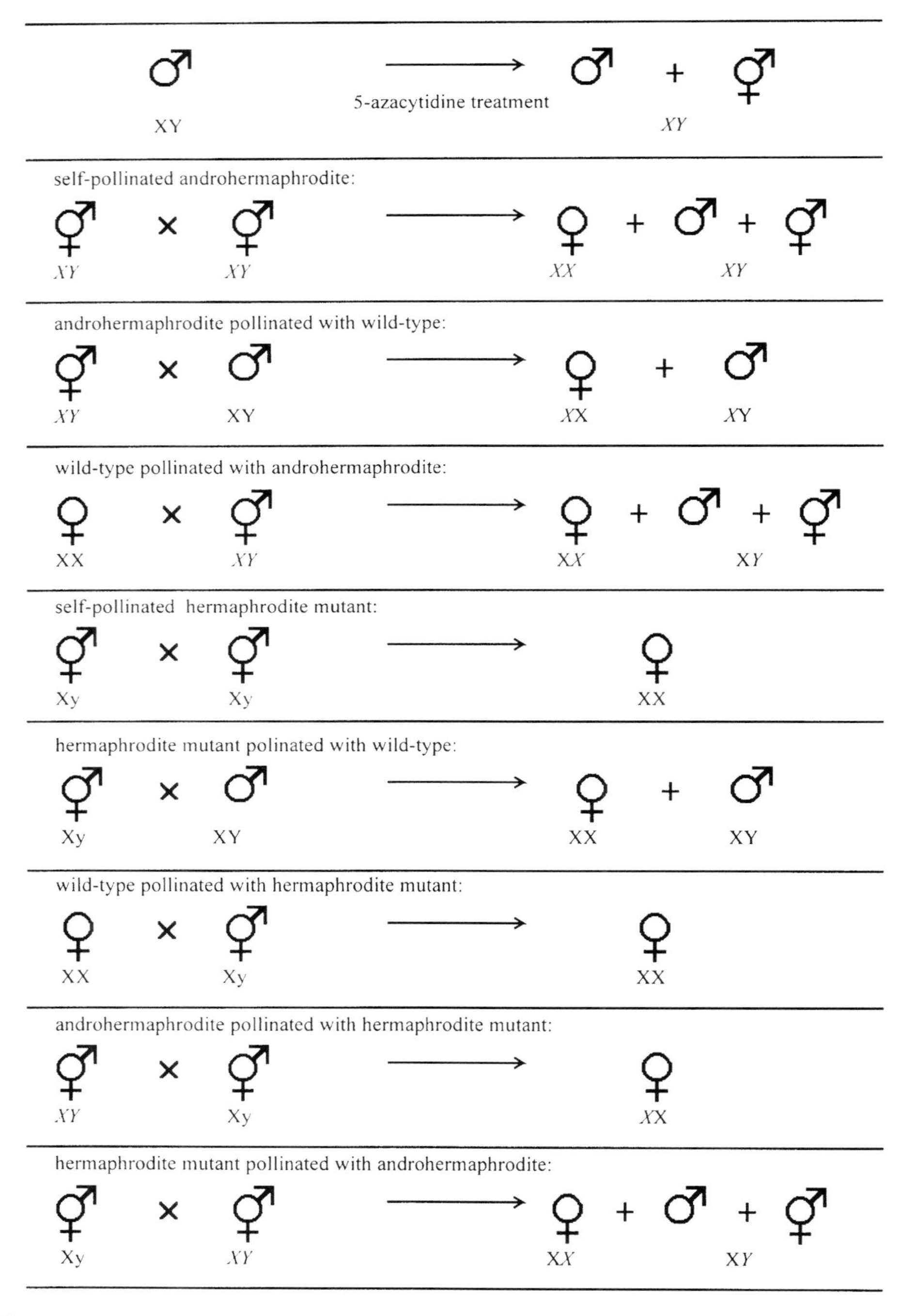

Figure 3. *General scheme of 5-azaC induced sex reversal (androhermaphroditism) in dioecious* M. album *and its inheritance by sexual progeny, including crosses with the hermaphrodite possessing an aberrant Y chromosome. The sex chromosomes, X and Y, of corresponding plants are indicated (italics mark the chromosomes possibly influenced by hypomethylation, y indicates the aberrant Y chromosome).*

expressivity of the epimutation. Genetic analyses have shown further that the androhermaphroditism is inherited through the male line (holandric inheritance) for at least four sexual generations with incomplete penetrance and varying expressivity. From these crosses it is clear that the gene(s) affected by epimutation cannot be located on the X chromosome, since the androhermaphrodites in the progeny of the wild-type parent pollinated with the androhermaphrodite could not obtain the X chromosome from the pollen donor. The incomplete dominant epimutation, leading from maleness to androhermaphroditism, if located on the Y chromosome, must cause partial inactivation of Y-linked gene(s), the products of which normally suppress pistil development, or if located on the autosome(s), the hypomethylation must activate gene(s) responsible for the female sex organ formation and which are normally repressed. In the latter case, the epimutation would be maintained through male gametogenesis, but restored when passing through the female gametophyte. If this is the case, it can be considered as a rare example of genomic imprinting at the embryonic level in plants, and has possibly co-evolved with the dioecious condition. Recent data indicate that a new methylation pattern is formed during plant gametogenesis as has been demonstrated in mammals (Oakeley and Jost, 1996).

Hermaphrodite *M. album* plants can also be isolated by mutation or deletion of gynoecium suppressor genes (reviewed by Westergaard, 1958). Such mutants have been obtained recently by pollinating wild-type female plants with X-ray mutagenized pollen (Grant *et al.*, 1994a). One of these hermaphroditic mutants (H8) possessed a large deletion on the non-homologous arm of the Y chromosome which led to its inability to pass this aberrant Y chromosome through both pollen and eggs (Donnison *et al.*, 1996). This implies that when this plant is self-fertilized or used as the pollen donor, only female plants appear in the progeny (*Figure 3*). In further experiments this mutant was used as a sexual partner to the 5-azaC-induced androhermaphrodite epimutant, to follow the meiotic transmission of the Y chromosome from the androhermaphrodite plant. Because of the monosporic origin of the *Melandrium* embryo sac, all the cells have the same chromosome constitution. Transmission of the Y chromosome through the female gametic line presumes the formation of megaspores of the $n = 11 + Y$ chromosome constitution (i.e. without any X chromosome) and their further development into the viable mature embryo sac. In evolutionary less advanced dioecious plant species, such as *Asparagus* and kiwi fruit (Testolin *et al.*, 1995), it is possible, after self-pollination of the AAXY hermaphrodite plants, to obtain progeny possessing the 'supermale' chromosome constitution, AAYY. However, self-pollination of spontaneous androhermaphrodite *M. rubrum* yielded no plants of the AAYY karyotype (van Nigtevecht, 1966). In accordance with this observation, only female haploid plants could be regenerated by androgenesis, indicating that the X-chromosome harbours some genes necessary for proper embryo development and plant survival (Veuskens *et al.*, 1992).

In our experiments, the control population (the hermaphrodite mutant as female × androhermaphrodite epimutant as male) contained, apart of females, both males and androhermaphrodites (possessing the Y chromosome from the androhermaphrodite pollen donor), while the progeny of the reciprocal cross (androhermaphrodite × hermaphrodite) consisted of females only (*Figure 3*). The genotypes of both parents and progeny were verified by karyological and RFLP analyses. The results show that the Y chromosome cannot pass through the female gametophyte, probably due to the fact that the absence of the X chromosome is incompatible with the proper development of the embryo sac (Janousek *et al.*, 1998). This can be caused by the absence of X-linked genes necessary for the egg cell (and/or embryo sac) survival. Our data on the non-transmissibility of the Y chromosome do not show that the Y chromosome can never pass through the female line. Westergaard (1946) showed that an aberrant Y chromosome (Y′) was transmitted from the tetraploid seed parent ($2n = 4n = 44 +$ XXXY′) to some male progeny when pollinated with wild-type pollen. However, in this case the karyotype of the megaspore and embryo sac included the X chromosome (AAXY′). Taken together, our results clearly demonstrate that the non-transmissibility of the Y chromosome through the female line in *M. album* is not caused by the inability of the Y chromosome to pass through the female gametophyte, but by the absence of the X chromosome at this particular stage of development.

3.2 *Different roles of X and Y chromosomes in plant development: when does the differentiation of the sexes begin?*

Many data indicate that the X and Y chromosomes differ substantially in size and have evolved relatively recently through replicated, independent events (Charlesworth, 1996). Attempts to isolate sex chromosome-specific DNA sequences by microdissection followed by DOP-PCR amplification yielded some repetitive clones which were localized both on the sex chromosomes (X and Y) and even on the autosomes (Buzek *et al.*, 1997; Grant *et al.*, 1994a; Scutt *et al.*, 1997). The first sex-specific DNA sequences (Donnison *et al.*, 1996) or even transcribed genes (Guttman and Charlesworth, 1998; Matsunaga *et al.*, 1996; Robertson *et al.*, 1997) have been isolated and characterized very recently. Although a close homology between the X and Y has been indicated (Guttman and Charlesworth, 1998), their meiotic pairing has been shown to be restricted to the pseudoautosomal regions which largely consist of telomeric and subtelomeric heterochromatin (Buzek *et al.*, 1997; Westergaard, 1958).

Melandrium album (as other dioecious plant species) comprises female and male individuals possessing pistillate and staminate flowers, respectively, but the differences between sexes are evident only upon flowering. A detailed analysis has shown that the first gender-specific differences in floral buds occur after the gynoecium and stamen primordia appear; the male gynoecium primordium then forms as an undifferentiated rod-like structure, while in later stages of female flower development, stamens are arrested after the anthers become distinguishable from the filaments (Grant *et al.*, 1994b). Whilst it is clear that genes on the Y chromosome are active during floral development, the question is raised as to whether Y chromosome genes are also active earlier, during plant embryonic and vegetative development. There is a clear evidence that the X chromosome is necessary for development of the female embryo sac (Janousek *et al.*, 1998) and sporophyte development (Veuskens *et al.*, 1992). The Y chromosome in *M. album* is largely euchromatic, but no data on its transcriptional activity, except its obvious function in the male flowering phase, are available (for review, see Grant *et al.*, 1994a). In *M. album* plant populations, a higher frequency of females than males is regularly found (the female bias). Genetic analyses have indicated that the female/male gender ratio is controlled by genes located on both the Y chromosome and the autosomes (Taylor, 1994). The female bias in *M. album* is a well-described, classical phenomenon (Correns, 1917), but no experimental evidence is available implicating a specific stage of plant development, in which a putative selective advantage of female plants is realized. It is obvious that the female bias, which also occurs in some animal species, has evolved to ensure the production of enhanced seed progeny; for successful fertilization of females a relatively small number of male plants is sufficient for both wind- and insect-pollinated species.

Segregation of X and Y chromosomes takes place during male meiosis; this is thus the first step in which a selective advantage of X-pollen grains could occur (meiotic drive). At least two pieces of evidence do not support this hypothesis: no morphological or size differences among pollen grains have been found (Caroll and Mulcahy, 1991), and since pollen fertility is very high (always higher than 90%; Janousek *et al.*, 1998), a substantial abortion of the Y-pollen grains is not indicated. The next step at which the female bias could be exerted is faster and more vigorous pollen germination and tube growth of the X-pollen leading to a higher probability of fertilization. This possibility could be checked by, for example, *in situ* hybridization on pollen tubes with X or Y chromosome specific DNA probes. Embryo and seed development represent highly complex processes in which the basic body pattern of the plant is established. It has been shown unambiguously that at least one copy of the X chromosome is a prerequisite for both haploid (Veuskens *et al.*, 1992) and diploid (Janousek *et al.*, 1998) sporophyte formation. Since some data indicate that one X chromosome is inactivated in homogametic females (Siroky *et al.*, 1994, 1998; Vyskot *et al.*, 1993), a negative role of some Y chromosome product(s) in the male embryo development cannot be excluded. The other possibility, still neglected, is the role of the early embryo feeding tissue, the endosperm, which comes from the second fertilization event. The embryo sac of *M. album* is of the *Polygonum* type (the endosperm is triploid), and could be subject to genomic imprinting, as described in many plant species (Kermicle and Alleman, 1990). Finally, female and male developing seeds differ in their genomes (as far as the X and Y chromosomes are concerned), in both embryo and endosperm (*Figure 4*). Taken together, we can speculate that the X chromosome dosage or the presence/absence of the Y chromosome could play an important role for the proper development of both the embryo and endosperm tissue. This difference would be manifested as a reduced production of male seeds or their lower germinability. Actually, the frequency of seed germination in our *M. album* material never exceeded 75% (Janousek *et al.*, 1998). The later stages of plant development, for example, the time and rate of seed germination, growth competition between female and male seedlings, as well as later phases of growth, do not seem to be responsible for the female bias. Individual seeds were grown on agar medium and plants cultured in individual pots till maturity: the female bias (usually between 60% and 70%) was similar to natural populations potentially subject to competition between sexes (Janousek and Vyskot, unpublished data). In summary, we conclude that the primary events leading to the female bias occur either during pollen tube growth or embryo and endosperm development, and are connected with the dosage of X chromosomes or with the presence/absence of the Y chromosome. The bias phenomenon in *M. album* populations, after more than 80 years of investigation, still represents a challenge for plant biologists; it is only part of the problem to identify the stage of development at which the differentiation between female and male sexes begins and whether the Y chromosome is transcriptionally active before flowering.

3.3 *Is there dosage compensation of X-linked genes?*

The inactivation of one X chromosome in mammalian females is the best described example of facultative heterochromatin formation in eukaryotes. On the inactive X

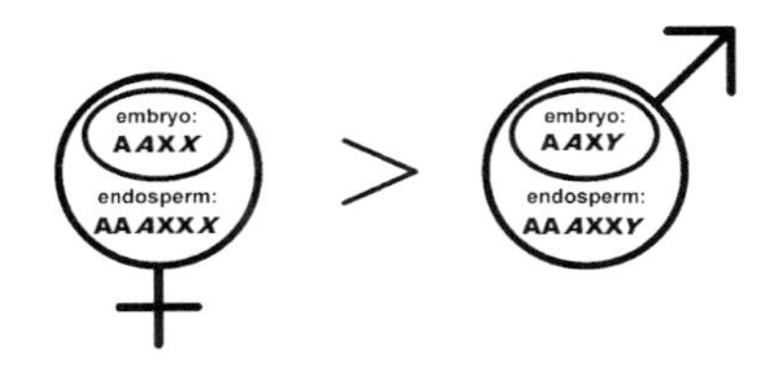

Figure 4. *Schematic comparison of the chromosome constitutions of female and male developing seeds of* M. album, *populations of which regularly display a more/less strong female bias. Sets of autosomes are indicated as the A, the chromosomes coming from the male parent (pollen donor) are shown in italics.*

chromosome, only a few genes, namely those contained in the X inactivation centre, escape from the transcriptional silencing. Recent studies have shown that a number of different mechanisms have evolved to maintain the X-inactivation, and are activated during early embryogenesis by RNA encoded by the *XIST* gene, in successive cell lineages (reviewed by, e.g., Jamieson *et al.*, 1996). While the positive correlation of DNA hypermethylation and the X chromosome inactivation is well documented by molecular data, the results of cytological and immunocytochemical experiments are ambiguous: some authors have observed a global condensation and hypermethylation of the inactive X chromosome (e.g. Haaf *et al.*, 1993; Prantera and Ferraro, 1990), while the others have not found any correlation (Bernardino *et al.*, 1996). Other mechanisms maintaining the X inactive status in mammalian females, late replication and core histone hypoacetylation, have also been demonstrated. In non-mammals, two different processes which regulate the levels of X chromosome-derived transcripts have been described: positive (up)regulation of genes on the single X in *Drosophila* XY males and negative (down)regulation in *Caenorhabditis* XX hermaphrodites (reviewed by Hodgkin, 1990). Finally, some animal species have no apparent dosage compensation mechanism (Chandra, 1991).

Melandrium album, in which female plants are homogametic (XX) and males are heterogametic (XY), uses the mammalian type of sex determination, as the male Y chromosome is strictly dominant and at least one X chromosome is necessary for survival. Provided that products encoded by the X chromosome are necessary for development of any sex, the question arises as to whether in this dioecious plant species (and as well as in others, e.g. *Rumex acetosa*) there is a mechanism compensating two doses of the X chromosome in females with the single X in males. To investigate the possibility of a dosage compensation mechanism in *M. album*, various experimental strategies have been applied. Some older data have indicated that there is late replicating sex chromatin in female nuclei (Choudhuri, 1969) and different banding patterns of the two X chromosomes resulted after quinacrine-mustard staining (Kampmeijer, 1972). In our first experiments, we estimated transcriptional activities of sex chromosomes prepared from *M. album* hairy root cultures (Vyskot *et al.*, 1993). Using *in situ* nick translation (RE/NT) driven by DNase or 5-mC-sensitive restriction endonucleases, we found a reproducible difference between the two X chromosomes in female metaphases (*Figure 5a*). This result indicates differences in chromatin condensation rather than in DNA methylation, since only transcriptionally active, relaxed chromatin regions are susceptible to nuclease treatment. However, chromatin structure is related to DNA methylation: 5-azaC treatment on cells before the RE/NT reaction led to a substantial reduction in the differences observed between the two X chromosomes (Vyskot *et al.*, 1993). These experiments also showed that the male Y chromosome is not largely heterochromatic, as has also been demonstrated by *in situ* hybridization with repetitive DNA clones and by chromosome C-banding (Grant *et al.*, 1994a).

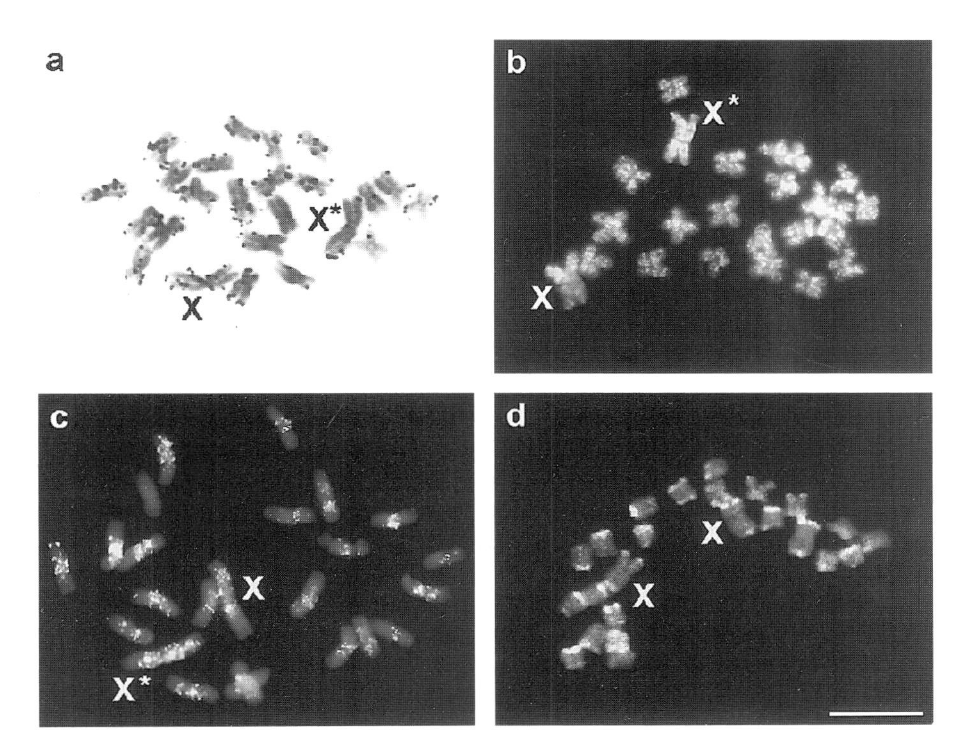

Figure 5. *Female metaphases from* M. album *root tips with special reference to their X chromosomes. (a)* in situ *nick translation driven by HpaII. Sites of cuts are visualized by incorporating [³H]dCTP followed by autoradiography; (b) DNA methylation patterns as detected by immunostaining with the mouse monoclonal antibody raised against 5-mC and the FITC-conjugated anti-mouse antibody; (c) DNA replication patterns after a short 5-bromodeoxyuridine pulse applied at late S phase and visualized with the mouse monoclonal antibody against 5-bromodeoxyuridine and the FITC-conjugated anti-mouse antibody; (d) histone H4 acetylated at lysine 5 detected with the specific rabbit antiserum against H4Ac5 and the FITC-conjugated anti-rabbit antibody. Chromosome slides were performed according to Hladilova* et al. *(1998) and counterstained with Giemsa (a) or propidium iodide (b–d). X chromosomes are indicated. Asterisks indicate hypermethylation (a, b) or late replication (c). Bar = 10 μm.*

DNA methylation patterns of *M. album* sex chromosomes have been studied using two experimental protocols: by *in vivo* labelling with *S*-adenosyl-L-[methyl-³H]methionine followed by autoradiography (Vyskot *et al.*, 1993) and by indirect immunofluorescence with a mouse monoclonal antibody raised against 5-mC (Siroky *et al.*, 1998). Both types of experiment showed reproducible differences between the two female X chromosomes and a similar level of methylation of the Y chromosome and the autosomes. The immunocytological data demonstrated that the 5-mC pattern of one of the two Xs strikingly resembles that of the single X chromosome present in males, while the other X chromosome in female cells displayed a global hypermethylation (*Figure 5b*). Moreover, pre-treatment of root tips with 5-azaC erased any specific 5-mC labelling patterns. We can conclude that the hypermethylation of one of the two X chromosomes in *M. album* female cells could be related to its inactivation, and interpreted as a specific dosage compensation mechanism (similar to mammals;

Lyon, 1961). This phenomenon could be unique in plants: it may have evolved with dioecy and must be linked with the fact that the X chromosome harbours genes necessary for development of both sexes. The 5-azaC-treated, undermethylated females did not display a sex change or any other disturbance in vegetative and reproductive development (Janousek *et al.*, 1996). However, our cytological and molecular analyses have shown that this treatment leads to a global genome hypomethylation, which includes the two X chromosomes. No differences between the two Xs were observed using *in situ* nick translation or anti-5-mC labelling and no shift in replication was demonstrated (Siroky *et al.*, 1998; Vyskot *et al.*, 1993). This indicates that even if the single X chromosome hypermethylation (and inactivation) occurs, it can be erased without any deleterious consequences.

Some recent data demonstrate that DNA methylation patterns are modified in mammalian tumour cells as a consequence of altered gene expression and that these changes could be monitored using 5-mC-antibody binding and digital imaging (Veilleux *et al.*, 1995). Similar data have also been described recently in plants (Lambe *et al.*, 1997). In our experiments, metaphase and interphase nuclei from the primary root meristem (root tips from germinating seeds after few cell divisions of the embryo) and from hairy root cultures resulting from *Agrobacterium rhizogenes* infection were compared. No differences in the methylation patterns of both sex chromosomes and autosomes were observed between the germinating seeds and the hairy root lines (Siroky *et al.*, 1998; Vyskot *et al.*, 1993). This result can be explained by at least two hypotheses. Firstly, hairy root transformation of plant cells is not accompanied by gross methylation changes in the nuclear genome, in contrast to mammalian tumours. Secondly, hairy root cell lines represent an unusual type of transformation which does not lead to non-differentiated tumour cells (in contrast to, e.g. crown gall tumours induced by *Agrobacterium tumefaciens*), but yields highly differentiated root tissues which may not necessarily differ from normal roots in their methylation patterns.

Other experiments have been focused on sex chromosome replication patterns. Autoradiographic analysis after [^{3}H]thymidine pulses showed that the X chromosome in male cells was relatively early replicating (similar to the autosomes and one of the two X chromosomes in females), while the second X chromosome in female cells was late replicating (Siroky *et al.*, 1994). The kinetics of DNA replication is clearly connected with DNA methylation since 5-azaC treatment led to a substantial reduction of the X replication difference. Similar experiments on chromosome kinetics have been also performed using an indirect immunofluorescence approach with a mouse anti-5-bromodeoxyuridine antibody after short *in vivo* pulses with 5-bromodeoxyuridine, and one of the two X chromosomes in female root tips has been shown to be very late replicating (*Figure 5c*). Some additional data obtained using [^{3}H]uridine pulses and precise measurement of chromosome lengths showed that one of the two female X chromosomes is prone to aberration, which could indicate a higher transcriptional activity, and less condensed state (Siroky *et al.*, 1994). It can be concluded that, as is the case with the inactive mammalian X chromosome, one of the two *M. album* female X chromosomes is late replicating, a fact which also supports its facultatively heterochromatic character.

Recent data have demonstrated that the inactive mammalian X chromosome is depleted in histone H4 acetylation (Jeppesen and Turner, 1993). Using the same antisera against various isoforms of acetylated H4 histones, we have analysed labelling patterns of *M. album* chromosomes. In these experiments any visible differences between the two female X chromosomes were detected. We observed that in *M. album*, all distal/subtelomeric chromosome regions, on both the sex chromosomes and autosomes, displayed strong signals of histone H4 acetylated at N-terminal lysine positions 5, 8 and 12 (*Figure 5d*), while the distribution of H4Ac16 was uniform along the chromosomes. Similar results were obtained in three other *Silene* species, but not in two non-related plant species tested (*Allium cepa* and *Nicotiana tabacum*), where H4Ac5, 8, 12 and 16 distributions were rather heterogeneous along the chromosomes (Vyskot *et al.*, 1999). Supported by our recent studies on the structure and early replication of *M. album* chromosome ends (Buzek *et al.*, 1997; Riha *et al.*, 1998; Vyskot *et al.*, 1999), we conclude that the chromosomes of this species (in common with other *Silene* species) have gene rich regions in distal/subtelomeric regions. Provided that the differences in replication timing and DNA methylation really reflect the inactivation of one of the two X chromosomes, one can speculate on why this chromosome inactivation is not accompanied by a global histone underacetylation. Mammalian species have evolved several different mechanisms to maintain and transmit the transcriptionally silenced status of the X chromosome. In marsupials, the inactive X chromosome (always paternal) is late-replicating (Sharman, 1971) and histone-hypoacetylated but not hypermethylated (Wakefield *et al.*, 1997), while in eutherian mammals, the X inactivation is random and the inactive X is late-replicating (Comings, 1967), hypoacetylated (Jeppesen and Turner, 1993) and hypermethylated. However, no H4 histone underacetylation of the transcriptionally inactive male X chromosome was found in meiotic cells of the mammalian male germline (Armstrong *et al.*, 1997). This last example may indicate that core histone underacetylation is not always essential for the transcriptional inactivation.

Histone acetylation and DNA methylation seem to be two evolutionary unique epigenetic mechanisms which could be responsible for long-term functional inertness or potential transcriptional activity of genes, chromosome regions, and whole chromosomes or genomes. Although they refer to two different basic components of chromatin, histones and DNA, both substantially influence the final chromatin structure and function. Histone acetylation and DNA methylation patterns could be simply modified using drugs which specifically inhibit histone deacetylases (e.g. trichostatin A; Yoshida *et al.*, 1990) or DNA-cytosine methyltransferases (e.g. 5-azaC; Santi *et al.*, 1984), respectively. These drugs have similar effects, both releasing a condensed chromatin structure, and could induce gene activity (Chen and Pikaard, 1997). Their mutual interrelationship is not yet clear, but recent data indicate that transcriptional repression by a methyl-CpG-binding protein in mouse involves a histone deacetylase complex (Nan *et al.*, 1998). In preliminary experiments, we have observed that both trichostatin A and 5-azaC induced the decondensation of plant constitutive heterochromatin and increased intensities of H4Ac5 chromatin labelling (Vyskot *et al.*, unpublished data).

4. Conclusions

During the last few years, data have accumulated which show that DNA methylation and core histone acetylation play important roles in the long-term control of gene expression and development. They are obviously responsible for (or at least accompany) mitotic (cell memory) and meiotic transmission (epigenetic inheritance and genomic imprinting) of gene expression. DNA demethylation and/or histone acetylation seem to be a necessary step toward potential gene activation (reviewed by Turner, 1998).

However, not all eukaryotic genes are regulated by methylation and these exceptions, which include *Drosophila*, nematodes, ciliate protozoans and yeasts, illustrate that DNA methylation is only one of several possible modifying controls used in epigenetic regulation. This suggests that methylation as a device for transcriptional regulation appeared relatively late in the evolution of animal and plant kingdoms. The single X inactivation in female marsupial cells indicates that histone underacetylation, unlike DNA methylation, is a feature of dosage compensation in a common mammalian ancestor (Wakefield *et al.*, 1997). The molecular effects of methylation in transcriptional regulation have not yet been elucidated fully. Recent data implicate transcriptional repressors of methylated DNA accompanied by a specific chromatin assembly (Kass *et al.*, 1997) and indicate that CpG methylation alters the chromatin structure by preventing the histone octamer from interaction with an otherwise high affinity positioning sequence in the promoter region (Davey *et al.*, 1997).

DNA methylation obviously plays a pleiotropic role in eukaryotic cells and organisms. It represents a defence mechanism used to inactivate mobile genetic elements and other types of intrusive DNA (e.g. transgenes). It is involved in the condensed, heterochromatic structure of repetitive DNA sequences. It ensures monoallelic expression of some genes depending on their parent-of-origin (genomic imprinting). It could accompany permanently or transiently inactive chromosomes or their complete sets (e.g. one X chromosome inactivation in female mammals), and it participates in a long-term inactivation of developmentally controlled genes. Recent data demonstrate that DNA hypermethylation and/or core histone underacetylation are chromatin modifications which characterize a transcriptional inertness in both constitutive and facultative heterochromatin. They do not seem to be causative processes of transcriptional inactivation as demonstrated by experiments on the mammalian inactive X chromosome; both DNA hypermethylation (Lock *et al.*, 1987) and histone H4 underacetylation (Keohane *et al.*, 1996) follow the silencing of X-linked genes. In flowering plants, both the most important mechanisms of epigenetic control, DNA methylation and histone acetylation, have been unambiguously demonstrated. The experimental data on *Melandrium* presented here show that sex expression is under DNA methylation control and indicate that in dioecious plants, due to the gender separation, epigenetic mechanisms could also play other roles (dosage compensation, genomic imprinting) as in mammals. The isolation of genes involved in *M. album* sex expression or genes located on the sex chromosomes will clearly help to shed light on these processes.

Acknowledgements

I am grateful to Dr Charles C. Ainsworth (Wye College, University of London) for critical reading and English revision of the manuscript, and to all colleagues in my laboratory, particularly Drs Jiri Siroky, Bohuslav Janousek and Karel Riha, for collaboration. This work was supported by the Grant Agency of the Czech Republic and the Grant Agency of the Czech Academy of Sciences.

References

Adams, R.L.P. (1996) DNA methylation. *Princ. Med. Biol.* **5**: 33–66.

Ainsworth, C., Parker, J. and Buchanan-Wollaston, V. (1998) Sex determination in plants. *Curr. Topics Develop. Biol.* **38**: 167–223.

Armstrong, S.J., Hulten, M.A., Keohane, A.M. and Turner, B.M. (1997) Different strategies of X-inactivation in germinal and somatic cells: histone H4 underacetylation does not mark the inactive X chromosome in the mouse male germline. *Exp. Cell Res.* **230**: 399–402.

Bernardino, J., Lamoliatte, E., Lombard, M., Niveleau, A., Malfoy, B., Dutrillaux, B. and Bourgeois, C.A. (1996) DNA methylation of the X chromosomes of the human female: an *in situ* semi-quantitative analysis. *Chromosoma* **104**: 528–535.

Burn, J.E., Bagnall, D.J., Metzger, J.D., Dennis, E.S. and Peacock, W.J. (1993) DNA methylation, vernalization, and the initiation of flowering. *Proc. Natl Acad. Sci. USA* **90**: 287–291.

Buzek, J., Koutnikova, H., Houben, A., Riha, K., Janousek, B., Siroky, J., Grant, S. and Vyskot, B. (1997) Isolation and characterization of X chromosome-derived DNA sequences from a dioecious plant *Melandrium album*. *Chromosome Res.* **5**: 57–65.

Buzek, J., Ebert, I., Ruffini-Castiglione, M., Siroky, J., Vyskot, B. and Greilhuber, J. (1998a) Structure and DNA methylation pattern of partially heterochromatinised endosperm nuclei in *Gagea lutea* (Liliaceae). *Planta* **204**: 506–514.

Buzek, J., Riha, K., Siroky, J., Ebert, I., Greilhuber, J. and Vyskot, B. (1998b) Histone H4 underacetylation in plant facultative heterochromatin. *Biol. Chem.*, **379**: 1235–1241.

Caroll, S.B. and Mulcahy, D.L. (1991) The relationship between pollen grain size and progeny gender in dioecious *Silene latifolia* (Caryophyllaceae). *Sex. Plant Reprod.* **4**: 203–207.

Chandra, H.S. (1991) How do heterogametic females survive without gene dosage compensation? *J. Genet.* **70**: 137–146.

Charlesworth, B. (1996) The evolution of sex determination and dosage compensation. *Curr. Biol.* **6**: 149–162.

Chen, Z.J. and Pikaard, C.S. (1997) Epigenetic silencing of RNA polymerase I transcription: a role for DNA methylation and histone modification in nucleolar dominance. *Genes Develop.* **11**: 2124–2136.

Choudhuri, H.C. (1969) Late DNA replication pattern in sex chromosomes of *Melandrium*. *Can. J. Genet. Cytol.* **11**: 192–198.

Comings, D.E. (1967) The duration of replication of the inactive X chromosome in humans based on the persistence of the heterochromatic sex chromatin body during DNA synthesis. *Cytogenetics* **6**: 20–37.

Correns, C. (1917) Ein Fall experimenteller Verschiebung des Geschlechtsverhaltnisses. *Sitzungsberg K. Preuss Akad. Wiss.* **51**: 685–717.

Davey, C., Pennings, S. and Allan, J. (1997) CpG methylation remodels chromatin structure *in vitro*. *J. Mol. Biol.* **267**: 276–288.

Donnison, I.S., Siroky, J., Vyskot, B., Saedler, H. and Grant, S.R. (1996). Isolation of Y chromosome-specific sequences from *Silene latifolia* and mapping of male sex-determining genes using representational difference analysis. *Genetics* **144**: 1893–1901.

Drozdenyuk, A.P., Sulimova, G.E. and Vanyushin, B.I. (1976) Changes in base composition and molecular population of wheat DNA on germination. *Mol. Biol.* **10**: 1378–1386.

Finnegan, E.J. and Dennis, E.S. (1993) Isolation and identification by sequence homology of a putative cytosine methyltransferase from *Arabidopsis thaliana*. *Nucl. Acids Res.* **21**: 2383–2388.

Finnegan, E.J., Peacock, W.J. and Dennis, E.S. (1996) Reduced DNA methylation in *Arabidopsis thaliana* results in abnormal plant development. *Proc. Natl Acad. Sci. USA* **93**: 8449–8454.

Finnegan, E.J., Genger, R.K., Kovac, K., Peacock, W.J. and Dennis, E.S. (1998) DNA methylation and the promotion of flowering by vernalization. *Proc. Natl Acad. Sci. USA* **95**: 5824–5829.

Follmann, H., Balzer, H.-J. and Schleicher, R. (1990) Biosynthesis and distribution of methylcytosine in wheat DNA. How different are plant DNA methyltransferases? In: *Nucleic Acid Methylation*. Alan R. Liss, pp. 199–210.

Grant, S., Houben, A., Vyskot, B., Siroky, J., Pan, W.-H., Macas, J. and Saedler, H. (1994a) Genetics of sex determination in flowering plants. *Develop. Genet.* **15**: 214–230.

Grant, S., Hunkirchen, B. and Saedler, H. (1994b) Developmental differences between male and female flowers in the dioecious *Silene latifolia*. *Plant J.* **6**: 471–480.

Gruenbaum, Y., Naveh-Many, T., Cedar, H. and Razin, A. (1981) Sequence specificity of methylation in higher plant DNA. *Nature* **292**: 860–862.

Guttman, D.S. and Charlesworth, D. (1998) An X-linked gene with a degenerate Y-linked homologue in a dioecious plant. *Nature* **394**: 263–266.

Haaf, T., Werner, P. and Schmid, M. (1993) 5-Azadeoxycytidine distinguishes between active and inactive X chromosome condensation. *Cytogenet. Cell Genet.* **63**: 160–168.

Haig, D. and Westoby, M. (1991) Genomic imprinting in endosperm: its effect on seed development in crosses between species, and between different ploidies of the same species, and its implications for the evolution of apomixis. *Phil. Trans. Roy. Soc. Lond. B* **333**: 1–13.

Hladilova, R., Siroky, J. and Vyskot, B. (1998) A cytospin technique for the spreading of plant metaphases suitable for immunofluorescence studies. *Biotechnic Histochem.* **73**: 150–156.

Hodgkin, J. (1990). Sex determination compared in *Drosophila* and *Caenorhabditis*. *Nature* **344**: 721–728.

Jacobsen, S.E. and Meyerowitz, E.M. (1997) Hypermethylated *SUPERMAN* epigenetic alleles in *Arabidopsis*. *Science* **277**: 1100–1003.

Jaenisch, R. (1997) DNA methylation and imprinting: why bother? *Trends Genet.* **13**: 323–329.

Jamieson, R.V., Tam, P.P.L. and Gardiner-Garden, M. (1996) X-chromosome activity: impact of imprinting and chromatin structure. *Int. J. Develop. Biol.* **40**: 1065–1080.

Janousek, B., Siroky, J. and Vyskot, B. (1996) Epigenetic control of sexual phenotype in a dioecious plant, *Melandrium album*. *Mol. Gen. Genet.* **250**: 483–490.

Janousek, B., Grant, S.R. and Vyskot, B. (1998) Non-transmissibility of the Y chromosome through the female line in androhermaphrodite plants of *Melandrium album*. *Heredity* **80**: 576–583.

Jeppesen, P. and Turner, B.M. (1993) The inactive X chromosome in female mammals is distinguished by a lack of histone H4 acetylation, a cytogenetic marker for gene expression. *Cell* **74**: 281–289.

Kafri, T., Ariel, M., Brandeis, M., Shemer, R., Urven, L., McCarrey, J., Cedar, H. and Razin, A. (1992) Developmental pattern of gene specific DNA methylation in the mouse embryo and germ line. *Genes Develop.* **6**: 705–714.

Kakutani, T., Jeddeloh, J.A. and Richards, E.J. (1995) Characterization of an *Arabidopsis thaliana* DNA hypomethylation mutant. *Nucl. Acids Res.* **23**: 130–137.

Kakutani, T., Jeddeloh, J.A., Flowers, S.K., Munakata, K. and Richards, E.J. (1996) Developmental abnormalities and epimutations associated with DNA hypomethylation mutations. *Proc. Natl Acad. Sci. USA* **93**: 12406–12411.

Kampmeijer, P. (1972) Fluorescence pattern of the sex chromosomes of *Melandrium dioicum*, stained with quinacrine-mustard. *Genetica* **43**: 201–206.

Kass, S.U., Pruss, D. and Wolffe, A.P. (1997) How does DNA methylation repress transcription? *Trends Genet.* **13**: 444–449.

Keohane, A.M., O'Neill, L.P., Belyaev, N.D., Lavender, J.S. and Turner, B.M. (1996) X-inactivation and histone H4 acetylation in embryonic stem cells. *Develop. Biol.* **180**: 618–630.

Kermicle, J.L. and Alleman, M. (1990) Gametic imprinting in maize in relation to the angiosperm life cycle. *Development* (suppl.): 9–14.

Koukalova, B., Kuhrova, V., Vyskot, B., Siroky, J. and Bezdek, M. (1994) Maintenance of induced hypomethylated state of tobacco nuclear repetitive DNA sequences in the course of protoplast and plant regeneration. *Planta* **194**: 306–310.

Lambe, P., Mutambel, H.S.N., Fouche, J.-G., Deltour, R., Foidart, J.-M. and Gaspar, T. (1997) DNA methylation as a key process in regulation of organogenic totipotency and plant neoplastic progression? *In Vitro Cell. Develop. Biol.* **33P**: 155–162.

Li, E., Bestor, T.H. and Jaenisch, R. (1992) Targeted mutation of the DNA methyltransferase gene results in embryonic lethality. *Cell* **69**: 915–926.

Lock, L.F., Takagi, N. and Martin, G.R. (1987) Methylation of the *Hprt* gene on the inactive X occurs after chromosome inactivation. *Cell* **48**: 39–46.

Lyon, M.F. (1961) Gene action in the X chromosome of the mouse. *Nature* **190**: 372–373.

Matsunaga, S., Kawano, S., Takano, H., Uchida, H., Sakai, A. and Kuroiwa, T. (1996) Isolation and developmental expression of male reproductive organ-specific genes in a dioecious campion, *Melandrium album* (*Silene latifolia*). *Plant J.* **10**: 679–689.

Matzke, M. and Matzke, A.J.M. (1993) Genomic imprinting in plants: parental effects and *trans*-inactivation phenomena. *Annu. Rev. Plant Physiol. Plant Mol. Biol.* **44**: 53–76.

Messeguer, R., Ganal, M.W., Steffens, J.C. and Tanksley, S.D. (1991) Characterization of the level, target sites and inheritance of cytosine methylation in tomato nuclear DNA. *Plant Mol. Biol.* **16**: 753–770.

Meyer, P., ed. (1995) *Gene Silencing in Higher Plants and Related Phenomena in Other Eukaryotes*. Springer-Verlag, Berlin.

Meyer, P., Niedenhof, I. and Lohuis, M.T. (1994) Evidence for cytosine methylation of non-symmetrical sequences in transgenic *Petunia hybrida*. *EMBO J.* **13**: 2084–2088.

Monk, M. (1990) Variation in epigenetic inheritance. *Trends Genet.* **6**: 110–114.

Monk, M., Boubelik, M. and Lehnert, S. (1987) Temporal and regional changes in DNA methylation in the embryonic, extraembryonic and germ cell lineages during mouse embryo development. *Development* **99**: 371–382.

Munksgaard, D., Mattsson, O. and Okkels, F.T. (1995) Somatic embryo development in carrot is associated with an increase in levels of S-adenosylmethionine, S-adenosylhomocysteine and DNA methylation. *Physiol. Plant.* **93**: 5–10.

Nan, X., Ng, H.-H., Johnson, C.A., Laherty, C.D., Turner, B.M., Eisenman, R.N. and Bird, A. (1998) Transcriptional repression by the methyl-CpG-binding protein MeCP2 involves a histone deacetylase complex. *Nature* **393**: 386–389.

Oakeley, E.J. and Jost, J.-P. (1996) Non-symmetrical cytosine methylation in tobacco pollen DNA. *Plant Mol. Biol.* **31**: 927–930.

Prantera, G. and Ferraro, M. (1990) Analysis of methylation and distribution of CpG sequences on human active and inactive X chromosomes by *in situ* nick translation. *Chromosoma* **99**: 18–23.

Razin, A. and Cedar, H. (1993) DNA methylation and embryogenesis. In: *DNA Methylation: Molecular Biology and Biological Significance* (eds J.P. Jost and H.P. Saluz). Birkhauser-Verlag, Basel, pp. 343–357.

Reik, W. (1996) Genetic imprinting: the battle of the sexes rages on. *Exp. Physiol.* **81**: 161–172.

Richards, E.J. (1997) DNA methylation and plant development. *Trends Genet.* **13**: 319–323.

Riha, K., Fajkus, J., Siroky, J. and Vyskot, B. (1998) Developmental control of telomere lengths and telomerase activity in plants. *Plant Cell* **10**: 1691–1698.

Robertson, S.E., Li, Y., Scutt, C.P., Willis, M.E. and Gilmartin, P.M. (1997) Spatial expression dynamics of *Men-9* delineate the third floral whorl in male and female flowers of dioecious Silene latifolia. *Plant J.* **12**: 155–168.

Ronemus, M.J., Galbiati, M., Ticknor, C., Chen, J. and Dellaporta, S.L. (1996) Demethylation-induced developmental pleiotropy in *Arabidopsis*. *Science* **273**: 654–657.

Sano, H., Kamada, I., Youssefian, S. and Wabiko, H. (1989) Correlation between DNA under-methylation and dwarfism in maize. *Biochim. Biophys. Acta* **1009**: 35–38.

Sano, H., Kamada, I., Youssefian, S., Katsumi, M. and Wabiko, H. (1990) A single treatment of rice seedlings with 5-azacytidine induces heritable dwarfism and undermethylation of genomic DNA. *Mol. Gen. Genet.* **220**: 441–447.

Santi, D.V., Norment, A. and Garrett, C.E. (1984) Covalent bond formation between a DNA-cytosine methyltransferase and DNA containing 5-azacytosine. *Proc. Natl Acad. Sci. USA* **81**: 6993–6997.

Scutt, C.P., Kamisugi, Y., Sakai, F. and Gilmartin, P.M. (1997) Laser isolation of plant sex chromosomes: studies on the DNA composition of the X and Y sex chromosomes of *Silene latifolia*. *Genome* **40**: 705–715.

Sharman, G.B. (1971) Late DNA replication in the paternally derived X chromosome of female kangaroos. *Nature* **230**: 231–232.

Siroky, J., Janousek, B., Mouras, A. and Vyskot, B. (1994). Replication pattern of sex chromosomes in *Melandrium album* female cells. *Hereditas* **120**: 175–181.

Siroky, J., Ruffini Castiglione, M. and Vyskot, B. (1998) DNA methylation patterns of *Melandrium album* chromosomes. *Chromosome Res.* **6**: (in press).

Taylor, D.R. (1994) The genetic basis of sex ratio in *Silene alba* (= *S. latifolia*). *Genetics* **136**: 641–651.

Testolin, R., Cipriani, G. and Costa, G. (1995) Sex segregation ratio and gender expression in the genus *Actinidia*. *Sex. Plant Reprod.* **8**: 129–132.

Theiss, G. and Follmann, H. (1980) 5-Methylcytosine formation in wheat embryo DNA. *Biochem. Biophys. Res. Commun.* **94**: 291–297.

Turner, B.M. (1998) Histone acetylation as an epigenetic determinant of long-term transcriptional competence. *Cell. Mol. Life Sci.* **54**: 21–31.

van Nigtevecht, G. (1966) Genetic studies in dioecious *Melandrium*: II. Sex determination in *Melandrium album* and *Melandrium dioicum*. *Genetica* **37**: 307–344.

Veilleux, C., Bernardino, J., Gibaud, A., Niveleau, A., Malfoy, B., Dutrillaux, B. and Bourgeois, C.A. (1995) Changes in methylation in tumor cells: a new *in situ* quantitative approach on interphase nuclei and chromosomes. *Bull. Cancer* **82**: 939–945.

Veuskens, J., Ye, D., Oliveira, M., Ciupercescu, D.D., Installe, P., Verhoven, H.A. and Negrutiu, I. (1992) Sex determination in the dioecious *Melandrium album*: androgenic embryogenesis requires the presence of the X chromosome. *Genome* **35**: 8–16.

Vongs, A., Kakutani, T., Martienssen, R.A. and Richards, E.J. (1993) *Arabidopsis thaliana* DNA methylation mutants. *Science* **260**: 1926–1928.

Vyskot, B., Araya, A., Veuskens, J., Negrutiu, I. and Mouras, A. (1993) DNA methylation of sex chromosomes in a dioecious plant, *Melandrium album*. *Mol. Gen. Genet.* **239**: 219–224.

Vyskot, B., Koukalova, B., Kovarik, A., Sachambula, L., Reynolds, D. and Bezdek, M. (1995) Meiotic transmission of a hypomethylated repetitive DNA family in tobacco. *Theor. Appl. Genet.* **91**: 659–664.

Vyskot, B., Siroky, J., Hladilova, R., Belyaev, N.D. and Turner, B.M. (1999) Euchromatic domains in plant chromosomes as revealed by H4 histone acetylation and early DNA replication. *Genome* (in press).

Wakefield, M.J., Keohane, A.M., Turner, B.M. and Graves, J.A.M. (1997) Histone underacetylation is an ancient component of mammalian X chromosome inactivation. *Proc. Natl Acad. Sci. USA* **94**: 9665–9668.

Westergaard, M. (1946) Structural changes of the Y chromosome in the offspring of polyploid *Melandrium*. *Hereditas* **32**: 60–64.

Westergaard, M. (1958) The mechanism of sex determination in dioecious plants. *Adv. Genet.* **9**: 217–281.

Yisraeli, J. and Szyf, M. (1984) Gene methylation patterns and expression. In: *DNA Methylation: Biochemistry and Biological Significance* (eds A. Razin, H. Cedar and A.D. Riggs). Springer-Verlag, New York, pp. 352–370.

Yoder, J.A., Walsh, C.P. and Bestor, T.H. (1997) Cytosine methylation and the ecology of intragenomic parasites. *Trends Genet.* **13**: 335–340.

Yoshida, M., Kijima, M., Akita, M. and Beppu, T. (1990) Potent and specific inhibition of mammalian histone deacetylase both *in vivo* and *in vitro* by trichostatin A. *J. Biol. Chem.* **255**: 17174–17179.

Zagris, N. and Podimatas, T. (1994) 5-Azacytidine changes gene expression and causes developmental arrest of early chick embryo. *Int. J. Develop. Biol.* **38**: 741–749.

7

Sex determination by X:autosome dosage: *Rumex acetosa* (sorrel)

Charles C. Ainsworth, Jianping Lu, Mark Winfield and John Parker

1. Sex chromosome-based sex determination systems: active Y versus X:autosome dosage

Dioecy which is accompanied by sex chromosome systems in flowering plants is rare. Despite this fact, the main sex chromosome systems characterizing sex determination in animals are represented in plants. The two basic systems are the active-Y system and the X:autosome dosage system.

The white campion (*Silene latifolia*; formerly known as *Melandrium album* and *Silene alba*), is a member of the family Caryophyllaceae and is one of a small group of dioecious plants which has an active-Y system of sex determination. Of the members of this group, white campion is undoubtedly the best characterized and has been the subject of considerable research efforts for many years since the first genetic studies in the 1930s and 1940s (Warmke and Blakeslee, 1939; Westergaard, 1940, 1946; Winge, 1931; for reviews see Grant *et al.*, 1994, 1996; Ye *et al.*, 1991). The active-Y system found in campion mirrors that of mammals in that the sex of the individual is determined by the presence or absence of the Y chromosome. However, the plant and animal systems are fundamentally different. In mammals, the default pathway, in the absence of the signal from the Y chromosome, is the female one in which the bipotential indifferent gonad develops into the female organs. In males, a single locus, *SRY* on the mammalian Y chromosome, encodes a protein which represses a negative regulator of male development and induces the testis determination pathway (McElreavey *et al.*, 1993; Sinclair *et al.*, 1990; reviewed in Capel, 1996). In campion, in contrast to the single primary sex-determining gene of mammals, genetic analysis of Y chromosome deletion mutants has shown that the Y chromosome carries dominant genes in three regions of the chromosome. These suppress carpel development, promote stamen development, and allow development of the stamens once initiated (Westergaard, 1946, 1958; Ye *et al.*, 1991; see also Chapters 4 and 5 in this volume for detailed accounts).

Sex Determination in Plants, edited by C.C. Ainsworth.

The alternative system of chromosomally based sex determination in plants, found only in some *Rumex* species, hop (*Humulus lupulus* and *H. japonicus*) and cannabis (*Cannabis sativa*), is X:autosome dosage or X: autosome balance which parallels that found in *Drosophila* and *Caenorhabditis* (reviewed in Mittwoch, 1996). Here, the Y chromosome is not an important signal (and is absent in *Caenorhabditis*). The chromosomal signal in these organisms is the ratio of the number of X chromosomes to the number of sets of autosomes (the X:autosome ratio); in females of both organisms, the X:A ratio is 1.0, and in males is 0.5. In *Drosophila*, the primary binary switching gene, *Sxl*, is transcribed in both sexes throughout development but is functionally active only in females, directing female development through its effect on at least two separate cascades of genes that regulate sexual differentiation and dosage compensation (Keyes *et al*., 1992). *Sxl* is activated by a system which assesses the X:A ratio, and includes X-located numerator genes and autosomal denominator genes. Several numerator genes, *sis-a*, *sis-b*, *sis-c* and *runt* have been identified, all of which encode transcription factors and which must increase the probability of activating the *Sxl* promoter.

We will not know the extent of the similarity between this system and those found in *Rumex*, hop and cannabis until some of the key genes are cloned. However, the similarity at the level of chromosomal involvement is striking. From the study of *Rumex* aneuploids we do know that several different autosomes, in addition to the X chromosome, are involved in sex determination and might carry genes analogous to the *Drosophila* autosomal denominator genes. However, this seems unlikely and will be discussed below.

Of the dioecious plant species which use X:autosome dosage, the most intensively studied has been *Rumex acetosa* (commonly named sorrel), a strictly dioecious perennial member of the Polygonaceae which is cultivated in some countries for its acidic leaves. The subgenus Acetosa of the *Rumex* genus comprises ten dioecious species *(R. acetosa, R. rothschildianus, R. thyrsiflorus, R. intermedius, R. papillaris, R. nivalis, R. alpestris, R. nebroides, R. thyrsoides and R. tuberosus*), which are all characterized by a distinctive sex chromosome system, and also a number of hermaphrodite species (for example, *R. scutatus* and *R. vesicarius*). In all species, females have the chromosome constitution $2n = 2x = 12 + XX$ whilst males have $2n = 2x = 12 + XY_1Y_2$. It is clear, therefore, that the evolution of dioecy and the XX/XY_1Y_2 sex chromosome system predated speciation in this genus. This group of species is remarkably diverse in habitat, form and geographical location, and includes an annual which grows in the dry coastal plains of Israel and North Africa (*R. rothschildianus*), a species which inhabits snowpatches at 2800 m in the Alps (*R. nivalis*) and a large woody perennial which grows in forests in the high European mountains and up to the Arctic circle (*R. alpestris*). Interestingly, two species which are outside the Acetosa subgenus, appear to have evolved dioecy independently of the members of the Acetosa subgenus. These are the North American species, *R. hastatulus*, which is distinct in that both the X and Y chromosomes are involved in an intermediate system (Smith, 1969) and *R. acetosella*, a dioecious member of the Acetosella subgenus, but which has no heteromorphic sex chromosomes. A recently cloned Y chromosome AFLP fragment (see below) will enable us to look in depth at the evolution of the sex chromosome system and sex determination in these species.

2. Morphology and development of *Rumex acetosa* flowers

Mature flowers of all species of *Rumex* are small (usually less than 4 mm in diameter) and non-showy, characters that are often associated with wind pollination. In dioecious species of *Rumex*, male and female flowers have only three whorls of developed organs. The male flowers consist of a whorl of stamens and two whorls of perianth segments; the female flower consists of three fused carpels with a single ovule and two whorls of perianth segments, one of which encloses the ovary. Male and female flowers from two dioecious species, *R. acetosa* and *R. rothschildianus*, are shown in *Figure 1.* However, in the basic *Rumex* flower as carried by hermaphrodite *Rumex* species, there are four whorls of organs. The two outer whorls belong to the perianth (organs in both whorls are small and sepaloid), then follows one stamen whorl, with the gynoecium occupying the central whorl. Mature flowers of *R. scutatus*, a hermaphrodite species, and a hermaphrodite mutant of *R. acetosa* are shown in *Figure 1.* In the vegetative phase, plants of

Figure 1. *Mature flowers from hermaphrodite and dioecious* Rumex *species from the subgenus Acetosa. (a)* R. scutatus *(normal hermaphrodite); (b)* R. acetosa *hermaphrodite (full hermaphrodite flower from a near tetraploid plant of chromosome constitution* $2n = 3x = 18 + XXY_1Y_2$*); (c)* R. acetosa *male* ($2n=2x=12+ XY_1Y_2$*); (d)* R. acetosa *female* ($2n= 2x= 12+ XX$*); (e)* R. rothschildianus *male* ($2n=2x=12+ XY_1Y_2$*); (f)* R. rothschildianus *female* ($2n=2x=12+ XX$*).*

the two different sexes of *R. acetosa* are indistinguishable from one another, as is also the case with the sexes of other dioecious species of *Rumex*.

The first differences between male and female are manifested very early during the development of the flowers, at a stage soon after initiation of the organ primordia. Analysis of floral development in *R. acetosa* by light microscopy and scanning electron microscopy has shown that, at least in the case of stamens in the female flower, initiation of the inappropriate set of organs does occur although there is very little development of them (Ainsworth *et al*., 1995). In the developing male flower, there is no significant proliferation of cells in the centre of the flower (*Figure 2a* and *d*), in the position normally occupied by the carpels of a hermaphrodite plant (*Figure 2c* and *f*).

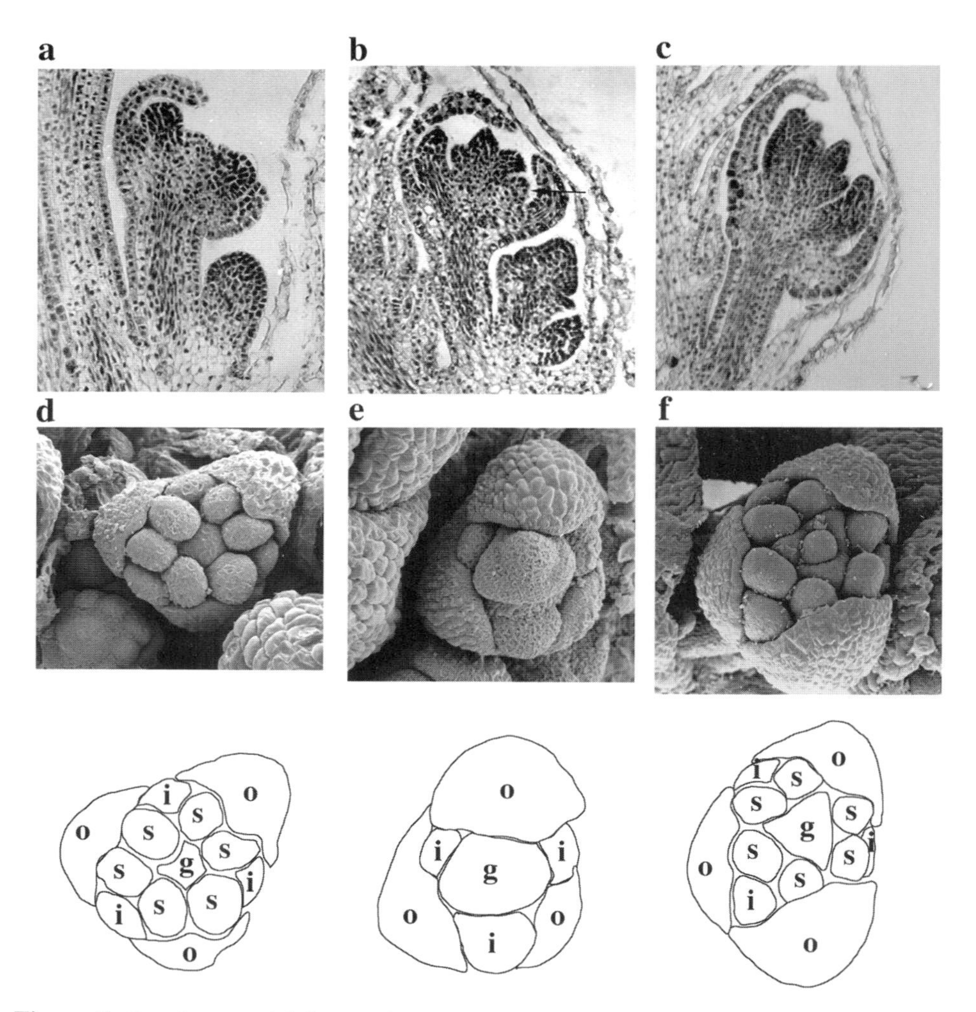

Figure 2. *Developmental differences between stage five floral primordia from male (a, d), female (b, e) and hermaphrodite (c, f) plants of* R. acetosa. *(a–c) are transverse sections viewed by light microscopy (an arrested stamen primordium in the female is arrowed in the female). (d–f) are scanning electron micrographs with annotations below (o, outer perianth segment; i, inner perianth segment; s, stamen; g, gynoecium).*

In the female flower, groups of cells in the third whorl proliferate to form stamen primordia which develop to a very limited extent, and are soon overgrown by the developing carpels (*Figure 2b* and *e*). In flowers of both sexes, DAPI staining indicates that the arrest in development does not appear to be accompanied by cell death and tissue degeneration (C.C. Ainsworth, unpublished data).

It is clear that in higher plants, the default developmental programme in flowers is that leading to a perfect or hermaphrodite flower, and that in both dioecious and monoecious plants, differentiation of the male and female occurs by modification of the hermaphrodite programme, the primary result being the suppression of one set of floral sex organs. Examination of the floral organs present in the unisexual flowers of the different monoecious and dioecious species reveals that there are many different ways in which a plant species can manipulate the normal hermaphrodite developmental programme. These differences relate to the timing during flower development of the action of sex determination genes. In *R. acetosa*, for example, the inappropriate sex organs are initiated in both male and female floral primordia, but their development is arrested at an early stage. This is also the case in *Zea mays* (maize; Veit *et al.*, 1993; for a recent review of sex determination in maize, see Chapter 12 in this volume) and *Pistacia vera* (pistachio; Hormaza and Pollito, 1996). In white campion (*Silene latifolia*), the inappropriate sex organs develop further; a rudimentary gynoecium is formed in male flowers-while in the female flower, rudimentary stamens are produced which abort and degenerate (Grant *et al.*, 1994). In some dioecious systems, the arrest in the development of the inappropriate sex organs occurs very late so that male and female flowers are superficially indistinguishable from each other and from perfect flowers. In *Actinidia deliciosa* (kiwi fruit) the inappropriate organs are of normal size, the difference being that the pollen produced by anthers in the female is sterile, as is the gynoecium in the male (Schmid, 1978). Similarly, in *Asparagus officinalis*, the gynoecium in the male flower can develop to the same size as that found in the female (Caporali *et al.*, 1994; Galli *et al.*, 1993).

At the other end of the developmental spectrum are plants such as *Mercurialis annua* (annual mercury; Durand and Durand, 1991), *Cannabis sativa* (cannabis or hemp; Heslop-Harrison, 1958; Mohan Ram and Nath, 1964), *Spinacia oleracea* (spinach; Sherry *et al.*, 1993), and *Humulus* species (hop; see Chapter 8 in this volume), where a hermaphrodite or bisexual stage is not apparent in flowers of either sex. Male and female flowers are strikingly different in these plant species and in all cases the male flowers resemble perfect flowers whilst the female flowers are quite different. In such plants, sex determining genes must act very early, before the initiation of organ development.

3. Genetic control of sex differentiation in *Rumex acetosa*

Rumex acetosa has a multiple sex system with $2n=12+XX$ in females and $2n=12+XY_1Y_2$ in males (*Figure 3a and b*). The regularity of the sex chromosome constitution in progeny is maintained by trivalent formation during male meiosis, between the single X chromosome and the two Y chromosomes, where, at zygotene, the Y chromosomes pair with the telomeric regions of opposite arms of the X (Parker and Clark, 1991). Gametes produced by males and females contain six autosomes and either the X chromosome or Y_1 plus Y_2.

When the ratio between X chromosomes and autosome sets is 1.0 or higher, the individuals are female, whereas X:autosome ratios of 0.5 and lower result in males.

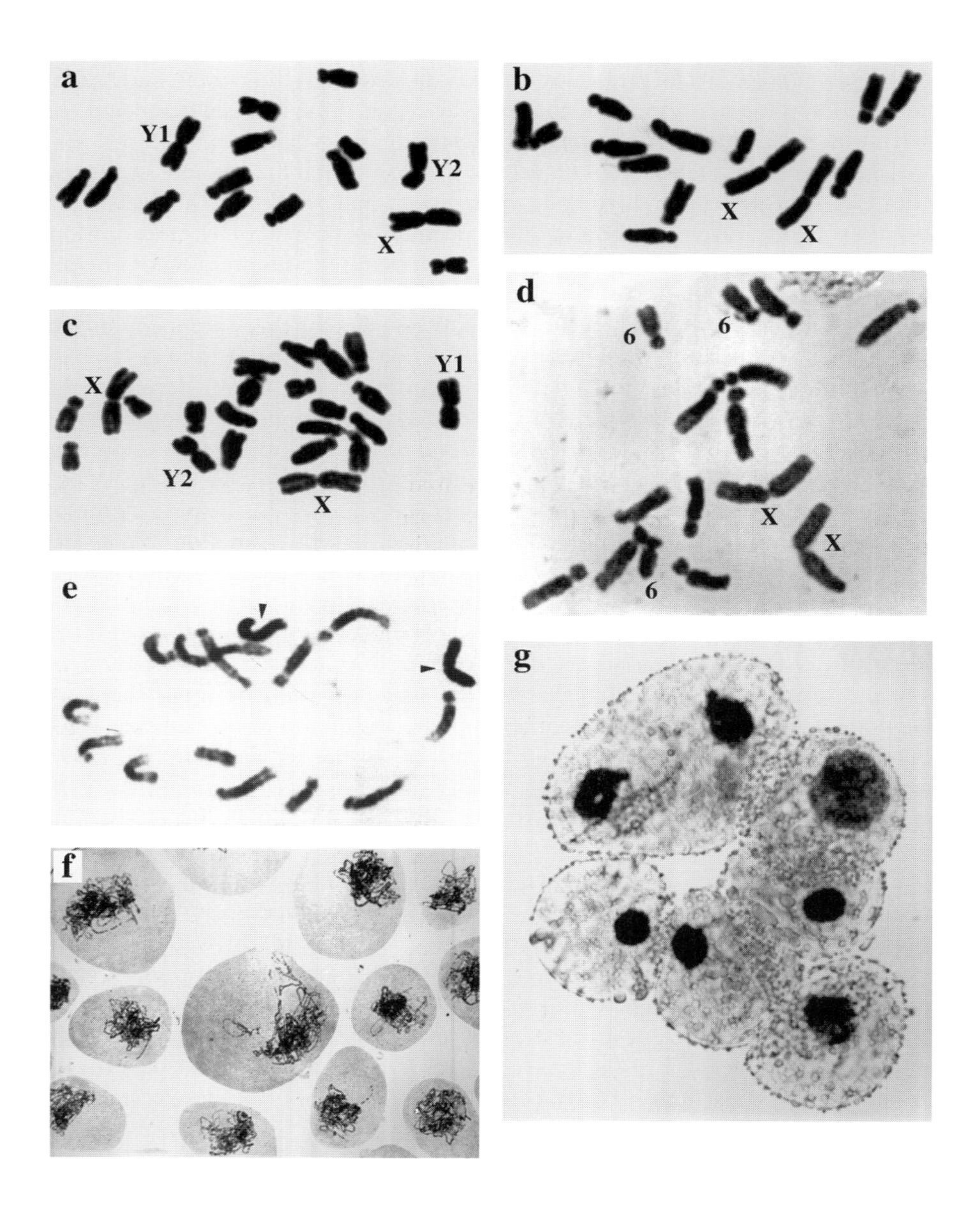

Figure 3. *Karyotypes of* R. acetosa *wild types and chromosomal mutants and the effect of deletions in the Y chromosome. (a) Mitotic metaphase in male* R. acetosa *($2n = 2x = 12+ XY_1Y_2$); (b) mitotic metaphase in female* R. acetosa *($2n=2x=12+ XX$); (c) mitotic metaphase in near triploid hermaphrodite* R. acetosa *($2n=3x=18+XXY_1Y_2$); (d) mitotic metaphase in female* R. acetosa, *trisomic for chromosome 6 ($2n=2x=13+ XX$); (e) mitotic prophase in a male of* R. acetosa *($2n=2x=12+ XY_1Y_2$) showing the heterochromatic Y chromosomes (arrowed); (f) zygotene in PMCs of a male of* R. thyrsoides *carrying a large deletion in chromosome Y_1; (g) aborted PMCs after meiosis in a male of* R. thyrsoides *carrying a large deletion in chromosome Y_1. Part (e) reprinted from* Plant Science, *vol. 80, Parker and Clark, Dosage sex-chromosome systems in plants, pp. 79–92, 1991, with permission from Elsevier Science.*

Ratios of between 0.5 and 1.0 can give an intermediate, hermaphrodite phenotype. The primary sex determination, therefore, is independent of the presence or absence of the Y chromosomes. The production of the male floral parts is also independent of the Y chromosomes because near tetraploids in which $2n = 24 + XXX$ (with a ratio of 0.75) are hermaphrodite, although they are sterile. The X chromosome clearly carries genes which are key to sex determination. Indeed, in plants that carry a translocation resulting in disomy for only a quarter of the X chromosome in addition to XY_1Y_2, hermaphrodite flowers are formed that are female-fertile but have shrivelled anthers (Parker, 1990). Plants lacking this X chromosome segment are male. The extra X chromatin is, therefore, able to give a functionally female phenotype, although it is insufficient to suppress the development of an androecium. Such an X:autosome system of determination must also involve genes on the autosomes. The role of the autosomes will be discussed later.

What is the role of the Y chromosomes? Although not required for the determination of sex, the two Y chromosomes are required for male fertility. One of the remarkable features, which is functionally related to the sequences that the Y chromosomes carry, is that they exhibit a novel form of heterochromatin. The heterochromatic Y chromosomes are clearly distinguishable from the rest of the complement at mitotic prophase (*Figure 3e*) where they are uniformly heteropycnotic with only a minute euchromatic region at one telomere, corresponding to the X/Y pairing region (Wilby and Parker, 1986). Males can be identified by massive chromosome-like chromocentres formed by the Y chromosomes in interphase nuclei of root tips and developing leaves. During mitotic prophase, the Ys are DAPI-positive but are unique in containing no C-band material (Ruiz Rejon *et al.*, 1994) which would be expected for heterochromatin. The Ys thus seem to represent massive blocks of non-C-banding heterochromatin. Evidence for the facultative nature of the heterochromatin has been given by Zuk (1969) and Parker (unpublished data) who observed uncoiling of Ys during pre-meiotic interphase, but not in roots, leaves or young stems. Deletions of Y chromosome material affect male meiotic progress, resulting in three identifiable changes. Firstly, pollen mother cells (PMCs) are normally extremely constant in size, but in deletion mutants, PMC diameter is extremely variable with a two-fold range – the control of cell development is clearly impaired. Secondly, meiotic pairing is arrested at mid-zygotene. Lastly, chromosome fragments are found, indicating replication problems (*Figure 3f*). The overall result is that the PMCs coalesce, nuclei abort, and total sterility ensues (*Figure 3g*). Therefore, transcription products from at least three genes carried on the Y chromosomes are required at pre-meiotic interphase, presumably for the regulation of entry into, and progression through, male meiosis.

A second remarkable feature of the Y chromosomes is that the Y_1 and Y_2 chromosomes of all species of the Acetosa group show massive structural variation (Wilby and Parker, 1986). The centromere can be located anywhere within the central 40% of the chromosome but is excluded from the distal segments. There are apparently no preferred sites for centromere location within the central region. The variation in location is hypervariable and maintained by an exceptionally high mutation rate: 1 in 80 mutant individuals per generation in controlled crosses (Parker 1990; Wilby and Parker, 1986; Wilby and Parker, 1987). The mechanism of centromeric transposition is not known but must involve the expression of sequences on the Y chromosomes during meiosis.

4. Genes involved in sex differentiation

Our understanding of the developmental biology of the flowers of higher plants and of the genes which control development has expanded hugely in the past decade, due largely to the work on *Arabidopsis* and *Antirrhinum*. This information can be of great value in studies aimed at understanding the genetic control of the development of monoecious and dioecious flowers. Although it is not possible to use this information to unravel sex determination *per se* (i.e. the nature of the primary determination event and the gene(s) responsible), we can look at the expression of the downstream genes involved in differentiation of the androecium and gynoecium. Obvious candidates in such studies are the organ identity genes which are discussed in Chapter 1 of this volume. In order to examine whether the differences between male and female *R. acetosa* flowers are the result of differences in expression of the organ identity genes, the putative B and C function homeotic genes were isolated and their patterns of gene expression were assessed by *in situ* hybridization (Ainsworth *et al.*, 1995).

In this study, the expression of the putative B function genes in *R. acetosa* was found to be coincident with the initiation and development of the stamens in the male and with the formation of the stamen primordia which abort after a short period of development in the female. B function gene expression, therefore, only differed between male and female flowers in the area of the domain of expression. By contrast, *RAP1*, the putative C function gene showed a sex-specific expression pattern. The gene was shown to be expressed in the young male flower primordia in whorls three and four, the stamen and carpel whorls, respectively, in both male and female floral primordia. However, the expression in the inappropriate organ primordia was transient. In female flowers, as the stamen primordia begin to enlarge significantly, *RAP1* expression in the centre of the flower (the carpel whorl) was undetectable. In young female flower primordia, *RAP1* is expressed in whorls three and four, the stamen and carpel whorls. Later than this, when the stamen primordia are visible as hemispherical structures, the expression is retained in the carpel whorl but is lost from the stamen primordia themselves. Therefore, arrest of the development of the inappropriate organs is associated with inactivation of the C function gene. Whether this is cause or effect remains to be demonstrated in transgenic experiments. What is clear, is that the block in development is not caused by cell death since B function expression continues in the stamen primordia in the female and there is no DNA degradation in cells of the arrested organs (C.C. Ainsworth, unpublished data).

An interesting finding, unrelated to sex determination, is that the expression of two putative B function genes, both *DEFICIENS* homologues, is confined to a single whorl – whorl 3, the stamen whorl – in both male and female flowers rather than in the stamen and petal whorls that would be expected for a hermaphrodite flower such as *Antirrhinum* or *Arabidopsis* (Ainsworth *et al.*, 1995). The fact that the second whorl organs in *R. acetosa* flowers are sepaloid may result from the lack of expression of the B function genes in this whorl. A parallel exists in petunia where the whorl 2 organs of the homeotic *gp* (green *p*etals) mutant have the same shape and colour as the sepals in whorl 1, and petaloid cells occur on the stamen filaments (van der Krol and Chua, 1993). Analysis of the expression of the petunia MADS-box genes shows that in the wild type, the *DEFICIENS* homologue, *pMADS1*, is expressed in the petal and stamen whorls (Angenent *et al.*, 1995), whereas, in the *gp* mutant, expression of *pMADS1* mRNA is not detectable in any whorl (Angenent *et al.*, 1995; van der Krol and Chua, 1993).

5. The role of the autosomes

Analysis of a hermaphrodite line of *R. acetosa*, a near triploid with the chromosome constitution $2n=3x=18+XXY_1Y_2$ (*Figure 2c*) has provided insights into the timing of sex determination and its control (C.C. Ainsworth, V. Buchanan-Wollaston and J.S. Parker, unpublished data). These plants have an X:autosome ratio of 0.66, between the ratio of 0.5 required for maleness and the ratio of 1.0 required for femaleness. All flowers produced on inflorescences of these hermaphrodite plants are staminate although there is variation between adjacent flowers in the degree of anther development. Some anthers produce pollen whilst others develop much less, are shrivelled and produce no pollen. The flowers, however, differ in the degree of development of the gynoecium, ranging from a simple rod-like structure or filament (varying in length from a few cells to the height of the normal gynoecium) to a gynoecium consisting of three separate carpel elements each developing its own stigma (*Figure 1b*). Thus, in the hermaphrodite plants, with an X:A ratio of 0.66, intermediate dosages of genes lead to molecular confusion; the levels of transcripts or their products from X located and autosomal genes must be critical in ensuring proper control of the developmental pathway. Interestingly, mixed organs are sometimes found in these hermaphrodite flowers, such that a carpel-like structure made up the upper part and a stamen the lower part. This may indicate that whatever gene product(s) are required for determination of sex, i.e. which organ sets are promoted and suppressed, it may also be needed for maintaining the correct development of the organs in the set. A key implication of this is that it is the sex of the individual flowers that is determined, rather than that of the whole inflorescence.

Sex determination in *R. acetosa* is influenced not only by the X:autosome ratio but also by the number and types of autosome. Yamamoto (1938) and Wilby (1987) analysed intersex triploid plants which were tetrasomic for specific autosomes. Association of floral phenotype with tetrasomy for particular autosomes enabled the autosomes to be classified as male promoting (chromosomes 1 and 4), female promoting (chromosomes 2 and 5), or sex neutral (chromosomes 3 and 6). The near triploid line described above has enabled the production, by crossing with a normal male, of a population of plants which are trisomic for one or more autosomes and which may have aberrant sex chromosome constitutions. Not only does this population allow us to confirm the effect of individual autosomes on gender, but it also allows us to assess the effects of combinations of trisomic autosomes and to look for interactions between the autosomes. *Figure 2d* shows the karyotype of a female plant which is trisomic for chromosome 6.

6. Molecular markers for sex chromosomes

We know little about the nature of the DNA sequences carried on plant sex chromosomes but evidence points to plant sex chromosomes being rich in repeated sequences and carrying genes which have degenerated so as to become non-functional. Representational difference analysis of Y-linked sex (deletion) mutants in *S. latifolia*, where the deletions are small, has enabled mapping of the sex determining loci and allowed the sequences from the Y chromosome to be isolated (Donnison *et al.*, 1996). Detailed analysis of some of these sequences suggests that much of the *S. latifolia* Y chromosome is composed of simple sequence repeats (Chapter 4). This is not a surprising finding given the increase in size of the *S. latifolia* Y chromosome (relative to the autosomes) and the lack of pairing over most of the arms of the sex chromosomes. Recently, Guttman and Charlesworth (1998) compared the sequence of a transcribed gene present in *S. latifolia* on both the X chromosome and the Y chromosome and found that the sequence on the Y chromosome was considerably degenerate relative to the homologue on the X chromosome.

The Y chromosomes of *R. acetosa* appear as chromocentres in interphase nuclei and are positively heteropycnotic during mitotic prophase. The Ys, however, consist of facultative heterochromatin and are transcriptionally active during a few cell cycles immediately prior to meiosis in pollen mother cells. Attempts by Clark *et al.* (1993) to isolate Y-specific repetitive sequences proved unsuccessful although seven non-homologous repeats were isolated. These sequences were all dispersed with no areas of localization within the genome. A repetitive sequence of 180 bp, found in tandem arrays, was isolated by Ruiz Rejon *et al.* (1994) from *R. acetosa* which mapped by *in situ* hybridization to the sex chromosomes alone. Both Y_1 and Y_2 were heavily labelled as were both arms of the X. There was no evidence of the presence of this sequence in the autosomes at this level of discrimination. The distribution of the sequence suggests a common origin of both Y chromosomes from the X chromosome or its evolutionary precursor.

Recently, we have begun microsatellite and AFLP analyses in *R. acetosa* in order to investigate the sequence composition of the sex chromosomes. A microsatellite library enriched for di-, tri-, and tetra-nucleotide repeat sequences has been developed, and enrichment of approximately 50% was achieved for microsatellite sequences in general. Dinucleotide sequences were the most frequently encountered, with GA microsatellites predominant. In preliminary AFLP studies, DNA was extracted from male and female plants, their sex being verified on the basis of mitotic spreads and floral phenotype. After digestion with the enzyme combination MseI/PstI, the samples were analysed both individually and as male and female pools, to facilitate the identification of male-specific, therefore Y chromosome-specific, fragments. A range of primer sets has been used, but only one male specific fragment, 94 bp long, has so far been identified (*Figure 4*). More primer sets and enzyme combinations are being tried in order to identify further Y-specific sequences. In this type of analysis, the X chromosome represents a more difficult target than the Y chromosome since it is present in cells from both sexes. X chromosome deletion lines will be valuable in this respect.

It has been proposed that the Ys have evolved from the X chromosome by centric fission followed by isochromosome formation (Ruiz Rejon *et al.*, 1994) and, therefore, at one time would have carried much the same sequences as the X chromosome. An alternative evolutionary route for the Y chromosome is by X/autosome interchange. We will be able to use Y chromosome specific AFLP sequences to distinguish between these alternatives.

7. Isolation of sex determination genes

It is clear that the identification of the key genes which determine sex in monoecious and dioecious plants will not be easy. The identification of downstream genes, the sex differentiation genes, represents an easier, but less significant, target. Attempts at cloning the sex determining genes have focused on the use of genes which are involved in the control of flower development in hermaphrodites. This approach is likely, in the main, to be unsuccessful since we cannot predict easily which genes will be important

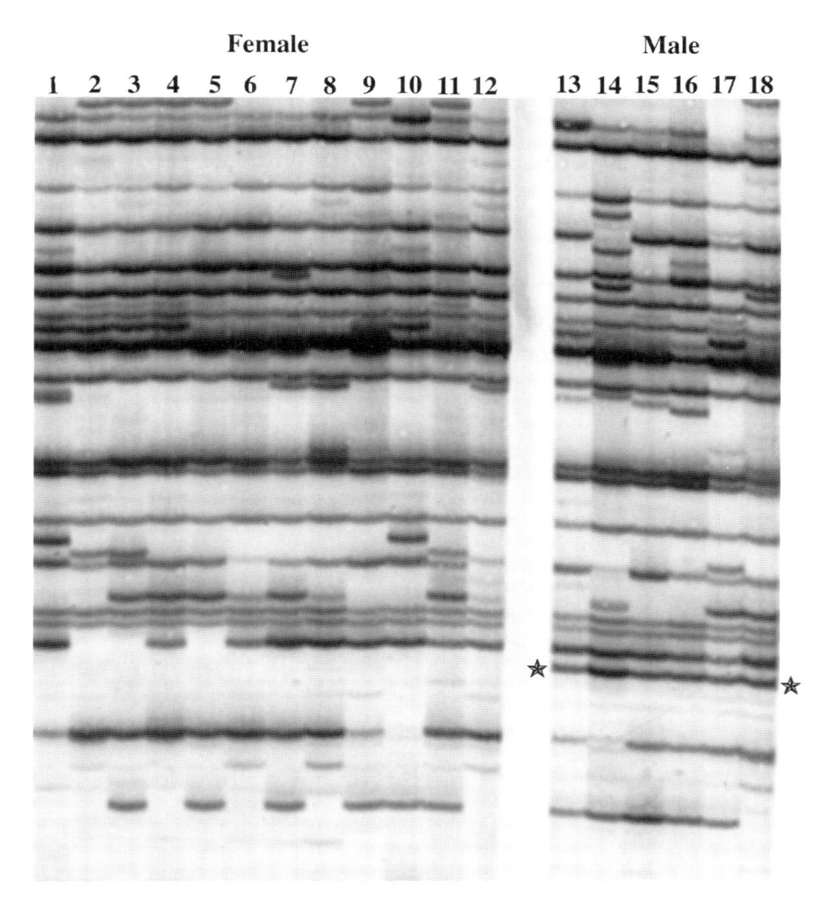

Figure 4. *AFLP analysis of* R. acetosa. *DNA isolated from 12 females (tracks 1–12) and six males (tracks 13–18) and digested with MseI and PstI. The band (marked with an asterisk) is present in all males but no females and must derive from a sequence on the Y chromosome.*

simply from expression patterns and their roles in hermaphrodites. Various different approaches are being taken in different laboratories to attempt to isolate sex-determining genes including chromosome microdissection methods, preparative *in situ* hybridization, differential display, and representational difference analysis.

Preparative *in situ* hybridization (Prep-ISH) is currently being used in *R. acetosa* to isolate genes from the X chromosome which are expressed during floral development. Prep-ISH has been previously used to clone genes from specific regions of chromosome 2 from mouse and man (Hozier *et al.*, 1994) but there are no reports of its use in plants. We have used Prep-ISH as follows: mitotic chromosome spreads from root tips of *R. acetosa* were hybridized *in situ* with linkered cDNA synthesized from RNA isolated from young male or female floral primordia of *R. acetosa*. After hybridization, the X chromosomes were microdissected and PCR-amplified using primers specific to the cDNA linkers. After restriction digestion of the sites in the linkers, the PCR products were cloned. A large number of clones resulted from these experiments. As we are interested in X chromosome-specific genes expressed in young floral primordia (which may be involved in sex determination), the clones have been screened by colony screening with shoot and root cDNA to eliminate clones not related to inflorescence development and then screened on virtual Northern blots carrying cDNA from different aged inflorescences and vegetative tissues. Clones for genes which show expression patterns specific to young inflorescences or where expression levels were too low to detect (as might be shown by regulatory genes) have been investigated further by isolation of full length clones and confirmation of X-location by Southern blotting (*Figure 5*). Expression patterns of

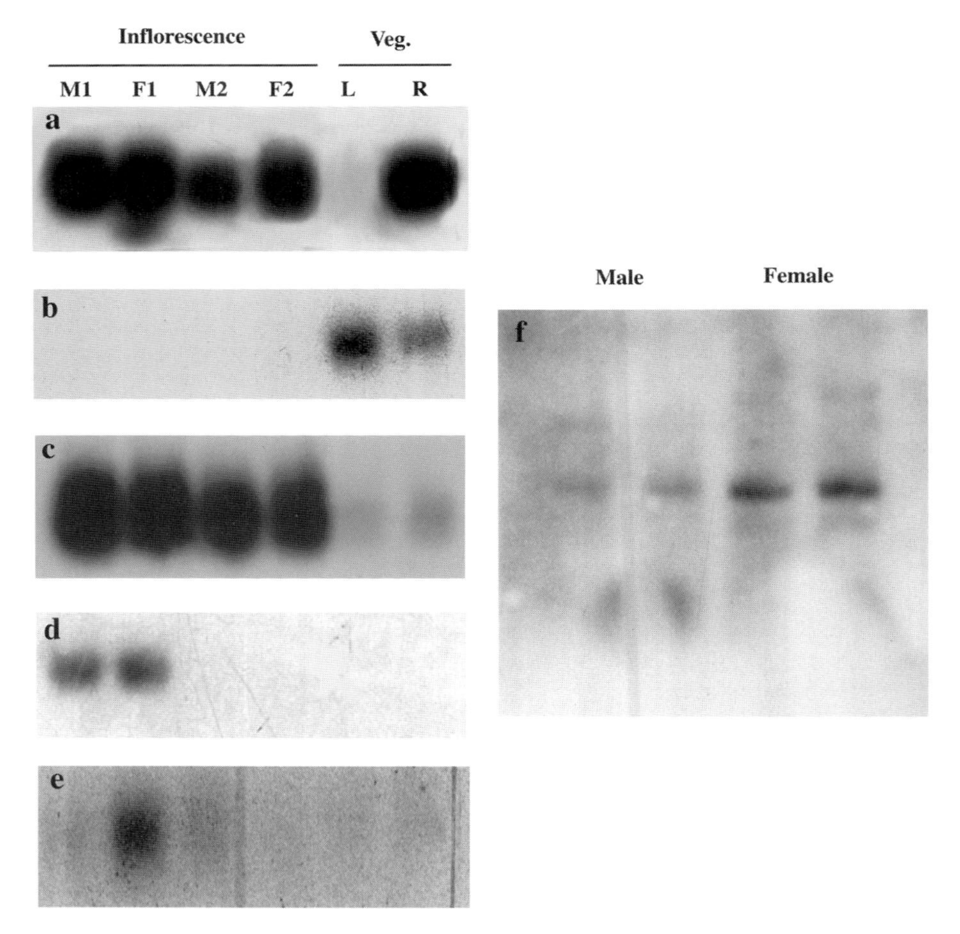

Figure 5. *Analysis of cDNA clones isolated from the X chromosome of* R. acetosa *by preparative in situ hybridization (Prep-ISH). Panels (a–e) are virtual Northerns (hybridization of cDNA inserts to blots carrying cDNA synthesized from mRNA isolated from tissues) of five different Prep-ISH clones which detect the expression of genes with quite different expression patterns. (a) Expression in all tissues with the exception of leaves; (b) expression in roots and leaves only; (c) expression in inflorescence; (d) expression in young inflorescences; (e) expression in young female inflorescences. cDNA was from very young male inflorescences (M1), very young female inflorescences (F1), older male inflorescences (M2), older female inflorescences (F2), leaves (L) and roots (R). (f) A Southern hybridization of genomic DNA from a male and a female digested with Mbo1 (left lanes) and Sau3A (right lanes), blotted and hybridized with a Prep-ISH cDNA insert. Note that the band intensity corresponds to X-chromosome dosage.*

some of the cDNA clones isolated by Prep-ISH are shown in *Figure 5*. A number of X chromosome genes are currently under study and it is hoped that some may have roles in sex determination.

8. Dosage compensation in *Rumex acetosa*

A question which is raised by the existence of systems of sex determination in plants which use sex chromosomes is whether systems of dosage compensation operate. In many eukaryote species, of both animals and plants, sex is determined by differences in the number of copies of a single chromosome. In animals, dosage compensation regulates the activity of genes carried on the X chromosomes; several different systems have evolved to allow the different number of copies of genes in the two sexes to produce the same amount of functional product (reviewed in Lyon, 1996). These include complete inactivation of one of the two X chromosomes in the female (as in mammals; Lyon, 1961), decreasing the rate of transcription of genes on both the X chromosomes in the female (as in *Caenorhabditis*; Villeneuve and Meyer, 1990) or increasing the rate of transcription of genes on the X chromosome in the male (as in *Drosophila*, reviewed by Jaffe and Laird, 1986; Kelley and Kuroda, 1995; Lucchesi and Manning, 1987). In some animals, such as birds and butterflies, there is no obvious mechanism of dosage compensation (Chandra, 1991).

Little information is available as to whether dioecious plant species have analogous systems. In the X chromosomes of mammals there is a marked difference in DNA replication pattern between the active and inactive X chromosome (Grant and Chapman, 1988; Rao and Padmaja, 1992). In *R. acetosa* and *R. thyrsiflorus* there is no evidence for differences in replication between the X chromosomes (J.S. Parker, unpublished data; Zuk, 1969), suggesting that, if a system of dosage compensation is used, it has a different basis. This contrasts with the situation in white campion, where one of the two X chromosomes carried by females was shown to be late replicating and was hypermethylated and inactive, suggesting that a system similar to that used in mammals may be employed (Chapter 6; Choudhuri, 1969; Siroky *et al*., 1994; Vyskot *et al*., 1993). The Prep-ISH cDNA clones from *R. acetosa* will enable us to test whether there is compensation for the dosage of specific genes on the X chromosomes.

9. Conclusions

Cytological investigations of the *Rumex* genus together with the study of intersex mutants have given us tantalizing pieces of information which allude to the nature of the X:autosome sex determination system. The default product in *R. acetosa* is clearly hermaphrodite where the pathways leading to male organ formation and female organ formation are functional. The system of sex determination is one of suppressing the progress of one or other developmental pathway (rather than its initiation) and, in this respect, is different from sex determination in animals, where the genetic decision is one initiating one or other of the alternative developmental programmes. At present, we know that there must be X-located genes and genes on several autosomes which are involved in sex determination and that there is interaction between the X-located genes and the autosomal genes. For this reason it seems unlikely that the X: autosome dosage system involves a *Drosophila*-like dosage sensing mechanism with X-located numerator genes and autosomal denominator genes, and more likely that the autosomal genes interact directly or indirectly with the X-located genes in the separate male and female developmental pathways.

In a simplistic model, one could envisage in a male (X:autosome ratio of 0.5 or less), there being an excess of a protein encoded by an autosomal gene which represses the female pathway relative to an X chromosome gene product which binds with it. In the female (X:autosome ratio of 1.0 or more) the repressor protein and its inhibitor are in equivalent amounts and the female pathway is operational. As an alternative to a protein–protein system, a protein transcribed from an X-located gene which binds with the promoter of a gene whose product represses female development could be involved. Similarly, in a situation of an autosomal gene product which is necessary for the male pathway and an inhibitor of its action, male development in the male would be allowed, but prevented in the female. It is clearly not this simple, given that several different autosomes are involved and, also, that expression of femaleness is so variable in intersex lines. The major challenge will be to identify the genes which are involved in the X:autosome interaction(s).

References

Ainsworth, C., Crossley, S., Buchanan-Wollaston, V., Thangavelu, M. and Parker, J. (1995) Male and female flowers of the dioecious plant sorrel show different patterns of MADS box gene expression. *Plant Cell* 7: 1583–1598.

Angenent, G.C., Busscher, M., Franken, J., Dons, H.J.M. and van Tunen, A.J. (1995) Functional interaction between the homeotic genes fbp1 and pMADS1 during Petunia floral organogenesis. *Plant Cell* 7: 507–516.

Capel, B. (1996) The role of Sry in cellular events underlying mammalian sex determination. *Curr. Topics. Dev. Biol.* **32**: 1–37.

Caporali, E., Carboni, A., Galli, M.G., Rossi, G., Spada, A. and Longo, G.P.M. (1994) Development of male and female flower in *Asparagus officinalis*. Search for point of transition from hermaphroditic to unisexual developmental pathway. *Sex. Plant Reprod.* 7: 239–249.

Chandra, H.S. (1991) How do heterogametic females survive without gene dosage compensation? *J.Genet.* 70: 137–146.

Choudhuri, H.C. (1969) Late replication pattern in sex chromosomes of Melandrium. *Can. J. Genet. Cytol.* **11**: 192–198.

Clark, M.S., Parker, J.S. and Ainsworth, C.C. (1993) Repeated DNA and heterochromatin structure in *Rumex acetosa. Heredity* 70: 527–536.

Donnison, I.S., Siroky, J., Vyskot, B., Saedler, H. and Grant, S. (1996) Isolation of Y chromosome-specific sequences from *Silene latifolia* and mapping of male sex determining genes using representational difference analysis. *Genetics* **144**: 1891–1899.

Durand, B. and Durand, R. (1991) Sex determination and reproductive organ differentiation in *Mercurialis. Plant Sci.* 80: 49–65.

Galli, M.G., Bracale, M., Falavigna, A., Raffaldi, F., Savini, C. and Vigo, A. (1993) Different kinds of male flowers in the dioecious plant *Asparagus officinalis* L. *Sexual Plant Reprod.* **6**: 16–21.

Grant, S.G. and Chapman, V.M. (1988) Mechanisms of X chromosome regulation. *Ann. Rev. Genet.* 22: 199.

Grant, S., Hunkirchen, B. and Saedler, H. (1994) Developmental differences between male and female flowers in the dioecious plant white campion. *Plant J.* **6**: 471–480.

Grant, S., Donnison, I.S., Hardenack, S. and Law, T.F. (1996) Studies of the genetics of sex determination in dioecious *Silene latifolia* by the front and the back doors. *Flowering Plant Newsletter* **21**: 21–26.

Guttman, D.S. and Charlesworth, D. (1998) An X-linked gene has a degenerate Y-linked homologue in the dioecious plant *Silene latifolia. Nature* **393**: 263–266.

Heslop-Harrison, J. (1958) Unisexual flower: a reply to criticism. *Phytomorphology* **8**: 177.

Hormaza, J.I. and Polito, V.S. (1996) Pistillate and staminate flower development in dioecious *Pistacia vera* (Anacardiaceae) *Am. J. Bot.* **83**: 759–766.

Hozier, J., Graham, G., Westfall, T., Siebert, P. and Davis, L. (1994) Preparative *in situ* hybridization of chromosome region-specific libraries on mitotic chromosomes. *Genomics* **19**: 441–447.

Jaffe, E. and Laird, C. (1986) Dosage compensation in *Drosophila*. *Trends Genet.* **2**: 316–321.

Kelley, R.L. and Kuroda, M.I. (1995) Equality for X- chromosomes. *Science* **270**: 1607–1610.

Keyes, L.N., Cline, T.W. and Schedl, P. (1992) The primary sex determination signal of *Drosophila* acts at the level of transcription. *Cell* **68**: 933–943.

Lucchesi, J.C. and Manning, J.E. (1987) Gene dosage compensation in *Drosophila melanogaster*. *Adv. Genet.* **24**: 371–429.

Lyon, M.F. (1961) Gene action in the X chromosome of the mouse. *Nature* **190**: 372–373.

Lyon, M.F. (1996) Molecular genetics of X chromosome inactivation. *Adv. Genome Biol.* **4**: 119–151.

McElreavey, K., Vilain, E., Abbas, N., Herskowitz, I. and Fellous, M. (1993) A regulatory cascade hypothesis for mammalian sex determination: SRY represses a negative regulator of male development. *Proc. Natl Acad. Sci. USA* **90**: 3368–3372.

Mittwoch, U. (1996) Genetics of sex determination. An overview. *Adv. Genet.* **4**: 1–28.

Mohan Ram, H.Y. and Nath, R. (1964) The morphology and embryology of *Cannabis sativa*. *Linn. Phytomorph.* **14**: 414–429.

Parker, J.S. (1990) Sex chromosomes and sexual differentiation in flowering plants. *Chromosomes Today* **10**: 187–198.

Parker, J.S. and Clark, M.S. (1991) Dosage sex-chromosome systems in plants. *Plant. Sci.* **80**: 79–92.

Rao, S.R.V. and Padmaja, M. (1992) Mammalian-type dosage compensation mechanism in an insect – *Gryllotalpa fossor* (Scudder) – Orthoptera. *J. Biosci.* **17**: 253–273.

Ruiz Rejon, C., Jamilena, M., Garrido Ramos, M., Parker, J.S. and Ruiz Rejon, M. (1994) Cytogenetic and molecular analysis of the multiple sex chromosome system of *Rumex acetosa*. *Heredity* **72**: 209–215.

Schmid, R. (1978) Reproductive anatomy of *Actinidia chinensis* (Actinidiaceae) *Bot. Jahrb. Syst. Pflanzengesch. Pflanzengeog.* **100**: 149–195.

Sherry, R.A., Eckard, K.J. and Lord, E.M. (1993) Flower development in dioecious *Spinacia oleracea* (Chenopodiaceae) *Am. J. Bot.* **80**: 283–291.

Sinclair, A.H., Berta, P., Palmer, M.S., Hawkins, J.R., Griffiths, B.L., Smith, M.J., Foster, J.W., Frischauf, A-M., Lovell-Badge, R. and Goodfellow, P.N. (1990) A gene from the human sex-determining region encodes a protein with homology to a conserved DNA-binding motif. *Nature* **346**: 240–244.

Siroky, J., Janousek, B., Mouras, A. and Vyskot, B. (1994) Replication patterns of sex chromosomes in *Melandrium album* female cells. *Hereditas* **120**: 175–181.

Smith, B.W. (1969) Evolution of sex-determining mechanisms in Rumex. *Chromosomes Today* **2**: 172–182.

van der Krol, A.R. and Chua, N.-H. (1993) Flower development in petunia. *Plant Cell* **5**: 1195–1203.

Veit, B., Schmidt, R.J., Hake, S. and Yanofsky, M. (1993) Maize floral development: new genes and old mutants. *Plant Cell* **5**: 1205–1215.

Villeneuve, A.M. and Meyer, B.J. (1990) The regulatory hierarchy controlling sex determination and dosage compensation in *Caenorhabditis elegans*. *Adv. Genet.* **27**: 117–188.

Vyskot, B., Araya, A., Veuskens, J., Negrutiu, I. and Mouras, A. (1993) DNA methylation of sex chromosomes in a dioecious plant, *Melandrium album*. *Mol. Gen. Genet.* **239**: 219–224.

Warmke, H.E. and Blakeslee, A.F. (1939) Sex mechanism in polyploids of *Melandrium album*. *Science* **89**: 391–392.

Westergaard, M. (1940) Studies on cytology and sex determination in polyploid forms of *Melandrium album*. *Dansk. Bot. Arkiv.* **5**: 1–131.

Westergaard, M. (1946) Aberrant Y chromosomes and sex expression in *Melandrium album*. *Hereditas* **32**: 419–443.

Westergaard, M. (1958) The mechanism of sex determination in dioecious flowering plants. *Adv. Genet.* **9**: 217–281.

Wilby, A.S. (1987) Ph.D. Thesis. University of London.

Wilby, A.S. and Parker, J.S. (1986) Continuous variation in the Y chromosome structure of *Rumex acetosa*. *Heredity* **57**: 247–254.

Wilby, A.S. and Parker, J.S. (1987) Population structure of hypervariable Y chromosomes in *Rumex acetosa*. *Heredity* **59**: 137–143.

Winge, O. (1931) X- and Y- linked inheritance in *Melandrium*. *Hereditas* **15**: 127–165.

Yamamoto, Y. (1938) Karyogenetische Untersuchungen bei der Gattung Rumex VI. Geschlechtsbestimmung bei En – und Aneuploiden Pflanzen von *Rumex acetosa* L. Kyoto Imp. Univ. Mem. Coll. Agr. **43**: 1–59.

Ye, D., Oliveira, M., Veuskens, J., Wu, Y., Installe, P., Hinnisdaels, S., Truong, A.T., Brown, S., Mouras, A. and Negrutiu, I. (1991) Sex determination in the dioecious *Melandrium*. The X/Y chromosome system allows complementary cloning strategies. *Plant Sci.* **80**: 93–106.

Zuk, J. (1969) Autoradiographic studies in *Rumex* with special reference to sex chromosomes. *Chromosomes Today* **2**: 183–188.

8

Sex expression in hop (*Humulus lupulus* L. and *H. japonicus* Sieb. et Zucc.): floral morphology and sex chromosomes

Helen Shephard, John Parker, Peter Darby and Charles C. Ainsworth

1. Introduction

The family Cannabidaceae consists of the two genera *Humulus* and *Cannabis*. The hop belongs to the genus *Humulus*, which comprises three species: *H. lupulus* (the cultivated or European hop), *H. japonicus* (the annual Japanese ornamental hop) and *H. yunnanensis* (a perennial Chinese hop) (Small, 1978). *H. lupulus*, which has been cultivated for over 1000 years in central Europe (Kohlmann and Kaster, 1976), is a dioecious, wind-pollinated perennial climbing plant. The commercially important strobiles or 'cones' are produced by the female plant and comprise the female inflorescence within enlarged bracts and bracteoles. Lupulin glands, located on the bases of the bracteoles and bracts of the cones, and on the pericarp of fertilized hops, contain a number of resins and essential oils of importance to the brewing industry (Hough *et al.*, 1982). Various essential oils contribute to the flavour and aroma of beer. Resins, of which the α-acids (principally humulone and cohumulone) comprise the most important fraction, impart the characteristic bitterness following their isomerization during the brewing process. The α-acids also act as natural preservatives and foaming agents (Burgess, 1964; Neve, 1991). Male flowers also contain a few resin glands on the locular grooves of the anthers and glands are also present in small numbers on leaves of both sexes, but in insufficient numbers to make their use in the brewing industry a commercially viable proposition.

Due to their dioecious nature, self-pollination in hops is not normally possible and thus the plants are genetically highly heterozygous. Plants raised from seed are therefore extremely variable and most are not commercially valuable. Hop gardens are established

Sex Determination in Plants, edited by C.C. Ainsworth.

with vegetatively propagated material to preserve clonal fidelity and elite brewing characteristics. In commercial production of conventional hop varieties, the entire aboveground part of the plant (bine) is cut down and the rootstock is left to overwinter. However, in the new practice of cultivating dwarf varieties on low trellis systems, the bines are left intact. After a period of dormancy, the plants regenerate in the spring from adventitious buds from the branched stem tissue which lies below ground level and forms the upper part of the rootstock (Williams, 1960). The plants then produce a number of vegetative shoots which rapidly extend and twine around any available support.

2. Vegetative development of *Humulus lupulus*

In both male and female hop plants, vegetative apical meristems appear to be almost identical in arrangement (*Figure 1a* and *b*). Pairs of oppositely-arranged leaf primordia subtend four bract primordia within each plane of phyllotaxy and successive alternate planes are arranged at 90° to each other. Each plane of vegetative organs enlarges and extends, whilst the tip of the apical meristem continues to differentiate to produce leaf and bract primordia on the flanks of the meristem. Within the leaf axils, small axillary vegetative meristems produce further leaf and bract primordia, eventually forming further sublaterals along each lateral branch.

3. Floral development of *Humulus lupulus*

The switch from vegetative to reproductive development in female hop plants is triggered by shortening daylength (Tournois, 1912; Thomas and Schwabe, 1970) and takes place below a critical daylength that is varietal-dependent. The influence of daylength on the transition to flowering in male hop plants is less clear. In a detailed study of the critical factors influencing floral initiation in female hop lines (*H. lupulus*), Thomas and Schwabe (1970) found that the once widely cultivated variety Fuggle had a critical daylength of between 21 and 24 hours under outdoor conditions and also showed that the plant must initiate a minimum number of nodes (30–32 for Fuggle) before flowering could be induced. Once these, and certain other environmental factors such as temperature, are met, the plant then switches from the vegetative to the reproductive phase of development.

When the hop plant enters the reproductive phase of development, the apical meristem is reduced in size by over one half compared to the diameter of a vegetative meristem. This decrease in size accompanying the transition to the reproductive phase is of the same order of magnitude for meristems of both male and female plants. The hop is one of only a few plant species known to exhibit a decrease in the diameter of the apical meristem on the transition to flowering. Conversely, most other species show an increase in the size of the apex on reaching the reproductive phase of growth. The hop inflorescence meristem also becomes more domed in appearance, compared to the flatter shape of the vegetative meristem. Male and female flowers both arise from differentiating floral meristems on the flanks of the inflorescence meristems.

4. Development of the male flower

Male flowers develop from clusters of small floral meristems located on the periphery of the inflorescence meristem. Five sepal primordia start to differentiate on the floral meri-

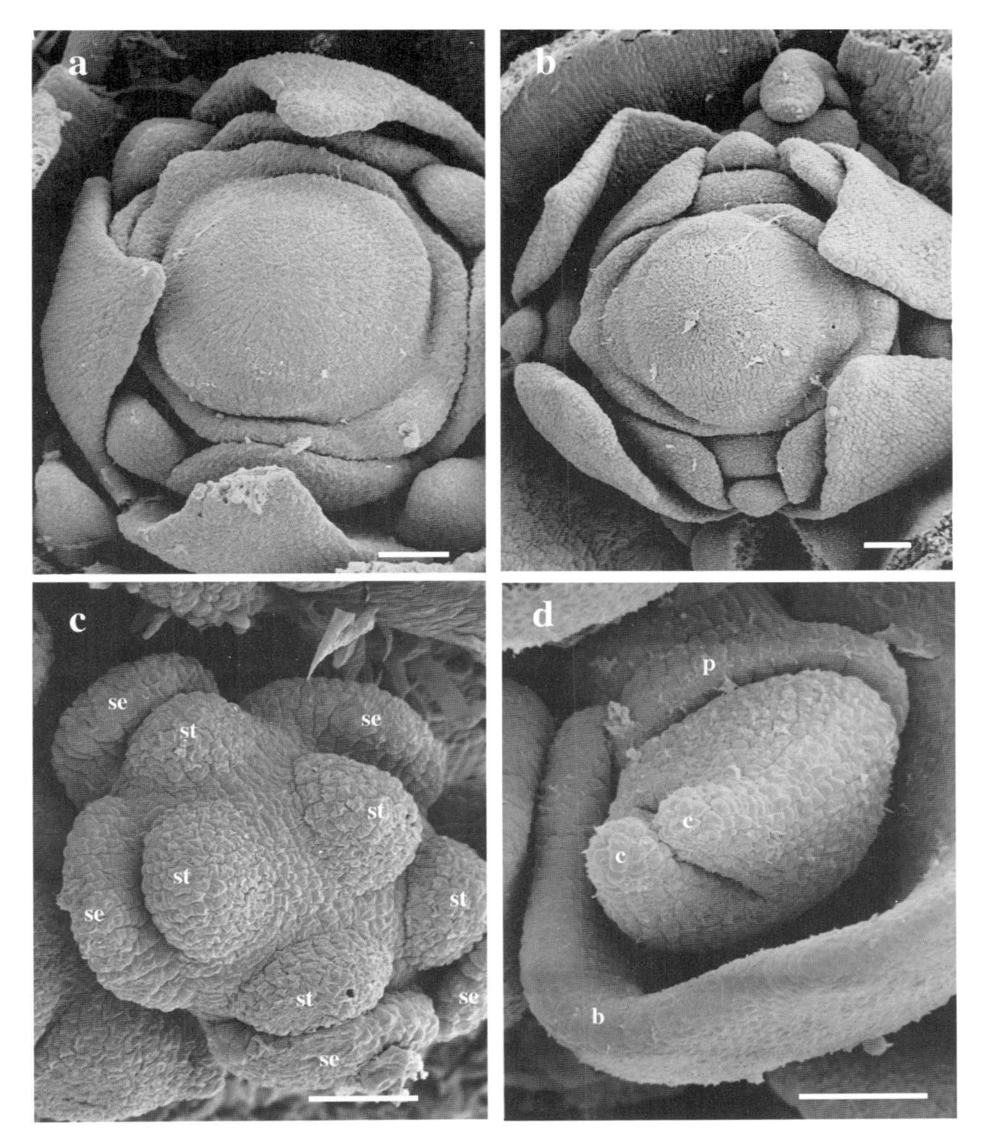

Figure 1. *Scanning electron micrographs of vegetative apices and developing flowers of dioecious* H. lupulus. *(a) vegetative male apical meristem; (b) vegetative female apical meristem; (c) initiation of sepal and stamen primordia in male flower; (d) initiation of perianth and carpel primordia in female flower, enclosed within a bracteole. b, bracteole; c, carpel; p, perianth; se, sepal; st, stamen. Bars = 50 µm. Parts (c) and (d) from Ainsworth* et al. *(1988) Sex determination in plants.* Current Topics in Developmental Biology, *vol. 38, pp. 167–223. Reprinted by permission of Academic Press, Inc.*

stem in a specific developmental pattern, in which three 'outer' sepal primordia are initiated, followed by the differentiation of two 'inner' sepal primordia on the flanks of the floral meristem. The development of five stamen primordia occurs in the same developmental pattern as the sepals, and takes place just after the initiation of the inner sepals (*Figure 1c*). Occasionally, a small number of male flowers arise that possess six sepals and either six or seven stamens. The sepals and stamens continue to develop and extend, such that each stamen primordium is enclosed within one sepal. The sepals continue to elongate and start to envelop the stamens. The formation of locular grooves on the stamens and the development of resin glands on the sepals precedes the complete enclosure of the stamens by the sepals. The anthers continue to develop on short thin filaments that aid wind dispersal of the pollen. Once the male flower is fully mature, the sepals open, the tapetum degenerates and wind-borne pollen is released (*Figure 2a*). The sepals then wither and the flowers become detached from the pedicel. Male flowers are borne in loose, highly branched cymose panicles, each cluster of which is subtended by a bract (*Figure 2b*). At no stage of development is there evidence of the formation of a central carpel whorl in male flowers.

5. Development of the female flower

In the female, the floral buttress emerging from the flanks of the reproductive apical meristem divides into two floral primordia, each of which is enclosed within a small bracteole. The bracteole extends to envelop the differentiating floral meristem, on which

Figure 2. *Differences in floral structure between male and female hop flowers of* H. lupulus. *(a) male flower, comprising five sepals and five anthers; (b) mature male flowers arranged in cymose panicles; (c) female inflorescence comprising many pairs of flowers arranged within bracts and bracteoles; (d) mature female inflorescences within papery bracteoles.Part (a) from Ainsworth* et al. *(1988) Sex determination in plants.* Current Topics in Developmental Biology, *vol. 38, pp. 167–223. Reprinted by permission of Academic Press, Inc.*

two carpel primordia begin to develop and extend, surrounded by a ring of vestigial perianth tissue at the base (*Figure 1d*). The bracteole and carpels continue to elongate, and the perianth, located at the base of the carpels, extends to enclose developing ovary tissues at the base of the flower. The bracteoles start to envelop the gynoecium and the carpels begin to develop stigmata on their surfaces. The ovary tissue at the base continues to develop and the carpels extend further. The stigmas at this stage are highly papillated and receptive to the wind-borne pollen. Female flowers are arranged in pairs, each of which is enclosed within a bracteole, and pairs of flowers are subtended by a bract. Two pairs of flowers are arranged in the same plane of phyllotaxy and each cluster of four flowers is arranged alternately along the central axis or 'strig' (*Figure 2c*). As the inflorescences mature, the central axis elongates, the stigmas degenerate and the bracteoles at the base of each flower begin to extend to envelop the remains of the stigmatic tissue. As the ovary tissue matures, the vestigial perianth part surrounding the ovary ceases further development, and the ovary, if pollinated, produces an achene enclosed within the perianth. The papery bracts and bracteoles extend rapidly to form the familiar hop strobiles or 'cones' (*Figure 2d*). These bract tissues turn brown (or 'ripen') as the α-acids and essential oils reach their highest levels within the lupulin glands located on the pericarp and on the bases of the bracts and bracteoles. No traces of stamen development are observed at any stage in female hop flowers.

6. Development of male and female flowers on monoecious lines

Occasionally, spontaneously arising monoecious hop plants occur which are often of predominately male phenotype, with many of the lateral branches terminating in female inflorescences (*Figure 3*). Monoecious lines of *H. lupulus* follow a similar developmental pathway to that of the dioecious lines before the formation of the reproductive organ primordia. After the onset of the reproductive phase, floral organs of both sexes are formed, either in separate male and female inflorescences or, very rarely, in 'bisexual' flowers containing both sets of sex organs. These 'bisexual' floral types are borne in loose cymose panicles or around the bases of the terminal female inflorescences.

Figure 3. *Mature monoecious flowering lateral of a triploid plant* ($2n = 27 + XXY$) *of* H. lupulus.

7. Sex chromosomes in *Humulus lupulus* and *H. japonicus*

The family Cannabidaceae is unique in the respect that all three species which have been chromosome-counted (*H. lupulus*, *H. japonicus* and *Cannabis sativa*) possess sex chromosomes. At least five different types of sex chromosomes have been identified in *H. lupulus*. The most common type is the 'Winge type' (Winge, 1923, 1929) which consists of a heteromorphic sex chromosome pair (where the X chromosome is almost twice the length of the Y) and nine autosomal bivalents forming at meiosis (*Figure 4a*). In addition, several other sex chromosome types have been identified from studies of meiotic events in male hop plants. These sex chromosome types are: the 'New Winge type', a heteromorphic sex chromosome pair with a proportionately longer Y chromosome than that of the 'Winge type' and nine bivalents; the 'Sinoto type' a quadrivalent with eight bivalents; the 'New Sinoto type', a quadrivalent with shorter Y chromosomes than those of the 'Sinoto type'; and a 'Homotype' with ten bivalents and no discernible sex chromosomes (Sinoto, 1929a, 1929b; Ono, 1937, 1955; Neve, 1958, 1961 and 1991). Furthermore, four- and six-chromosome meiotic complexes have been reported in some wild Japanese and American hops (Ono and Suzuki, 1962). Neve (1961), however, disputed the existence of sex-related quadrivalents in these plants and interpreted them as being autosomal in origin. This is despite strong evidence for the formation of sex-quadrivalents. Ono (1955) demonstrated that all the male offspring of Sinoto type males were themselves quadrivalent-formers during meiosis.

The common hop, *H. lupulus*, has a diploid chromosome number of 20 such that, in the simplest system, females are 2n = 18 plus sex chromosomes of constitution XX, whilst males are 2n = 18 + XY. The karyotype of a diploid *H. lupulus* male with XY sex chromosomes is shown in *Figure 4b*. Sex expression is determined by an X to autosome balance system (Neve, 1961) so that a ratio of 0.5 of X chromosomes to sets of autosomes gives a male phenotype, whilst a ratio of 1.0 gives rise to a female phenotype. Intermediate ratios are associated with intersexual, monoecious individuals, such that a ratio of 0.67, as exemplified by a triploid of chromosome constitution 2n = 27 + XXY, gives rise to a predominately male phenotype, with the plant bearing a small number of female inflorescences. The first report of a naturally occurring triploid of *H. lupulus* was by Tournois in 1914. These plants were predominately male, but carried occasional female inflorescences at the ends of laterals. An X to autosome ratio of 0.75, as in a tetraploid of chromosome constitution 2n = 36 + XXXY, gives plants which carry approximately equal numbers of male and female inflorescences. However, all of these phenotypes are prone to variation, probably due, in part, to the influence of background genotypic effects (Parker and Clark, 1991).

Wild hops originating in Japan (*H. lupulus* subsp. *cordifolius*) and most of their cultivated derivatives have a complex multiple sex chromosome system. Females have a chromosome constitution of 2n = 16 + $X_1X_1X_2X_2$, whilst males are 2n = 16 + $X_1Y_1X_2Y_2$. Males have eight autosomal bivalents and a chain of four chromosomes at metaphase I of meiosis (*Figure 4c*), while females form ten bivalents of Sinoto type (Parker and Clark, 1991). This sex chromosome system has arisen by a translocation between a sex chromosome and an autosome. Male fertility is dependent on the regular occurrence of alternate disjunction of the quadrivalent. Studies by Parker and Clark (1991) on a male plant of this type have shown that during meiosis, segregation efficiency of the quadrivalent is low (70%). Chains of four in alternate, adjacent

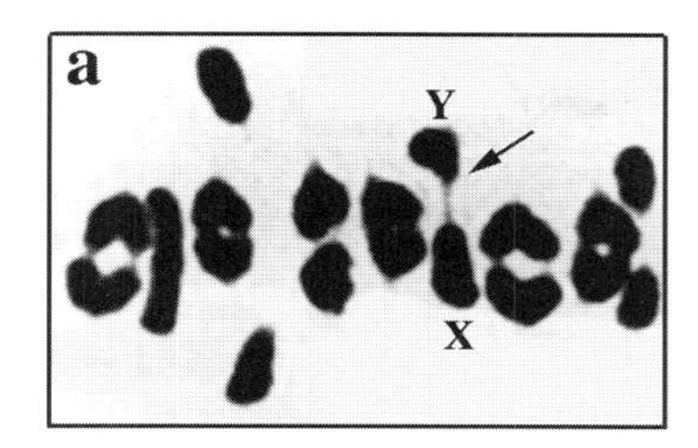

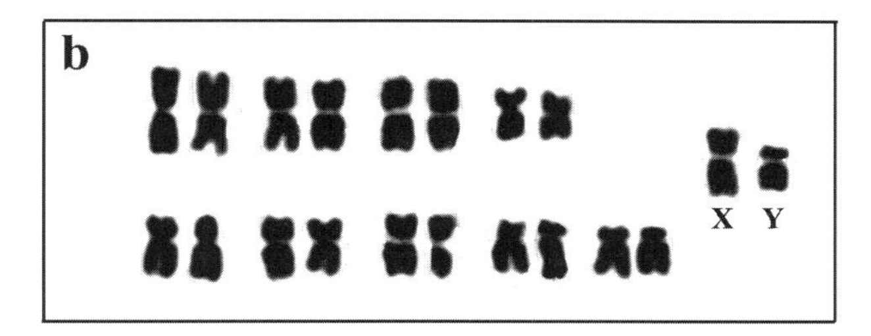

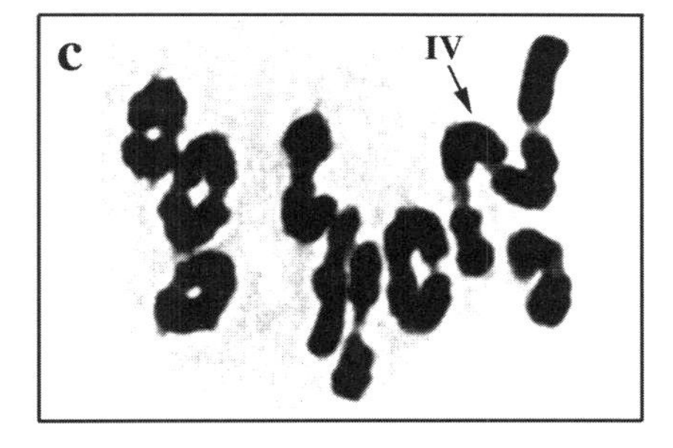

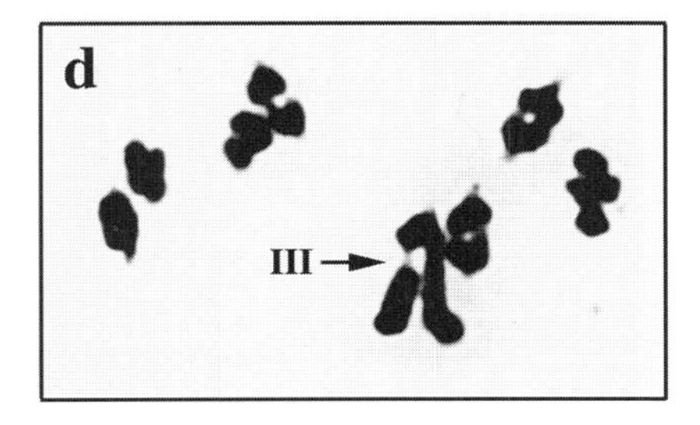

Figure 4. *(a) Metaphase I in pollen mother cell (PMC) of male* H. lupulus, *English variety, 2n = 18 + XY with heteromorphic XY bivalent (arrow). (b) Karyotype of male* H. lupulus, *English variety, with 2n = 18 + XY. Note that the Y is smaller than the X. (c) Metaphase I in PMC of the Japanese subspecies* H. lupulus *subsp.* cordifolius *male,* $2n = 16 + X_1X_2Y_1Y_2$. *Note the sex-quadrivalent in alternate orientation (IV). (d) Metaphase I in PMC of* H. japonicus *male,* $2n = 14 + XY_1Y_2$ *with sex-trivalent (III). Parts (a)–(d) from Ainsworth* et al. *(1988) Sex determination in plants.* Current Topics in Developmental Biology, *vol. 38, pp. 167–223. Reprinted by permission of Academic Press, Inc.*

and indifferent orientations can be produced, as well as trivalent plus univalent combinations.

The annual Japanese hop, *H. japonicus*, is also dioecious, although male plants occasionally carry terminal female inflorescences on their laterals. *H. japonicus* has a sex chromosome system similar to that of *Rumex acetosa*, with females of chromosome constitution 2n = 14 + XX and males of 2n = 14 + XY_1Y_2. In the male, the sex chromosomes form a linear trivalent Y_1-X-Y_2 with terminal chiasmata during meiosis which regularly undergoes convergent orientation (Parker and Clark, 1991; *Figure 4d*).

The Y chromosome of both *H. lupulus* and *H. japonicus* is unusual amongst plant sex chromosomes in that it is smaller than the X. In most plant species carrying sex chromosomes, the Y is larger than the X (e.g. in *Silene latifolia*, *R. acetosa*, *Coccinia indica*). The hop Y chromosome, like that of *R. acetosa*, is not necessary for the production of a male phenotype as Y deletion mutants still produce male flowers. However, the resultant plant lacking the Y is infertile. Sequences on the Y are necessary for directing the latter stages of pollen mother cell differentiation. Thus, sex expression in hops, like that of *R. acetosa*, may be governed by genes located on the autosomes or on the X chromosomes (Neve, 1961).

8. Sensitivity of sex expression in hop to plant hormones

Diploid hop plants tend not to show any lability in their sex expression and therefore are usually of strictly male or female phenotype. However, a relatively 'plastic' response in sex expression has been observed in some monoecious hop lines, manifested in changes in the numbers of terminal female inflorescences from one growing season to the next (P. Darby, unpublished data). These changes in sex expression suggest that the hop could be sensitive to changes in the local concentrations of endogenous plant growth hormones brought about by environmental fluctuations from one growing season to the next. Several experiments have been previously carried out to attempt to induce sex changes in dioecious diploid hop lines by the exogenous application of plant growth substances, including auxins and gibberellins.

In the cultivated hop, certain auxins can induce maleness on female plants. Weston (1960) reported the induction of male flowers on female plants of cv. Fuggle when they were sprayed with the weak synthetic auxin α (2-chlorophenylthio) propionic acid (2-CPTPA). This work suggests the possibility of a plant growth substance component to sex determination in hop. The timing of exogenous application of 2-CPTPA was critical for the induction of a 'monoecious' phenotype (Weston, 1960). It was not reported whether the male flowers were fertile, although it is likely that they were not, since the Y sex chromosome, which is required for pollen maturation, would be absent in these genetically female plants. The sterile nature of auxin-induced male flowers was confirmed when Weston's work was repeated on the female hop variety Yeoman (P. Darby, unpublished data). In *C. sativa* (hemp), a close relative of hop, the exogenous application of auxin was found to have the opposite effect, resulting in the feminization of male plants (Heslop-Harrison, 1956; Chailakhyan, 1979). Thus, even in closely related plant species, plant growth substances can have profoundly different effects on the resulting sexual phenotype. The effects of the exogenous application of gibberellin (GA) on female hop plants was investigated by Hartley and Neve (1966). They found that when 5 ppm GA was sprayed onto plants at the end of May, this increased the amount of lateral branching and the number of inflorescences that formed, but resulted in a reduction in cone size and no change in the sex of the plant (Hartley and Neve, 1966).

9. Conclusions and current work

Plant species bearing unisexual flowers have been divided into two groups according to their mode of reproductive organ initiation (Heslop-Harrison, 1963). In those species in the first group, both stamen and carpel primordia are initiated and the inappropriate set of sex organs is suppressed or aborted at a later stage. Most monoecious and dioecious species follow this floral development pathway, e.g. maize (Cheng *et al.*, 1983; Stevens *et al.*, 1986), cucumber (Atsmon and Galun, 1960; Malepszy and Niemirowicz-Szczytt, 1991), white campion (Grant *et al.*, 1994), pistachio (Hormaza and Polito, 1996), asparagus (Lazarte and Palser, 1979; Bracale *et al.*, 1991), kiwi fruit (McNeilage, 1991a, 1991b) and *Fragaria* spp. (Ahmadi and Bringhurst, 1991). Those species in the second group produce flowers that are unisexual at inception; they initiate only one set of reproductive organs and there are no obvious signs of development of the other sex organ set. This leads to gross developmental differences between staminate and pistillate flowers very early in floral ontogeny. This alternative mode of unisexual floral development is shared by a small number of plant species, including: *C. sativa* (Mohan Ram and Nath, 1964), spinach (Sherry *et al.*, 1993) and *Mercurialis annua* (Durand and Durand, 1991a,b).

Our work has revealed that both sexes of *H. lupulus* and *H. japonicus* fall into the more unusual second group as they show extreme divergence in their floral morphologies because the inappropriate set of sex organs is not initiated in the strictly unisexual flowers. These striking differences between male and female hop flower development, which are apparent very early in floral ontogeny, suggest that genes which could control the gender differences in hops probably act well in advance of floral organogenesis. The hop therefore appears to be well suited to sex determination studies, since it has the benefit of initiating only one set of sex organs. This means that there should be few or none of the compounding factors involved in the initiation of the other reproductive organ set. This is in contrast to most of the monoecious and dioecious plant species being used to investigate the molecular basis of sex determination (e.g. cucumber, white campion, asparagus, kiwi fruit), which initiate both stamens and carpels prior to the suppression of the inappropriate set. In addition, the auxin-induced flexibility of sex expression in hop could be useful in future work directed towards cloning sex determining genes, particularly via subtractive cloning techniques.

Our work is focusing on the possible role of homeotic floral MADS box genes in sex determination of hop. We predict that the B and C function organ identity genes could be expressed in a sex-specific manner in developing hop flowers since the inappropriate set of sex organs is never initiated in flowers of either sex. In the dioecious species *R. acetosa*, the expression of the B function homologues *RAD1* and *RAD2* and the C function *RAP1* show sex-specific expression patterns which coincide with the suppression of the 'wrong' sex organ set (Ainsworth *et al.*, 1995). In male hop flowers, which only initiate a sepal and stamen whorl, we predict that the B function homologue(s) may be restricted to the developing stamens and that the C function gene(s) should be absent from the centre of flower which would develop into the carpel whorl in a hermaphrodite flower. Female hop flowers, which comprise two whorls (perianth, carpel), appear to partially resemble severe *deficiens* mutants (sepal, sepal, carpel; Sommer *et al.*, 1990; Schwarz-Sommer *et al.*, 1992). If this comparison is valid, we would not anticipate B function expression in female flowers and C function expression should be restricted to the carpel whorl.

We are currently analysing the expression patterns of a putative *DEFICIENS/APETALA3* (Schwarz-Sommer *et al.*, 1992; Jack *et al.*, 1992, 1994) B function hop homologue in male, female and 'bisexual' hop flowers using *in situ* hybridization techniques. We are also analysing the expression of a putative *FLORAL BINDING PROTEIN 2 (FBP2)/TM5/AGL2* (Angenent *et al.*, 1992; Pnueli *et al.*, 1994; Ma *et al.*, 1991, respectively) hop homologue which may share similar mediatory roles between the activation of the floral meristem genes and floral organ identity genes that are exhibited by *FBP2*, *TM5* and *AGL2*. In addition, we are developing callus regeneration and transformation systems for hop in order to analyse the functions of the cloned hop homeotic genes using antisense strategies.

References

Ahmadi, H. and Bringhurst, R.S. (1991) Genetics of sex expression in Fragaria species. *Am. J. Bot.* **78**: 504–514.

Ainsworth, C., Crossley, S., Buchanan-Wollaston, V., Thangavelu, M. and Parker, J. (1995) Male and female flowers of the dioecious plant sorrel show different patterns of MADS box gene expression. *Plant Cell* **7**: 1583–1598.

Angenent, G.C., Busscher, M., Franken, J., Mol, J.N.M. and van Tunen, A.J. (1992) Differential expression of two MADS box genes in wild-type and mutant petunia flowers. *Plant Cell* **4**: 983–993.

Atsmon, D. and Galun, E. (1960) A morphogenetic study of staminate, pistillate and hermaphrodite flowers in *Cucumis sativus* (L.) *Phytomorphology* **10**: 110–115.

Bracale, M., Caporali, E., Galli, M.G. *et al.* (1991) Sex determination and differentiation in *Asparagus officinalis* L. *Plant Sci.* **80**: 67–77.

Burgess, A.H. (1964) Hops, botany, cultivation and utilisation. In: *Hops* (ed. N. Polumin). World Crops Books, London.

Chailakhyan, M.Kh. (1979) Genetic and hormonal regulation of growth, flowering and sex expression in plants. *Am. J. Bot.* **66**: 717–736.

Cheng, P.C., Greyson, R.I. and Walden, D.B. (1983) Organ initiation and the development of unisexual flowers in the tassel and ear of *Zea mays*. *Am. J. Bot.* **70**: 450–462.

Durand, B. and Durand, R. (1991a) Sex determination and reproductive organ differentiation in *Mercurialis*. *Plant Sci.* **80**: 49–65.

Durand, B. and Durand, R. (1991b) Male sterility and restored fertility in annual mercuries, relations with sex differentiation. *Plant Sci.* **80**: 107–118.

Grant, S., Hunkirchen, B. and Saedler, H. (1994) Developmental differences between male and female flowers in the dioecious plant *Silene latifolia*. *Plant J.* **6**: 471–480.

Hartley, R.G. and Neve, R.A. (1966) The effect of gibberellic acid on development and yield of Fuggle hops. *J. Hort. Sci.* **41**: 53–56.

Heslop-Harrison, J. (1956) Auxin and sexuality in *Cannabis sativa*. *Physiol. Plant.* **4**: 588–597.

Heslop-Harrison, J. (1963) Sex expression in flowering plants. In: *Brookhaven Symposia in Biology* **16**. Brookhaven National Laboratory, New York, NY, pp. 109–125.

Hormaza, J.I. and Polito, V.S. (1996) Pistillate and staminate flower development in dioecious *Pistacia vera* (Anacardiaceae) *Am. J. Bot.* **83**: 759–766.

Hough, J.S., Briggs, D.E., Stevens, R. and Young, T.W. (1982) *Malting and Brewing Science, Vol. 2. Hopped Wort and Beer*, 2nd Edn. Chapman and Hall, London.

Jack, T., Brockman, L.L. and Meyerowitz, E.M. (1992) The homeotic gene *APETALA3* of *Arabidopsis thaliana* encodes a MADS box and is expressed in petals and stamens. *Cell* **68**: 683–697.

Jack, T., Fox, G.L. and Meyerowitz, E.M. (1994) Arabidopsis homeotic gene *apetala3* ectopic expression: transcriptional and post-transcriptional regulation determine floral organ identity. *Cell* **76**: 703–716.

Kohlmann, H. and Kastner, A. (1976) *Der Hopfen*. Hopfen-Verlag, Wolnzach.

Lazarte, J.E. and Palser, B.F. (1979) Morphology, vascular anatomy and embryology of pistillate and staminate flowers of *Asparagus officinalis*. *Am. J. Bot.* **66**: 753–764.

Ma, H., Yanofsky, M.F. and Meyerowitz, E.M. (1991) *AGL1-AGL6*, an *Arabidopsis* gene family with similarity to floral homeotic and transcription factor genes. *Genes Develop.* **5**: 484–495.

Malepszy, S. and Niemirowicz-Szczytt, K. (1991) Sex determination in cucumber (*Cucumis sativus*) as a model system for molecular biology. *Plant Sci.* **80**: 39–47.

McNeilage, M.A. (1991a) Gender variation in *Actinidia deliciosa*, the kiwifruit. *Sex. Plant Reprod.* **4**: 267–273.

McNeilage, M.A. (1991b) Sex expression in fruiting male vines of kiwifruit. *Sex. Plant Reprod.* **4**: 274–278.

Mohan Ram, H.Y. and Nath, R. (1964) The morphology and embryology of *Cannabis sativa* Linn. *Phytomorphology* **14**: 414–429.

Neve, R.A. (1958) Sex chromosomes in the cultivated hop *Humulus lupulus*. *Nature* **181**: 1084–1085.

Neve, R.A. (1961) Sex determination in the cultivated hop *Humulus lupulus*. Ph.D. thesis, Wye College, University of London.

Neve, R.A. (1991) *Hops*. Chapman and Hall, London.

Ono, T. (1937) On the sex chromosomes in wild hops. *Bot. Mag., Tokyo* **51**: 110–115.

Ono, T. (1955) Studies in hop. I. Chromosomes of common hop and its relatives. *Bull. Brew. Sci.* **2**: 1–65.

Ono, T. and Suzuki, H. (1962) The wild hop native to Japan. In: *Cytological Studies* (ed. T. Ono). Sasaki Printing Company, Sendai, pp. 71–110.

Parker, J.S. and Clark, M.S. (1991) Dosage sex-chromosome systems in plants. *Plant Sci.* **80**: 79–92.

Pnueli, L., Hareven, D., Broday, L., Hurwitz, C. and Lifschitz, E. (1994) The *TM5* MADS box gene mediates organ differentiation in the three inner whorls of tomato flowers. *Plant Cell* **6**: 175–186.

Schwarz-Sommer, Z., Hue, I., Huijser, P., Flor, P.J., Hansen, R., Tetens, F., Lönnig, W.-E., Saedler, H. and Sommer, H. (1992) Characterisation of the *Antirrhinum* homeotic MADS-box gene deficiens: Evidence for DNA binding and autoregulation of its persistent expression throughout flower development. *EMBO J.* **11**: 251–263.

Sherry, R.A., Eckard, K.J. and Lord, E.M. (1993) Flower development in dioecious *Spinacia oleracea* (Chenopodiaceae) *Am. J. Bot.* **80**: 283–291.

Sinoto, Y. (1929a) On the tetrapartite chromosomes in *Humulus lupulus*. *Proc. Imp. Acad.* **5**: 46–47.

Sinoto, Y. (1929b) Chromosome studies in some dioecious plants with special reference to the allosomes. *Cytologia* **1**: 109–191.

Small, E. (1978) A numerical and nomenclatural analysis of morpho-geographic taxa of *Humulus*. *Syst. Bot.* **3**: 37–76.

Sommer, H., Beltrán, J.P., Huijser, P., Pape, H., Lönnig, W.-E. Saedler, H., Sommer, H. and Schwarz-Sommer, Z. (1990) *Deficiens*, a homeotic gene involved in the control of flower morphogenesis in *Antirrhinum majus*: the protein shows homology to transcription factors. *EMBO J.* **9**: 605–613.

Stevens, S.J., Stevens, E.J., Lee, K.W., Flowerday, A.D. and Gardner, C.O. (1986) Organogenesis of the staminate and pistillate inflorescences of pop and dent corns: relationship to leaf stages. *Crop Sci.* **26**: 712–718.

Thomas, G.G. and Schwabe, W.W. (1970) Apical morphology in the hop (*Humulus lupulus*) during flower initiation. *Ann. Bot.* **34**: 849–859.

Tournois, M.J. (1912) Influence de la lumiére sur la floraison du Houblon japonais et du chanvre. *C.r. hebd. Séanc. Acad. Sci., Paris* **155**: 297–300.

Tournois, M.J. (1914) Études sur la sexualité du Houblon. *Annales des Sci. Nat.* **XIX**: 49–91.

Weston, E.W. (1960) Changes in sex in the hop caused by plant growth substances. *Nature* **188**: 81–82.

Williams, I.H. (1960) Changes in the carbohydrate balance of the hop (*Humulus lupulus* L.) in relation to the annual growth cycle in young plants. Report of the Department of Hop Research, Wye College, 1959, pp. 98–106.

Winge, Ø. (1923) On sex chromosomes, sex determination and preponderance of females in some dioecious plants. *C.R. Lab. Carlsberg* **15**: 1–26.

Winge, Ø. (1929) On the nature of sex chromosomes in *Humulus*. *Hereditas* **12**: 53–63.

9

Search for genes involved in asparagus sex determination

Giovanna Marziani, Elisabetta Caporali and Alberto Spada

1. *Asparagus officinalis*: an important crop plant

Asparagus officinalis, a monocot which belongs to the family of Asparagaceae, is an important crop plant and is considered to be one of the most refined vegetables. It originated from the Mediterranean area and was known to Egyptians, Greeks and Romans. At present, several varieties of asparagus are cultivated world-wide, particularly in mild and sunny climates to which the plant is expecially well adapted.

Asparagus is a perennial species and is productive for 8–10 years. In spring the young stems, called spears, emerge through the ground; they represent the edible part of the plant and are collected for 40–60 days. Spears are produced annually and have a complex apical structure, bearing the primordia of leaves, lateral branches and flowers. The stems are covered with needle-like assimilatory structures, the phylloclades (Arber, 1925), while leaves are reduced to small scales subtending lateral branches. The young branches bear numerous small solitary flowers. The perennial part of the plant is a short sympodial rhizome where, during the vegetative season, carbohydrates are accumulated. The plant has a rest period during the winter.

Asparagus officinalis is a dioecious species and average populations exibit the expected 1: 1 ratio of males and females, bearing, respectively, staminate or pistillate flowers. Plants of the two sexes have the same external morphology and the diagnosis of gender may be performed only at flowering. The development of flowering branches is somewhat different in males and females. In male plants each branch bears a large number of flowers, most of which mature before the appearance of the phylloclades. In female plants, flowers are less numerous and develop together with phylloclades. After flowering, the only difference between sexes is the presence, on female plants, of several small globose berries which are red at maturity.

From an agronomical point of view, male plants are superior to females for longevity, growth precocity and productivity (Benson, 1982). These traits are probably the result of the lower energy investment required for reproduction in male plants.

Sex Determination in Plants, edited by C.C. Ainsworth.

2. Genetic mechanism of sex determination in asparagus

Asparagus officinalis has a haploid chromosome number of 10 and a 2C DNA content of 3.76 ± 0.26 pg (Galli *et al.*, 1988). This value is independent of sex and is rather low if compared with other members of the Liliflorae. Sex determination has been proposed to be under the control of two tightly associated regulatory genes of the type 'male activator' (M) and 'female suppressor' (F) as has also been proposed for *Silene latifolia* (Franken, 1970; Marks, 1973; Westergaard, 1958). These genes have been located on the homomorphic chromosome pair 5 (Loptien, 1979). The male is heterozygous (MF/mf) but homozygous males generated by anther culture are also viable (see Section 5). The two-gene hypothesis is supported by the appearance of rare hermaphrodite and sterile plants among asparagus populations.

3. The development of male and female flowers

Male and female asparagus flowers have six tepals of similar shape, which are arranged in two whorls. The tepals are free and spreading, are green in the young flower and are yellowish in the mature flower. Male flowers have six stamens, which in the mature female flower, are collapsed (*Figure 1*). The female flowers contain a tricarpellate and trilocular ovary with a short style with a lobate stigma. In males, the ovary remains very small and the style generally does not develop.

We have followed the development of male and female flowers from the appearance of flower primordia by SEM (Caporali *et al.*, 1994). The shoot apex produces the leaf primordia which develop as scales in an indeterminate spiral. Inside each scale, in an axillary position, five primordia arise, arranged in two rows. The internal row gives rise to two flowers and one lateral branch in the central position; the two primordia of the external row develop as small scales at the base of the flower stalk. During the first stages of development all flowers are hermaphrodite and it is impossible to distinguish male from female flowers (*Figure 1c*). The first visible difference between the sexes appears when the upper parts of the carpels in the female flower start to elongate and fuse to form the stylar tube (*Figure 1d*). In the male flower, this structure does not appear and the carpels curve inward, giving rise to a closed structure (*Figure 1e*). At this stage, the stamens are still growing in the female flower; their degeneration starts at a later stage and is visible by SEM as a progressive shrinkage leaving at anthesis small vestigial organs.

Histological observations conducted on flowers of both sexes through development give a clear picture of the events which lead to the abortion of the pistil in the male flower and the stamen in the female flower (Bracale *et al.*, 1990; Caporali *et al.*, 1994; Galli *et al.*, 1993; Lazarte and Palser, 1979). In the male flower, the ovary resembles the female ovary (except for its small size) and contains healthy tissue. The ovules, however, degenerate starting from the micropylar edge; this phenomenon leads to the appearance of empty dead cells with thickened walls. The internal structurs of the male and female anthers start to diverge with the appearance of the tapetum, at a time which is coincident with the appearance of the first differences at the pistil level. In female anthers, tapetum cells are very short-lived and soon begin to degenerate. The death of tapetum cells is followed by the degeneration of the sporogenous tissue, a very rapid phenomenon which leads to the coalescence of the contents of adjacent cells into a dense mass which is eventually reabsorbed, leaving

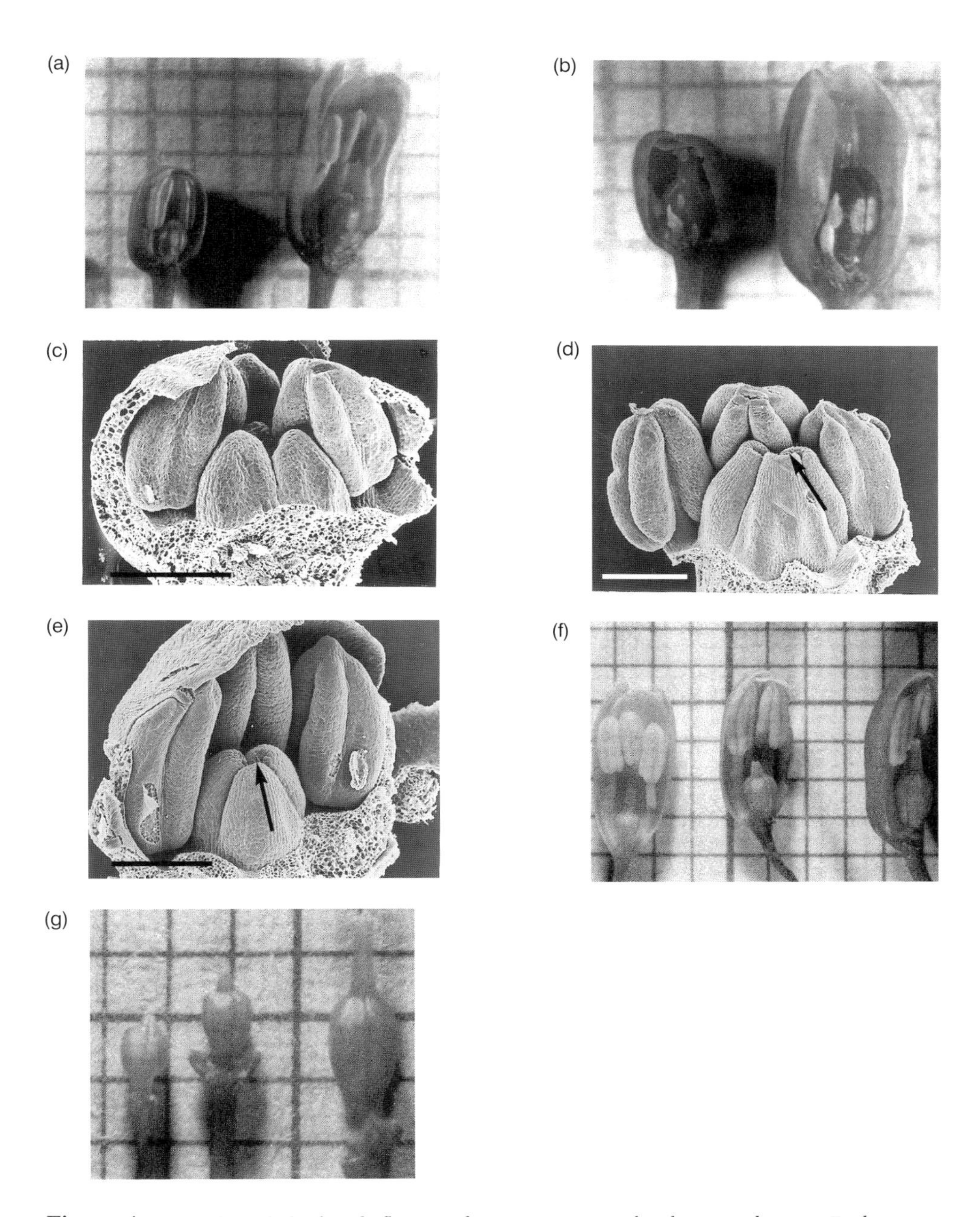

Figure 1. *(a) Male and (b) female flowers of asparagus at two developmental stages. Each square is one square millimetre in size. (c–e) Scanning electron micrographs of young asparagus flowers. (c) Hermaphrodite stage: male and female flowers up to this stage are identical. (d) First unisexual stage of the female flower; arrow indicates the appearance of stylar tube. (e) First unisexual stage of the male flower; arrow indicates the top of carpels that curve inwards. Bars: 200 μm. (f) Male flowers with different type of pistil. (g) Detail of the pistils. Each square is one square millimetre.*

empty pollen sacs. The degeneration of the external tissues of the anther is a slower phenomenon. The morphological events which lead to the degeneration of female anthers in asparagus are very similar to those described for anthers of *Brassica* transformed with a chimaeric ribonuclease gene specifically expressed in the tapetum cells (Mariani *et al.*, 1990).

Taken together, the morphological data indicate that the programmed abortion of developing organs which results in sex differentiation in asparagus is realized through different mechanisms in male and female flowers. In the male pistil, style development is inhibited and the cells of the ovules degenerates probably through a phenomenon of apoptosis, while the ovary itself remains viable. In the female anthers the target of the degenerative process seems to be the tapetum. The death of the tapetum is sufficient to induce the degeneration of the sporogenous tissue and the desiccation of the whole anther (Mariani *et al.*, 1990).

It is of interest to note that in some asparagus genotypes the inhibition of style development may be delayed leading to male flowers with a short style. Among the progeny derived from a cross where the male parent showed this characteristic, male plants had different types of pistil with styles of different length (*Figure 1f* and g) and some males bore a few hermaphrodite flowers with well developed styles. These flowers produce fertile seeds indicating that there is a relationship between style formation and ovule development (Galli *et al.*, 1993). From these data, it can be concluded that for sex differentiation of male flowers, the time of activation of the phenomenon which leads to style growth inhibition is critical. If the phenomenon is controlled by the sex-determining gene 'female suppressor', its expression can apparently be induced at different stages of development in different genotypes, giving rise, in some cases, to hermaphrodite flowers.

4. Physiological and biochemical changes during sex differentiation

The morphological changes which characterize the switch from the hermaphrodite developmental pathway to the unisexual pathway, are accompanied by physiological and biochemical changes. When the growth of stamens starts to slow in females and to accelerate in males, the endogenous level of auxin is 3.5 times higher in the male relative to the level in female flowers (Marziani Longo *et al.*, 1990) indicating that this growth factor is involved in stamen development. To control the effect of auxin on female anther development, we tried to administer exogenous auxin to developing flowers, but the treatments had no appreciable effect, probably due to the complex structure of the apex which bears the young flower buds (unpublished data). With the same aim, we are presently transforming female asparagus plants with genes involved in auxin synthesis, under the control of inducible promoters.

In female flowers, the activity of RNase, which remains constant in male flowers, increases steadily during the first stages of unisexual development, reaching a peak when anther degeneration starts to be visible by SEM. This finding suggests that the programmed degeneration of the anther could be induced through the activation of this hydrolytic enzyme. The development and arrest of the different reproductive organs in males and females is accompanied by a change in their polypeptide patterns (Caporali *et al.*, 1994). These data indicate that sex-determining genes are involved in the regulation of the different biochemical phenomena which led to the development or to the degeneration of the anthers in male or in female flowers. The block of pistil development in the male flower and the stimulation of its development in the female flower is not accompanied by biochemical changes which are detectable by the approaches described.

5. The doubled-haploid clones of asparagus

For many years, haploid and doubled-haploid plants of several different species have been obtainable through anther culture (Heberle-Bors, 1985; Sunderland, 1974; Sunderland and Roberts, 1979). In asparagus, pure lines cannot be obtained through conventional breeding methods, since dioecy makes selfing impossible. Anther culture has been applied to this species to obtain completely homozygous doubled-haploid clones (Doré, 1990; Falavigna *et al.*, 1983a; Qiao and Falavigna, 1990). Starting from male flower buds of a size range of 1.5–2 mm, whose anthers contain microspores at the uninucleate stages, both male and female plants can be obtained. The average frequency of male and female anther-regenerated plants is 60% and 40% respectively (Falavigna *et al.*, 1983b, 1983c). Male plants are of particular interest as they are homozygous for sex determinants (MF/MF). These plants, called 'supermales', are phenotypically identical to heterozygous males (MF/mf), but produce only MF pollen and an all-male progeny.

In another dioecious species, *Melandrium album* (*Silene latifolia*) which has well-characterized sex chromosomes, it is impossible to obtain males by anther culture. Starting from anthers isolated at the middle-to-late mononucleate stage, haploid plantlets develop from embryos which originate from microspores and, at the torpedo stage, protrude through the anther wall. All these plants are females suggesting that the absence of X chromosome during embryo development is lethal or, alternatively, that Y chromosome-bearing microspores are not competent for embryonic development (Ye *et al.*, 1990). If compared with *Melandrium* the viability of homozygous males of asparagus may have two possible explanations: (i) the sex chromosomes (conventionally called X and Y, even if homomorphic) are not involved in the development of plantlets from microspores; (ii) both X- and Y-bearing microspores are competent for this process.

The different behaviour of asparagus and *Melandrium* may depend on their different evolutionary origins. Most of the dioecious species are in taxa that also include hermaphrodite species, and probably originated independently through different modification of a primitive hermaphrodite condition. A simple evolutionary model for the transformation from the hermaphrodite to the dioecious condition assumes that the hermaphroditic ancestors had two master genes, one controlling the development of the male organs, the other the development of the female organs. Two mutations are necessary for transforming a hermaphrodite species into a dioecious one. A single recessive mutation in the 'M' master gene (M→m) leading to male sterility, which gives rise to functionally female plants, and a dominant mutation (f→F) leading to female sterility, which gives rise to functionally male plants (Bawa, 1980; Charlesworth and Charlesworth, 1978; Charnov *et al.*, 1976; Kohn, 1988). Thus, the hermaphrodite genotype Mf/Mf will have become MF/mf for male and mf/mf for females, with the male heterogamentic sex.

For maintenance of a stable dioecious population, recombination between male and female determining loci should be prevented as much as possible in the male. In the homomorphic chromosome system crossing-over is possible throughout the sex chromosome, and the sex determining loci should be tightly linked. In the heteromorphic chromosome system, recombination is suppressed in the whole chromosome or, at least, in a large part of it. This evolutionary model implies that sex chromosomes were originally morphologically identical, the only differences being the sex-determining genes. Therefore, the most recently evolved dioecious species are characterized by homomorphic sex chromosomes, while the appearance of sexual heterochromosomes should be a second step in the evolution of dioecy (Frankel and Galun, 1977). In the first case, no differences, besides the sex determinants, are accumulated between the sex chromosomes, and males and females derived from microspores are both viable. In the second case, deleterious mutations tend to accumulate in the Y chromosome, since they cannot be eliminated by homozygosity, and the YY genotype is not viable. This hypothesis has been recently confirmed by Guttman and Charlesworth (1998) who have isolated from *Melandrium album*, an X-linked gene with a degenerate Y-linked homologue. They observed an X-linked transmission pattern for the male reproductive organ-specific gene *MROS3* (Matsunaga *et al.*, 1996) and cloned the Y-linked homologue. Sequence comparison of *MROS3*-X and *MROS3*-Y showed that the first 235 bases of *MRSO3*-Y are not found in the X-gene and constitute a region of repetitive DNA. The gene on the Y-chromosome therefore appears to have degenerated to a non-functional state.

6. Search for DNA sequences associated with sex chromosomes

The molecular mechanisms which induce the morphological and biochemical changes characteristic of the point of transition from the hermaphrodite to unisexsual developmental pathways during asparagus flower development are still unknown and the sex-determining genes have not been isolated. The identification of DNA sequences tightly linked to a gene of interest is the first step for attempting the positional cloning. With the aim of identifying DNA sequences associated with asparagus sex-determining genes we have followed two different approaches: segregation analysis of polymorphic molecular markers in the progenies of asparagus crosses and bulked segregant analysis. For both approaches we have utilized doubled-haploid clones derived from anther culture at the Research Institute for Vegetable Crops, Section of Montanaso Lombardo.

6.1 *Segregation analysis of polymorphic markers*

The asparagus genotypes utilized for identifying molecular markers linked to sex determinants have been described previously (Maestri *et al.*, 1991; Restivo *et al.*, 1995). Parental lines were doubled-haploid clones, generated through anther culture, from male plants with different genetic backgrounds. Since male parents are homozygous for the sex determinants and the F1s are all male, linkage analysis has been carried out on the first backcross (BC1) progeny; 40–80 plants from five different BC1s have been utilized. A scheme of the genetic system utilized for the segregation analysis is shown in *Figure 2.* We have analysed different types of molecular markers: isoenzymes, restriction fragment length polymorphism (RFLP), random amplified polymorphic DNA (RAPD), amplified fragment length polymorphism (AFLP) and one morphological marker, sex. This research allowed the construction of an integrated asparagus genetic map (Spada *et al.*, 1998); one of the linkage groups of the map is

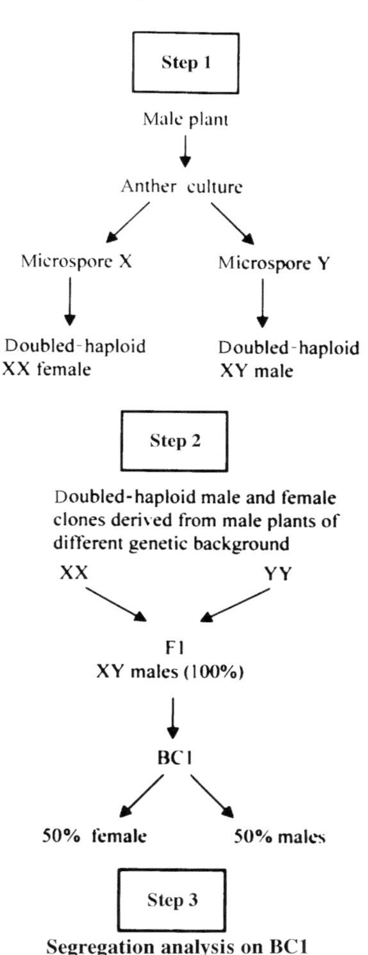

Figure 2. Genetic system utilized for segregation analysis.

chromosome 5, where the sex-determining factors are located. As shown in *Figure 3*, we have mapped a total of 33 markers (beside sex) on chromosome 5: 13 RFLP, 18 AFLP, 2 RAPD and one isoenzyme, malate dehydrogenase (MDH). The map of chromosome 5 covers 94.5 cM, with an average distance between markers of 2.8 cM; among these markers the nearest that maps to the sex determinant is the AFLP SV at 3.5 cM. It is interesting to note that the isoenzyme marker MDH, which in our map has been located at 25.7 cM from the sex determinant, is also present in the asparagus map presented by Jiang *et al.* (1997) and at a similar distance (22 cM).

6.2 *Bulked segregant analysis*

Bulked segregant analysis (Michelmore *et al.*, 1991) allows a rapid identification of molecular markers linked to a particular gene or genomic region. The technique is based on the comparison of two DNA samples, each derived from a mixture of the DNAs from several different individuals. The two pools are characterized by having a very similar genetic background but a different phenotype concerning a particular

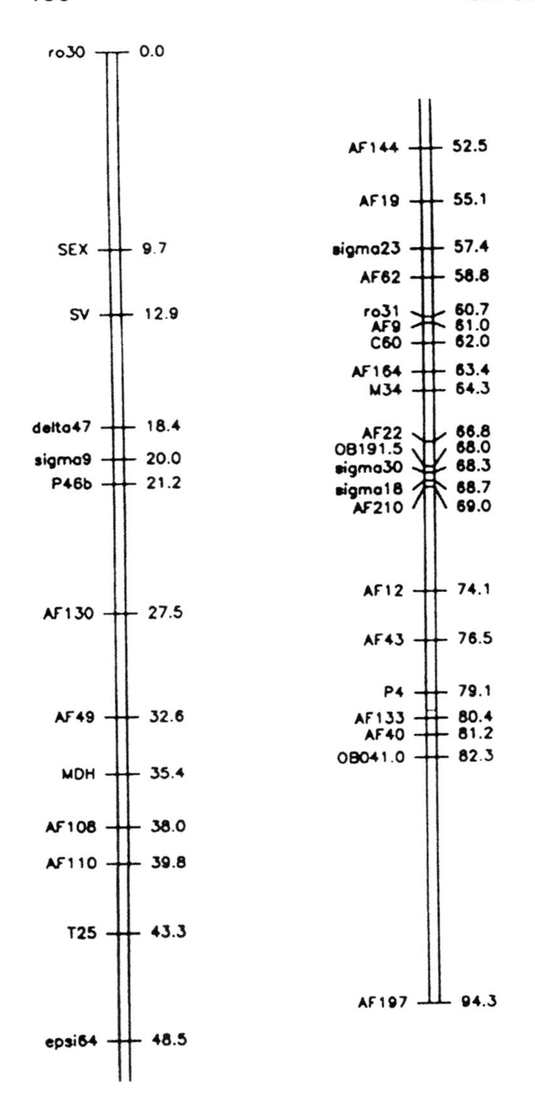

Figure 3. *The map of chromosome 5, where sex determinants are located. Names of markers are on the left, map distances (cM) on the right. AFLP markers, except SV, are indicated as AF and a number; OB119.5 and OB041.0 are RAPD markers; MDH is the malate dehydrogenase isoenzyme; all others are RFLP.*

character (i.e. resistant/susceptible to a disease or male/female) and are analysed to identify polymorphic markers. The markers which show a polymorphism between the two pools shoud be associated with the locus of interest.

For the search of molecular markers linked to sex through the bulked segregant analysis we have utilized doubled-haploid clones derived from a single male plant characterized by a high androgenetic capacity. The DNAs of 10 male and 10 female plants were pooled and analysed by the RAPD technique. The genetic system utilized is shown in *Figure 4*. We have analysed the amplification products obtained with 900 different primers: with an average of four products per primer this totals 3600 products. With this extensive analysis of asparagus DNA we have identified only two polymorphic products: one present in the male pool, the other in the female pool. Moreover, among the single individuals of the pools some recombinants were present, indicating

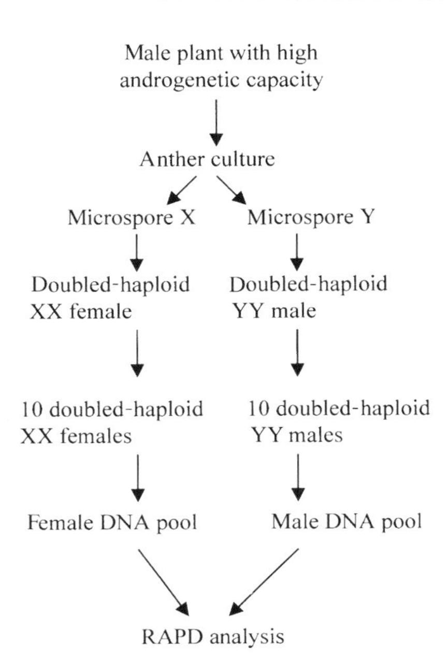

Figure 4. *Genetic system utilized for bulked segregant analysis.*

that the polymorphic sequences identified are quite distant from the sex locus. The product specifically amplified in the female pool is shown in *Figure 5*.

The results of the bulked analysis, together with those of the segregation analysis, contribute to our knowledge of asparagus sex chromosomes which appear to be highly homologous with recombination possible throughout their length. This finding, which supports the hypothesis of a recent origin of dioecy in asparagus, may also explain the viability of the homozygous male: besides the sex determinants, X and Y chromosomes are equivalent and all the genes present on the X chromosome are also present on the Y. The described characteristic of asparagus sex chromosomes make the

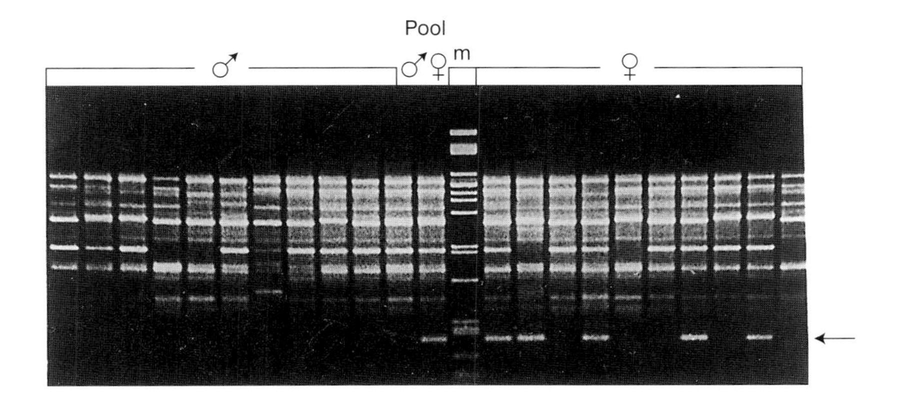

Figure 5. *RAPD amplification run on male and female DNA pools and single individual with the primer OF13 (Operon Technologies Inc., Alimeda, CA). The arrow indicates the product specifically amplified in the female pool; m: molecular weight marker.*

identification of the sex locus very difficult: dioecy in this species may be the result of two successive point mutations which could have introduced a stop codon or eliminated a splicing site.

7. Sex chromosome labelling

The molecular markers mapped to the sex chromosome to date are located too far from the sex determinants to attempt chromosome walking experiments but will be utilized for labelling the sex chromosomes in order to separate them by magnetic sorting. Starting from male and female doubled-haploid clones it is possible to utilize any of these markers to label and separate X and Y chromosomes for the construction of specific libraries. In a preliminary experiment, the RFLP marker Ro30 was utilized as a probe for *in situ* hybridization on isolated nuclei. The nuclei were prepared from root apices of 1-week-old asparagus seedlings and hybridized with a digoxigenin-labelled probe. The bound probe was immunodetected after a cascade reaction with anti-DIG antibodies and antibodies coupled to a fluorescent dye. As shown in *Figure 6*, only one or two hybridization signals per nucleus were visible (unpublished data).

8. Early diagnosis of gender

For perennial cultivated species which are dioecious and which do not produce flowers for several years, an early diagnosis of gender, not based on flower morphology, is very important. In fruit plants like kiwi or date palms few males are needed for pollen production and most of the plants are female. In asparagus, however, male plants are superior to female for some agronomically important traits (see Section 1). For dioecious crop plants, any marker linked to sex and visible at the seedling stage are of great importance. In date palm, the early diagnosis of gender is possible from analysis of the structure of the heterochromatin (Siljak-Yakovlev *et al.*, 1996). In asparagus, the polymorphic molecular markers linked to sex may be utilized for this purpose. The usefulness of the RFLP marker delta47 (8.25 cM from the sex determinant) has been tested in a previous work (Biffi *et al.*, 1995). A more recently identified AFLP marker, SV, whose distance from the sex determinant is 3.5 cM, should be more efficient.

Using bulked segregant analysis and AFLP markers, Reamon-Büttner *et al.* (1998) have identified three markers tightly linked to sex determinants (at 0.2, 0.3 and 0.5 cM).

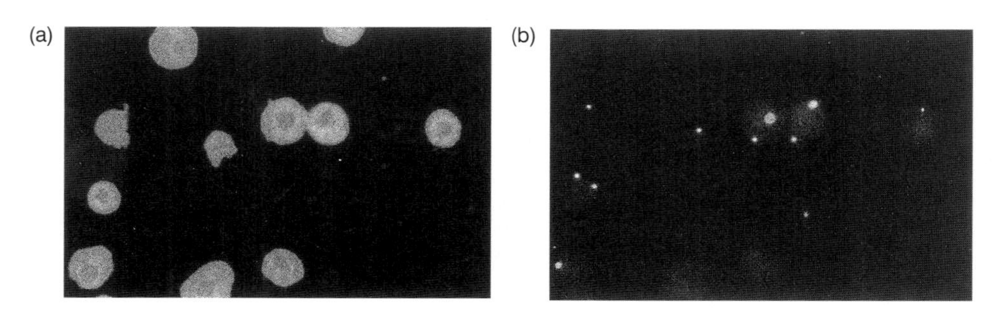

Figure 6. In situ *hybridization on isolated nuclei with the sex-linked RFLP probe Ro30. (a) DAPI staining; (b) hybridization signal (fluorescence).*

These markers, which did not show any recombination with the sex determinants in three different populations analysed, are considered by the authors as potential markers in screening for male and female plants in asparagus populations. Utilizing our genetic system, AFLP analysis with some of the primers used by Reamon-Büttner and co-workers, did not reveal any polymorphic markers linked to sex.

9. Discussion

The accumulating genetic and developmental data on sex differentiation in *Asparagus officinalis* are providing a detailed picture of the phenomenon.

9.1 *Genetic data*

The homomorphic sex chromosomes of asparagus appear to be highly homologous, with the possibility of recombination throughout their lengths; to maintain dioecy, the sex-determining genes must be tightly linked. Both X and Y chromosomes are functional; doubled-haploid YY males are perfectly viable, offering the possibility of utilizing male and female homozygous clones both for breeding programmes and for research projects. Extensive genetic analysis performed by ourselves and by other groups, however, has shown the difficulty of identifying the locus responsible for determining the sex of the individual and where the differences between the X and Y chromosome are maintained. These differences might be point mutations whose presence is difficult to demonstrate with the standard techniques of DNA analysis (RFLP, RAPD, AFLP).

9.2 *Developmental data*

Studies on asparagus male and female flower development have shown that the hermaphrodite phase is relatively long and it might be expected that the expression patterns of the organ identity genes involved in stamen and pistil development – the MADS-box genes of classes B and C – are identical in male and female flowers during the hermaphrodite stages. Their expression probably declines in the stamens of the female flower and in the pistil of the male flower after the switch from the hermaphrodite to the unisexual developmental pathway, as a consequence of the programmed abortion of these organs. We are currently isolating the asparagus homologues of the reproductive organ identity genes in order to define their expression patterns in male and female flowers. It will be particularly interesting to determine the expression pattern of the C-class gene in the male ovary, whose tissues remain viable.

A possible approach to the isolation of the sex-determining genes in asparagus is the analysis of mRNA specifically expressed in male or female flowers when the first event(s) leading to sex differentiation occur. As shown by histological analysis, the target cells of this event are very few: the cells of the tapetum in female anthers and a few cells in the ovule of the male pistil. These cells very rapidly degenerate. If the expression of the sex-determining genes is limited to these cells for a very short time period, the isolation of their product will be difficult. Indeed, we have tried to isolate stage- and sex-specific messengers through different approaches (differential display of mRNAs expressed during flower development and subtractive libraries) but with little success.

10. Future perspective

A possible experimental system for studying the transition from hermaphrodite to unisexual development is provided by the male plants whose flowers have different types of pistil with styles of different length and which bear some fertile bisexual flowers (see Section 3). The apparent correlation between style elongation and ovule development could be demonstrated by morphological and molecular analysis. In petunia, two MADS genes, *FBP7* and *FBP11*, are involved in ovule development. In wild-type flowers, the two genes are expressed in the centre of the gynoecium before ovule primordia are visible, and, in later stages, are exclusively expressed in ovules. Ectopic expression experiments indicate that *FBP11* represents an ovule identity gene (Angenet *et al.*, 1995; Colombo *et al.*, 1995). If ovule identity genes are present in asparagus also, their expression patterns, in the different types of male flower could be correlated with the viability of the ovules, while the expression of the C-class MADS-box genes might persist in the ovary after ovule abortion. The time course of the expression of these genes in different types of male flowers is probably correlated with the time of activation of the female suppressor gene in the male flower.

Acknowledgements

We thank Mr Antonio Grippo for careful preparation of illustrations and Prof. Marisa Levi for allowing utilization of unpublished data concerning *in situ* hybridization on isolated nuclei. We also thank Prof. Claudio Longo for reading the manuscript.

References

Angenent, G.C., Franken, J., Busscher, M., van Dijken, A., van Went, J., Dons, H.J.M. and van Tunen, A.J. (1995) A novel class of MADS box genes is involved in ovule development in petunia. *Plant Cell* 7: 1869–1582.

Arber, A. (1925) *Monocotyledons. A Morphological Study*. Cambridge University Press, Cambridge.

Bawa, K.S. (1980) Evolution of dioecy in flowering plants. *Annu. Rev. Ecol. Syst.* **11**: 15–39.

Benson, B.L. (1982) Sex influence on foliar trait morphology in *Asparagus*. *Hort. Sci.* **17**: 625–627.

Biffi, R., Restivo, F.M., Tassi, F., Caporali, E., Carboni. A., Marziani, G.P., Spada, A. and Falavigna, A. (1995) An RFLP probe for early diagnosis of gender in *Asparagus officinalis*. *Hort. Sci.* **30**: 1463–1464.

Bracale, M., Galli, M.G., Falavigna, A. and Soave, C. (1990) Sexual differentiation in *Asparagus officinalis* L. II. Total and newly synthesized proteins in male and female flowers. *Sex. Plant Reprod.* **3**: 23–30.

Caporali, E., Carboni, A., Galli, M.G., Rossi, G., Spada, A. and Marziani Longo, G.P. (1994) Development of male and female flowers in *Asparagus officinalis* . Search of point of transition from hermaphroditic to unisexual developmental pathway. *Sex. Plant Reprod.* **7**: 239–249.

Charlesworth, B. and Charlesworth, D. (1978) A model for the evolution of dioecy and gynodioecy. *Am. Nat.* **112**: 975–997.

Charnov, E.L., Smith, J.M. and Bull, I.J. (1976) Why be an hermaphrodite? *Nature* 263: 125–126.

Colombo, L., Franken, J., Koetje E., van Went, J., Dons, H.J.M., Angenent, G.C. and van Tunen, A.J. (1995) The petunia MADS box gene *FBP11* determines ovule identity. *Plant Cell* **7**: 1859–1868.

Dorè, C. (1990) Asparagus anther culture and field trials of dihaploids and F1 hybrids. In: *Biotechnology in Agriculture and Forestry. Vol.12. Haploid in Crop Improvement I* (ed. Y.P.S. Bajaj) Springer-Verlag, Berlin, pp. 322–345.

Falavigna, A., Tacconi, M.G. and Soressi, G.P. (1983a) Progress in the synthesis of F1 hybrids asparagus following in vitro androgenesis. *Genet. Agr.* **37**: 164–165.

Falavigna, A., Tacconi, M.G. and Soressi, G.P. (1983b) Asparagus breeding in Italy. *Asparagus Res. Newslett.* **1**: 14.

Falavigna, A., Tacconi, M.G. and Soressi, G.P. (1983c) Recent progress in asparagus breeding by anther culture. *Acta Hortic.* **131**: 215–222.

Frankel, R. and Galun, E. (1977) *Pollination Mechanisms, Reproduction and Plant Breeding.* Springer-Verlag, Berlin.

Franken, A.A. (1970) Sex characteristic and inheritance of sex in *Asparagus officinalis L. Euphytica* **19**: 277–287.

Galli, M.G., Bracale, M., Falavigna, A. and Soave, C. (1988) Sex differentiation in *Asparagus officinalis* L. I. DNA characterization and mRNA activities in male and female flowers. *Sex. Plant Reprod.* **1**: 202–207.

Galli, M.G., Bracale, M., Falavigna, A., Raffaldi, F., Savini, C. and Vigo, A. (1993) Different kinds of male flowers in the dioecious plant *Asparagus officinalis L. Sex. Plant Reprod.* **6**: 16–21.

Guttman, D.S. and Charlesworth, D. (1998) An X-linked gene with a degenerate Y-linked homologue in a dioecious plant. *Nature* **393**: 263–266.

Heberle-Bors, E. (1985) In vitro haploid formation from pollen: a critical review. *Theor. Appl. Genet.* **71**: 361–374.

Jiang, C., Lewis, M.E. and Sink, K.C. (1997) Combined RAPD and RFLP molecular linkage map of asparagus. *Genome* **40**: 69–76.

Kohn, J.R. (1988) Why be female? *Nature* **335**: 431–433.

Lazarte, J.I. and Palser, B.F. (1979) Morphology, vascular anatomy and embryology of pistillate and staminate flowers of *asparagus officinalis* L. *Am. J. Bot.* **66**: 753–764.

Loptien, D. (1979) Identification of the sex chromosome pair in asparagus (*Asparagus officinalis* L.). *Zeitschr. für Pflanzenzüchtung* **82**: 162–173.

Maestri, E., Restivo, F.M., Marziani Longo, G.P., Falavigna, A. and Tassi, F. (1991) Isozyme gene markers in the dioecious species *Asparagus officinalis* L. *Theor. Appl. Genet.* **81**: 613–618.

Mariani, C., De Beukeleer, M., Truettner, J., Leemans, J. and Goldberg, R.B. (1990) Induction of male sterility in plants by a chimaeric ribonuclease gene. *Nature* **347**: 737–741.

Marks, M. (1973) A reconsideration of the genetic mechanism for sex determination in *Asparagus officinalis.* In: *Proc. Eucarpia Meeting on Asparagus.* Versailles, France, pp. 122–128.

Marziani Longo, G.P., Rossi, G., Scaglione, G., Longo, C.P. and Soave, C. (1990) Sexual differentiation in *Asparagus officinalis* L. Hormonal content and peroxidase isoenzymes in female and male plants. *Sex. Plant Reprod.* **3**: 236–243.

Matsunaga, S., Kawano, S., Takano, H., Uchida, H., Sakai, A. and Kuroiwa T. (1996) Isolation and developmental expression of male reproductive organ-specific genes in a dioecious campion, *Melandrium album* (*Silene latifolia*). *Plant J.* **10**: 679–589.

Michelmore, R.W., Paran, I. and Kesselli, R.V. (1991) Identification of markers linked to disease-resistance genes by bulked segregant analysis: a rapid method to detect markers in specific regions by using segregating populations. *Proc. Natl Acad. Sci. USA* **88**: 9829–9832.

Qiao, Y. and Falavigna, A. (1990) An improved *in vitro* anther culture method for obtaining doubled haploid clones of asparagus. *Acta Hortic.* **271**: 145–150.

Reamon-Büttner, S.M., Schondelmaier, J. and Jung, C. (1998) AFLP markers tightly linked to the sex locus in *Asparagus officinalis* L. *Mol. Breeding* **4**: 91–98.

Restivo, F., Tassi, F., Biffi, R., Falavigna, A., Caporali, E., Carboni, A., Doldi, M.L., Spada, A. and Marziani, G.P. (1995) Linkage arrangement of RFLP loci in progenies from crosses between doubled haploid *Asparagus officinalis* clones. *Theor. Appl. Genet.* **90**: 124–128.

Siljak-Yakovlev, S., Benmalek, S., Cerbah, M., Coba de la Peña, T., Bounaga, N., Brown, S. C. and Sarr, A. (1966) Chromosomal sex determination and heterochromatin structure in date palm. *Sex. Plant Reprod.* **9**: 127–132.

Spada, A., Caporali, E., Marziani, G. and Portaluppi, P. (1998) An integrated genetic map of *Asparagus officinalis* based on different molecular markers. *Theor. Appl. Genet.* (in press).

Sunderland, N. (1974) Anther culture as a means of haploid induction. In: *Haploid in Higher Plants, Advance and Potential* (ed. K.J. Kasha) University of Guelph, Guelph, Ontario, pp. 89–106.

Sunderland, N. and Roberts, M. (1979) Cold pretreatment of excised flower buds in flot culture of tobacco anthers. *Ann. Bot. (Lond.)* **43**: 405–414.

Westergaard, M. (1958) The mechanism of sex determination in dioecious flowering plants. *Adv. Genet.* **9**: 217–281.

Ye, D., Installé, P., Ciuperescu, D., Veuskens, J., Wu, Y., Salesses G., Jacobs, M. and Negrutiu, I. (1990) Sex determination in the dioecious *Melandrium.* I. First lessons from androgenic haploid. *Sex. Plant Reprod.* **3**: 179–186.

10

Sex determination in *Dioscorea tokoro*, a wild yam species

Ryohei Terauchi and Günter Kahl

1. Introduction

1.1 *The genus* Dioscorea

The genus *Dioscorea* belongs to the monocotyledonous family Dioscoreaceae (Burkill, 1960; Dahlgren and Clifford, 1982; Knuth, 1930). It is a large genus comprising some 500 (Miège, 1982) to 850 (Al-Shehbaz and Schubert, 1989) species, mainly distributed in humid tropical and sub-tropical regions of the world. Most species are herbaceous perennial climbers and some have been cultivated for their starchy tubers, the 'yams'. World production of yam crops has doubled during the last two decades, and was 30 million metric tons in 1997 (FAO, 1998), which is about one-quarter of that of sweet potato, and one-fifth of that of cassava, two other important tropical tuber crops. More than 90% of world production of yams comes from West Africa (FAO, 1998; Hahn *et al.*, 1987) where economic and cultural relationships between yams and people have developed to such an extent that Miège (1954a) and Coursey (1972) described it as the 'civilization of the yams'. Other major yam-growing regions are Oceania and Southeast Asia. Economically important yam species are: *D. rotundata* and *D. cayenensis* in West Africa, *D. alata* and *D. esculenta* in the tropical regions of the world, and *D. opposita* in temperate East Asia.

Dioecy is one of the major characteristics of the genus *Dioscorea*. No hermaphrodite or monoecious species has been reported so far. In a few cases, hermaphrodite or monoecious individuals were reported, but they should be regarded as exceptions (Burkill, 1960; Sadik and Okereke, 1975). Dioecy is also found among other genera of the family Dioscoreaceae; *Tamus* and *Rajania*. These data indicate that dioecy is an ancient character which appeared at an early stage of the differentiation of the genus *Dioscorea* from other plant groups. Flowers of *Dioscorea* species are numerous and are usually borne on racemes, panicles or spikes. They are actinomorphic, small and unconspicuous, and are pollinated by small insects, including thrips, that are attracted by floral scent. Male plants

Sex Determination in Plants, edited by C.C. Ainsworth.

have six (rarely three) stamens and lack a gynoecium. Female plants lack fertile stamens, and have three carpels each containing two ovules (*Figure 1*). After fertilization, the inferior ovary develops into a capsule with three wings each containing two seeds. Within species of Liliflorae, dioecy is known in *Asparagus* and *Smilax* and a few other Liliaceae (Dahlgren and Clifford, 1982). Phylogenetic studies on the chloroplast *rbcL* gene suggest that *Dioscorea, Asparagus* and *Smilax* are not monophyletic (Terauchi, Shinwari, Kato and Kawano, unpublished data), indicating that dioecy has evolved independently at least three times in these plant groups. As Burkill (1960) noticed, most of the dioecious groups in Liliflorae are climbers which form shrubs around their supports. Flowers of these dioecious species are altogether unconspicuous. There is a possibility that this growth habit and inefficient pollinators forced the evolution of dioecy as a means to avoid inbreeding depression (Barrett and Harder, 1996).

To date, no major trials in yam breeding have been carried out. This is partly because of the difficulty of controlling the flowering of yams (Akoroda, 1983, 1985). In addition it is very common that all the plants of one cultivar express only one sex and even if two sexes are available, the synchronization of flowering time is very difficult. For the efficient improvement of yam crops through conventional breeding, we have to understand better the genetic mechanisms of dioecy in *Dioscorea*.

1.2. *Sex determination in* Dioscorea*: an overview*

In the past, much effort was invested to identify morphologically distinct sex chromosomes by cytological observations of meiosis and mitosis (*Table 1*). There are reports of extra chromosomes in females (male XO, female XX), heteromorphic sex chromosomes in males (male XY, female XX), heteromorphic sex chromoromes in females (male ZZ, female ZW), or no heteromorphic chromosomes at all. Obviously, the small sizes and large numbers of *Dioscorea* chromosomes made critical observations difficult (see Martin, 1966; Martin and Ortiz, 1963 for reviews). In *D. tokoro*, we could not find any evidence for the presence of heteromorphic chromosomes, either in mitotic or meiotic chromosome preparations (Terauchi, unpublished data).

The only source of data on the genetic analysis of sex determination in *Dioscorea* is

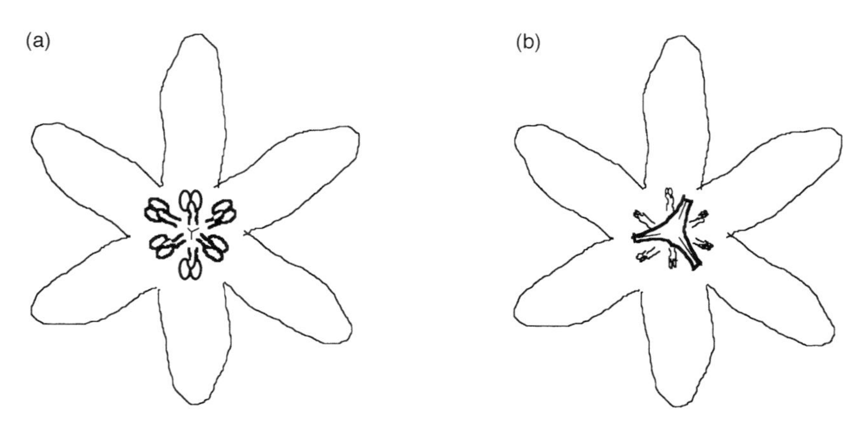

Figure 1. *Flowers of* Dioscorea tokoro: *(a) male and (b) female flowers. In female flowers, vestigial male organs develop, but are sterile.*

the work of Martin (1966). In controlled crosses in the tetraploid species, *D. floribunda*, ($2n = 4x = 36$) it has been observed that some families segregated in a 1:1 ratio and others in a 3:1 ratio, for the male and female progeny. When a common female was used for the crosses with different males, the segregations in their progeny were either 1:1 or 3:1, depending on the male parents used. By contrast, if a common male was crossed to different females, the segregation of sex in the progeny was consistent over families. On the basis of these observations, it was proposed that the male, but not the female determines the sex of the progeny, and that male is the heterogametic sex (XXYY or XXXY) with the female (XXXX) as the homogametic sex. Subsequent to Martin's work no further genetic analysis of *Dioscorea* has been carried out.

2. Genetic analysis of sex determination in *D. tokoro*

2.1. *Background*

Aiming at the isolation of sex determination genes from *Dioscorea*, we started a linkage analysis of the sex locus of *D. tokoro* using AFLP markers (Vos *et al.*, 1995). Our prospective strategy is the reverse genetics approach: (i) identification of molecular markers tightly linked to the sex locus, (ii) screening of a genomic library with the identified markers, (iii) isolation of candidate gene(s) responsible for sex determination from the genomic library, and (iv) confirmation of gene function by complementation analysis.

D. tokoro Makino is a dioecious, diploid ($2n = 2x = 20$) species, widely distributed in East Asia. In Japan, this species occurs commonly at the fringe of forests. We have used *D. tokoro* for studies on the population genetics of allozyme polymorphisms (Terauchi, 1990), microsatellite polymorphisms (Terauchi and Konuma, 1994), and DNA sequence variation in the phosphoglucose isomerase (PGI) gene (Terauchi *et al.*, 1997). The sex ratio in natural populations is close to 1:1. A controlled cross of *D. tokoro* is relatively easy and under suitable conditions, a plant can grow from seed to maturity in a single year. These favourable features made *D. tokoro* an attractive and potentially representative model species for genetic studies on yams generally.

2.2 *Crossing, molecular markers, and mapping strategy*

In 1995, a cross was made between a female plant (DT5 collected in Wakayama, Japan) and a male plant (DT7, Wakayama Japan), which gave rise to 200 seeds. Fifty-two progeny were grown in a nursery and their sex was determined once the flowers appeared. Total DNA was extracted from their leaves.

We chose the AFLP technique developed by Vos *et al.* (1995) to generate molecular markers to map the sex determination gene. AFLP is a combination of RFLP and PCR techniques. Genomic DNA is first digested with two restriction enzymes (*Eco*RI and *Mse*I in the original protocol), and double-stranded oligonucleotide adapters are ligated to the restriction sites. PCR primers complementary to the adaptors and restriction sites are then used for the amplification of fragments that are flanked by the adaptors. A subset of fragments is selectively amplified by PCR primers that have two- or three-base extensions into the restriction fragments. One of the primers is radioactively labelled. Amplified fragments are separated in denaturing polyacrylamide gels, and visualized by autoradiography. Usually 50–100 restriction fragments

Table 1. *Cytological observations on chromosomes of* Dioscorea *species with reference to sex chromosomes*

Species	Meiosis (Me)/ Mitosis (Mi)	Chromosome number	Sex or heteromorphic chromosomes	Reference
D. abyssinica	Mi	$2n = 40$	No information	Miège (1954b)
	Mi	$2n = 40$	No information	Martin and Ortiz (1963)
D. alata	Mi	$2n = 81$	Male XO	Smith (1937)
	Mi	$2n = 60$	No information	Miège (1954b)
	Mi	$2n = 40, 60$	No heteromorphic chromosome	Martin and Ortiz (1963)
	Me and Mi	$n = 20, 2n = 40$ $2n = 60$	Heteromorphic chromosome in male	Ramachandran (1968)
D. bulbifera	Mi	$2n = 80$	No heteromorphic chromosome	Smith (1937)
	Mi	$2n = 36, 40, 54, 60$	No information	Miège (1954b)
	Mi	$2n = 40, 80$	No heteromorphic chromosome	Martin and Ortiz (1963)
	Mi	$2n = 80, 100$	Male XY	Ramachandran (1962)
	Me and Mi	$2n = 80, 100$	No information	Ramachandran (1968)
D. caucasica	Me	$n = 10$	No heteromorphic chromosome	Meurman (1925)
	Mi	$2n = 20$	No heteromorphic chromosome	Smith (1937)
	Mi	$2n = 20$	No information	Miège (1954b)
D. cayenensis-rotundata	Mi	$2n = 36, 54$	No information	Miège (1954b)
	Mi	$2n = 40$	No information	Martin and Ortiz (1963)
		$2n = 40, 60, 80$	No information	Zoundjihekpon *et al.* (1990)
D. composita	Me and Mi	$2n = 36$	No heteromorphic chromosome	Martin and Ortiz (1963)
D. deltoidea	Me and Mi	$n = 10, 2n = 20$	Male ZZ female ZW	Bhat and Bindroo (1980)
D. discolor	Mi	$2n = 40$	Male XY	Smith (1937)
D. dumetorum	Mi	$2n = 36, 45, 54$	No information	Miège (1954b)
D. esculenta	Mi	$2n = 90, 100$	No information	Miège (1954b)
D. fargessii	Mi	$2n = 64$	Male XY	Smith (1937)
D. floribunda	Me and Mi	$2n = 36, 54$	No heteromorphic chromosome	Martin and Ortiz (1963)
D. friedrichsthalii	Me and Mi	$2n = 36$	No heteromorphic chromosome	Martin and Ortiz (1963)
D. japonica	Me	$n = 20$	Male XY	Nakajima (1942)
D. macroura	Mi	$2n = 40$	Male XY	Smith (1937)
D. opposita	Mi	$2n = 140$	No information	Nakajima (1937)

	Mi	$2n = 140$	No information	Teppner (1992)
D. pentaphylla	Mi	$2n = 144$	No heteromorphic chromosome	Smith (1937)
	Mi	$2n = 40, 80$	Male XY	Ramachandran (1962)
D. quarternata	Me	$n = 27$	No heteromorphic chromosome	Jensen (1937)
D. reticulata	Mi	$2n = 61$	Male XO	Smith (1937)
D. sinuata	Me	$n = 17$ (male) $n = 18$ (female)	Male XO	Meurman (1925)
D. spiculiflora	Me and Mi	$2n = 36$	No heteromorphic chromosome	Martin and Ortiz (1963)
D. spinosa	Mi	$2n = 90$	Male XY	Ramachandran (1962)
D. tokoro	Mi	$2n = 20$	No information	Nakajima (1933)
	Mi	$2n = 20$	No information	Takeuchi *et al.* (1970)
D. tomentosa	Mi	$2n = 40$	Male XY	Ramachandran (1962)

can be detected per gel, which makes AFLP a method with the highest throughput available at the moment. As the AFLP polymorphisms appear as the presence or absence of the amplified fragments, they are dominant markers in contrast to the co-dominant markers such as allozymes, RFLPs and sequence-tagged microsatellite sites.

The pseudo-testcross strategy (Grattapaglia and Sederoff, 1994) was employed for linkage analysis of the *D. tokoro* genome. This strategy was first developed for the mapping of tree genomes in species like *Eucalyptus*, using the RAPD technique (Williams *et al.*, 1990). In a cross between highly heterozygous parents, which is the case for *D. tokoro*, many dominant marker loci tend to be heterozygous in one parent (+/–), null in the other (–/–) and segregating among progeny in a 1: 1 ratio for +/– and –/– genotypes. This is similar to a testcross (backcross) to an F1 heterozygote in a usual crossing scheme of inbred lines. If we have two of these loci (+/- +/- for one parent and –/––/– for the other), their linkage can be tested by examining whether the segregation follows a 1:1:1:1 ratio (for genotypes +/– +/–, +/––/–, –/––/+ and –/– –/–, respectively) or not. With this mapping strategy, two maps are generated corresponding to the two parents. However, with the help of AFLP loci that are heterozygous in both male and female parents, it is possible to align the two linkage maps to each other (Terauchi and Kahl, unpublished data).

2.3 *AFLPs linked to sex*

After 28 primer combinations were tested for AFLP analysis, more than 2000 fragments were observed, among which 113 polymorphic fragments were used for mapping the genome of the female parent, DT5, and 103 polymorphic fragments for mapping the genome of the male parent, DT7 (Terauchi and Kahl, unpublished result). With the threshold of Log-Odds (LOD) = 3.5 and θ (recombination rate) = 0.3, 13 linkage groups and 12 linkage groups were generated for the female (DT5) and male (DT7) parent, respectively. These numbers are close to the haploid chromosome number of *D. tokoro* (n = 10). Ten AFLPs heterozygous in the male parent (DT7) were tightly linked (LOD > 5) to the sex of the progeny individuals (*Figure 2*). Among these 10 markers, four markers most closely linked to the sex locus showed an identical segregation pattern; the AFLP fragment present in all the male progeny (24 individuals) and absent in all the female progeny (18 individuals), with only one individual as the exception (male plant without the fragment). The distance between the sex locus and these four marker loci were calculated to be 2.4 cM. No AFLP markers heterozygous in the female parent (DT5) showed any linkage with sex of the progeny. This observation is only compatible with the hypothesis that male is the heterogametic sex (XY) and female is the homogametic sex (XX, *Figure 3*). This system of sex determination is similar to *Silene* and *Asparagus* (Bracale *et al.*, 1991; Westergaard, 1958) and mammals (see Dellaporta and Calderon-Urrea, 1993; Lebel-Hardenack and Grant, 1997 for reviews).

There was no marker completely linked to the sex. Even the most tightly linked markers were still separated from the sex locus at a 2.4 cM distance. This result indicates, that the physical difference between X and Y chromosomes is not large. If the region where the sex locus is located is extensively different between the X and Y chromosomes, then there should be a large chromosomal area around the sex locus, where no recombination can occur. Thus, we should expect at least some marker loci that are completely linked to the sex locus. This is, however, not the case and a small

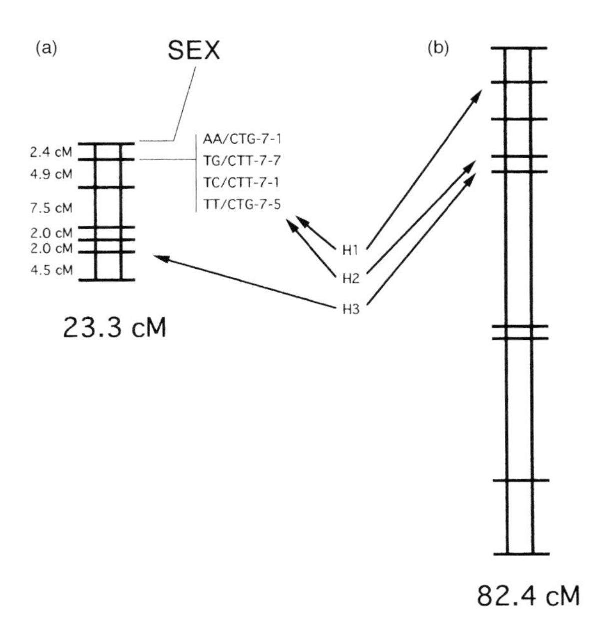

Figure 2. *Linkages of sex with AFLP markers. (a) Ten AFLP markers heterozygous in the male parent showed linkage with the sex of the progeny. This linkage group was aligned with a linkage group generated by the markers heterozygous in the female parent (b), with the help of three AFLP markers that are heterozygous in both the male and female parents (H1, H2 and H3).*

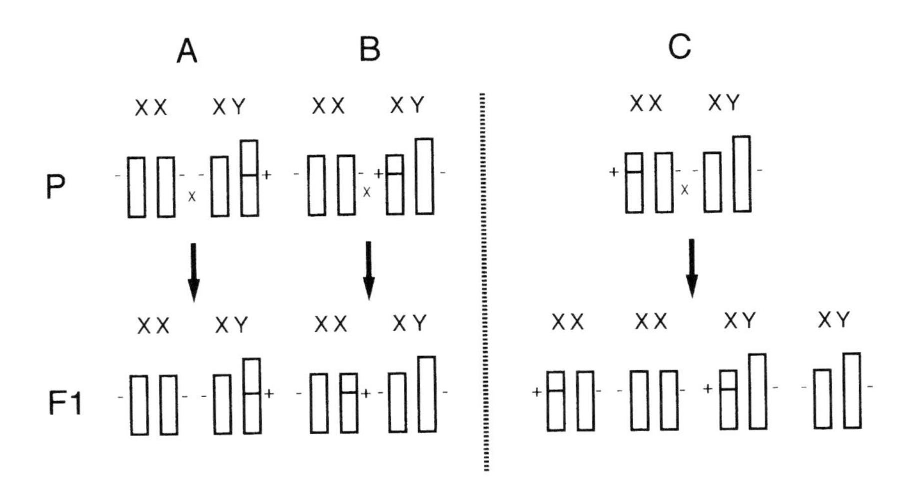

Figure 3. *Diagrammatic schemes of the inheritance of sex-linked AFLP markers. AFLP markers (indicated by +) heterozygous in heterogametic sex (A, B) are inherited by the heterogametic progeny only (A), or homogametic progeny only (B). Markers heterozygous in homogametic sex (C) are inherited by both the homogametic and heterogametic sex. Actually, 10 AFLP markers heterozygous in the male parent showed linkage with sex, whereas no marker heterozygous in the female parent showed any linkage with sex. This observation can be explained only when the heterogametic sex (XY) is male and the homogametic sex (XX) is female.*

amount of differentiation between the X and Y chromosomes is in agreement with the cytological observation that a heteromorphic sex chromosome is not observable in *D. tokoro*.

3. Towards cloning the sex gene of *Dioscorea*

The relatively small genome size of *Dioscorea* (555 Mbp reported for polyploid *D. alata*; Arumuganathan and Earle, 1991) is an attractive feature for the reverse genetics approach. The obvious future strategy is to isolate markers completely linked to sex by adding more molecular markers to the map, and to screen a bacterial artificial chromosome (BAC) library (Shizuya *et al.*, 1992) with them. To make a final confirmation of the function of the candidate genes by complementation analysis, we need to establish a reliable transformation system for *Dioscorea*.

Acknowledgements

RT appreciates an Alexander von Humboldt fellowship. Research of the authors was supported by GTZ (Grant 95.2072.7–001.00).

References

Akoroda, M.O. (1983) Floral biology in relation to hand pollination of white yam. *Euphytica* **32**: 931–838.

Akoroda, M.O. (1985) Pollination management for controlled hybridization of white yam. *Scientia Horticulturae* **25**: 201–209.

Al-Shehbaz, I.A. and Schubert, B.G. (1989) The Dioscoreaceae in the southeastern United States. *J. Arnold Arboretum* **70**: 57–95.

Arumuganathan, K. and Earle, E.D. (1991) Nuclear DNA content of some important plant species. *Plant Mol. Biol. Rep.* **9**: 208–218.

Barrett, S.C.H. and Harder, L.D. (1996) Ecology and evolution of plant mating. *Trends Ecol. Evolution* **11:** 73–79.

Bhat, B.K. and Bindroo, B.B. (1980) Sex chromosomes in *Dioscorea deltoidea* Wall. *Cytologia* **45**: 739–742.

Bracale, I., Caporall, E., Galli, M.G. *et al.* (1991) Sex determination and differentiation in *Asparagus officinalis* L. *Plant Sci.* **80**: 67–77.

Burkill, I.H. (1960) The organography and evolution of Dioscoreaceae, the family of yams. *J. Linn. Soc. Lond. Bot.* **56**: 319–412.

Coursey, D.G. (1972) The civilizations of the yam: interrelationships of man and yams in Africa and the Indo-pacific region. *Archaeol. Phys. Anthropol. Oceanea.* **7**: 215–233.

Dahlgren, R.M. and Clifford, H.T. (1982) *The Monocotyledons – A Comparative Study.* Academic Press, London.

Dellaporta, S.L. and Calderon-Urrea, A. (1993) Sex determination in flowering plants. *Plant Cell* **5**: 1241–1251.

FAO (1998) FAOSTAT, FAO Statistics Database, FAO, Rome.

Grattapaglia, D. and Sederoff, R. (1994) Genetic linkage maps of *Eucalyptus grandis* and *Eucalyptus urophylla* using a pseudo-testcross: mapping strategy and RAPD markers. *Genetics* **137**: 1121–1137.

Hahn, S.K., Osiru, S.O., Akoroda, M. and Otoo, J.A. (1987) Yam production and its future prospects. *Outlook on Agriculture* **16**: 105–110.

Jensen, H.W. (1937) Meiosis in several species of dioecious Monocotyledoneae: 1. The possibility of sex chromosome. *Cytologia* Fujii Jubilee Vol.: 96–103.
Knuth, R. (1930) Dioscoreaceae. In: *Nat. Pflanzenfam.* 2nd Edn (eds. A. Engler and K. Prantl). **15a**: 438–462.
Lebel-Hardenack, S. and Grant, S.R. (1997) Genetics of sex determination in flowering plants. *Trends Plant Sci.* **2**: 130–136.
Martin, F.W. (1966) Sex ratio and sex determination in *Dioscorea. J. Heredity* **57**: 95–99.
Martin, F.W. and Ortiz, S. (1963) Chromosome numbers and behavior in some species of *Dioscorea*. *Cytologia* **28**: 96–101.
Meurman, O. (1925) The chromosomal behavior of some dioecious plants and their relatives, with special reference to the sex chromosomes. *Soc. Sci. Fenn. Comm. Biol.* **2**: 1–105.
Miège, J. (1954a) Les cultures vivrières en Afrique Occidentale. *Cah. d'outre-Mer* **7**: 25–50.
Miège, J. (1954b) Nombres chromosomiques et répartition géographique de quelques plantes tropicales et équatoriales. *Rev. Cytol. Paris* **15**: 312–348.
Miège, J. (1982) Preface. In: *Yams Ignames.* (eds. J. Miège and S.N. Lyonga). Oxford University Press, Oxford, pp. ix–xi.
Nakajima, G. (1933) Chromosome numbers in some angiosperms. *Jap. J. Genet.* **9**: 1–5.
Nakajima, G. (1942) Cytological studies in some flowering dioecious plants, with specieal reference to the sex chromosomes. *Cytologia* **12**: 262–270.
Ramachandran, K. (1962) Studies on the cytology and sex determination of the Dioscoreaceae. *J. Indian Bot. Soc.* **41**: 93–98.
Ramachandran, K. (1968) Cytological studies in Dioscoreaceae. *Cytologia* **33**: 401–410.
Sadik, S. and Okereke, O.U. (1975) A new approach to improvement of yam *Dios*corea rotundata. *Nature* **254**: 134–135.
Shizuya, H., Birren, B., Kim, U.-J., Mancino, V., Slepak, T., Tahiiri, Y. and Simon, M. (1992) Cloning and stable maintenance of 300-kilobase-pair fragments of human DNA in *Escherichia coli* using an F-factor-based vector. *Proc. Natl Acad. Sci. USA* **89**: 8794–8797.
Smith, B.W. (1937) Notes on the cytology and distribution of the Dioscoreaceae. *Bull. Torr. Bot. Club* **64**: 189–197.
Takeuchi, Y., Iwao, T. and Akahori, A. (1970) Chromosome numbers of some Japanese Dioscorea species. *Acta Phytotax. Geobot.* **24**: 168–173.
Teppner, H. (1992) Notizen Über Igname de Chine (*Dioscorea opposita, Dioscoreaceae*). *Samentauschverzeichnis* 1992. (Bot. Garten. Univ. Graz) 41–46.
Terauchi, R. (1990) Genetic diversity and population structure of *Dioscorea tokoro* Makino, a dioecious climber. *Plant Species Biol.* **5**: 243–253.
Terauchi, R. and Konuma, A. (1994) Microsatellite polymorphism in *Dioscorea tokoro*, a wild yam species. *Genome* **37**: 794–801.
Terauchi, R., Terachi, T. and Miyashita, N. T. (1997) DNA polymorphism at the *Pgi* locus of a wild yam, *Dioscorea tokoro. Genetics* **147**: 1899–1914.
Vos, P., Hodgers, R., Bleeker, M. *et al.* (1995) AFLP: a new technique for DNA fingerprinting. *Nucl. Acids Res.* **23**: 4407–4414.
Westergaard, M. (1958) The mechanism of sex determination in dioecious flowering plants. *Adv. Genet.* **9**: 217–281.
Williams, J.G.K., Kubelik, A.R., Livak, K.J., Rafalski, J.A. and Tingey, S.V. (1990) DNA polymorphisms amplified by arbitrary primers are useful as genetic markers. *Nucl. Acids. Res.* **18**: 6531–6535.
Zoundjihekpon, J., Essad, S. and Toure, B. (1990) Dénombrement chromosomique dans dix groupes variétoux du complexe Dioscorea cayenensis-rotundata. *Cytologia* **55**: 115–120.

11

Sex control in *Actinidia* is monofactorial and remains so in polyploids

Raffaele Testolin, Guido Cipriani and Rachele Messina

1. Introduction

The kiwi fruit (*Actinidia deliciosa* (A. Chev.) C.-F. Liang and A.R. Ferguson) is probably the most important of the few fruit crops to have been domesticated in the 20th century. Commercial kiwi fruit cultivation began in New Zealand in about 1930 and spread world-wide throughout temperate and subtropical zones from around 1970 onwards (Ferguson *et al.*, 1996). Total world production, currently about one million tonnes of fruit per year, gives the kiwi fruit industry a prominent place in the fresh fruit market.

The kiwi fruit belongs to the genus *Actinidia*, which has been variously grouped with related genera such as *Clematoclethra*, *Saurauia*, and formerly *Sladenia*, to form the family *Actinidia*ceae (Ferguson, 1984). The position of the family Actinidiaceae is even more controversial, having been placed first in the Dilleniales, then in the Theales, subsequently in the Ericales, and once again back in the Theales (Morton *et al.*, 1996). The genus consists of more than 60 species and about 100 taxa, grouped into four sections on the basis of the type and degree of tomentum formation and the presence or absence of lenticels on the fruit surface (Liang, 1984). Most species are native to south-western China, with only a few found in the bordering countries. Plants are typically found climbing over small trees and spreading from tree to tree on forest edges or forming large straggling clumps on hillsides or in high grassland areas. Most species are deciduous, but a few that occur in more subtropical regions, are evergreen.

Actinidia species are usually diploid ($2n = 58$) or tetraploid ($2n = 4x = 116$), with a few being hexaploid and one, apparently, octoploid, but chromosome races with different ploidy levels are not uncommon within an individual taxon (reviewed in Ferguson *et al.*, 1996, 1997).

Sex Determination in Plants, edited by C.C. Ainsworth.

One of the most characteristic features of the genus is dioecy. All *Actinidia* species are dioecious (Ferguson, 1990; Ferguson *et al*., 1996), with male and female flowers carried on different plants (*Figure 1*). Such sexual dimorphism must have evolved before speciation and it must therefore be considered of ancient origin (Harvey *et al*., 1997a).

2. Phenotypic gender variants naturally occurring in *Actinidia* species

Male plants carry staminate flowers with numerous stamens producing viable pollen, and rudimentary ovaries which lack styles and do not form ovules (*Figure 1*). Female plants carry pistillate flowers with both well-developed ovaries containing numerous ovules and what appear to be fully-developed stamens, which, however, produce non-viable pollen grains (Ferguson, 1984) because the microspores have degenerated at the tetrad stage. Comparative observations carried out on the ultrastructure of anthers in male and female flowers have shown that microsporogenesis is not completed in female flowers because the intine is not formed in microspores (*Figure 2*) and the cytoplasm then degenerates before pollen grains are shed (Messina, 1993).

Sex determination is somewhat weak in *Actinidia* and phenotypically inconstant plants occasionally occur amongst males (Schmid, 1978, cited in Harvey and Fraser, 1988; Ferguson, 1984). Such plants, called 'inconstant' or 'fruiting males', carry, beside staminate flowers, bisexual flowers which show varying degrees of differentiation of ovary and styles (*Figure 3*) and which sometimes develop into small fruit with viable, albeit fewer, seeds. These gender variants are well known in kiwi fruit (*A. deliciosa*)

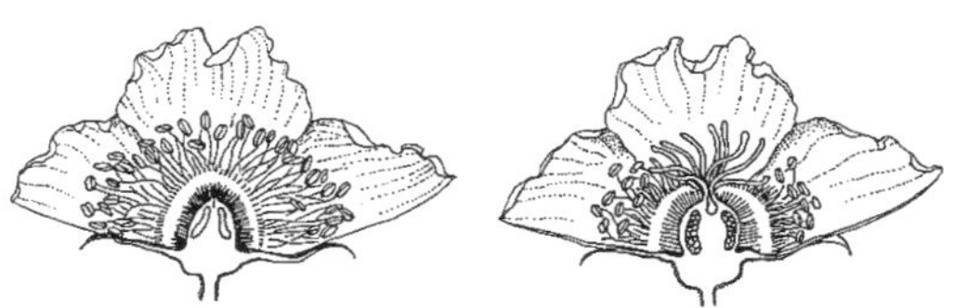

Figure 1. *Photographs (top) and diagrams (bottom) of staminate (left) and pistillate (right) flowers of* Actinidia deliciosa, *the kiwi fruit (drawing by A. Sensidoni). Note the rudimentary ovary and the lack of styles in the staminate flower. The pistillate flower has a hermaphroditic appearance, but stamens produce non-viable pollen grains because microspores degenerate at the tetrad stage.*

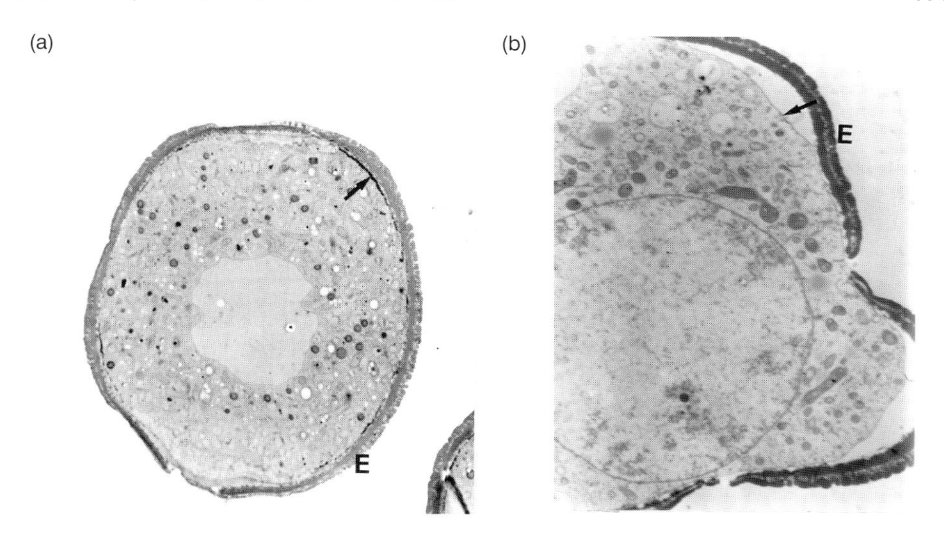

Figure 2. *Microsporogenesis in (a) male and (b) female flowers of* Actinidia deliciosa. *The microspore from the male flower shows a thick exine layer (E) and intine (arrow) which is being laid down (3060 ×). The microspore from the female flower shows a thick exine (E) layer as in male flower, but an intine is not formed and the plasma membrane (arrow) shrinks away from the wall (4200 ×). The absence of intine deposition is the first anomaly which occurs during microsporogenesis in female flowers and is followed by the collapse of the cytoplasm and the maturation of empty pollen grains.*

Figure 3. *Flowers from an 'inconstant male'. Note the presence of few developed pistils. 'Inconstant males' are occasionally found in orchards. Such plants carry flowers which show varying degrees of differentiation of ovary and styles and which develop into small fruit with viable seeds. The tendency to produce bisexual flowers varies greatly among genotypes, within vine, and from year to year.*

(Harvey and Fraser, 1988; Hirsch *et al.*, 1990; McNeilage, 1988; Messina *et al.*, 1990; Seal and McNeilage, 1989) and they have also been observed in *A. arguta* (Hirsch *et al.*, 1990), possibly in *A. melanandra* (Ferguson *et al.*, 1996), and in *A. chinensis* (Tang and Jiang, 1995; Ferguson, personal communication cited in Harvey *et al.*, 1997a; Testolin, unpublished). The tendency to produce bisexual flowers varies greatly among genotypes, within an individual vine, and erratically from year to year both in the proportion of bisexual flowers and in the degree of enhanced pistil and ovary development (McNeilage, 1991b; Messina *et al.*, 1990). Surprisingly, no sub-gynoecious individual has been found to date (Testolin *et al.*, 1995).

3. The genetic control of sex as inferred from gender segregation in progeny

Most seedling families produced by crossing male and female phenotypes have male and female offspring in a ratio which approaches 1:1. This has now been observed in both diploid and hexaploid taxa (*Table 1*), suggesting that a single genetic determinant controls sex in diploid taxa and that the same disomic segregation is conserved in polyploids (McNeilage, 1997; Testolin *et al.*, 1995). Significant deviation from a 1:1 ratio towards an excess of males has sometimes been reported, but in most cases this is due to immaturity of the families observed (*Figure 4*), since males are more precocious in overcoming the juvenility (Beatson, 1991; Blanchet and Chartier, 1991; Testolin *et al.*, 1995).

Seedling families derived from crosses involving fruiting males as male parents, segregate into apparently normal males and females and only a very small proportion (0 – *ca.* 4%) show enhanced pistil and ovary development (Testolin *et al.*, 1995) with as few as 1 or 2% being fruiting males (Ferguson *et al.*, 1996). It is therefore possible that the main sex determinant is not involved in the appearance of such phenotypes. It is more likely that inconstancy is determined by minor genes, unlinked to the primary sex determinant, which should act as quantitative trait loci (QTLs), and that the phenotypic expression of these genes is greatly influenced by environmental conditions.

When inconstant males are selfed or are crossed one with another, they always produce male and female seedlings (*Table 1*). The male is therefore the heterogametic sex (McNeilage, 1997; Testolin *et al.*, 1995). Moreover, the male to female ratio closely approaches 3:1 (*Table 1*), confirming the view of the apparently monogenic control of sex.

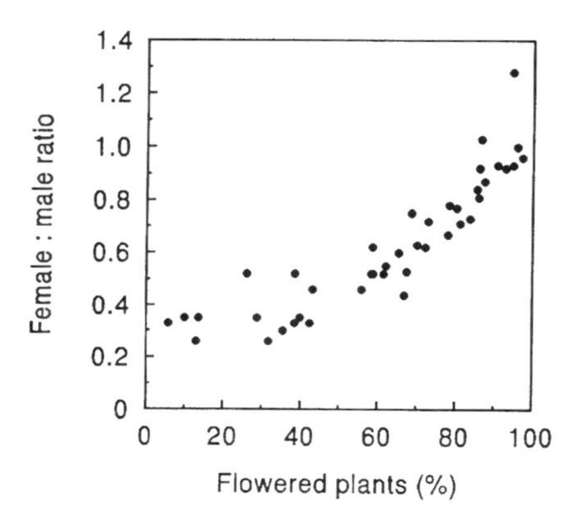

Figure 4. *Variation of gender ratio in* Actinidia *seedling families with the percentage of plants which overcame the juvenility. Each point represents one biparental family in a given year. The size of families varied from 283 to 593 individuals. Reprinted with the permission of Springer-Verlag from* Sexual Plant Reproduction, *Sex segregation ratio and gender expression in the genus* Actinidia, *R. Testolin, G. Cipriani and G. Costa, Vol. 8, pp. 129–132, Figure 1. © Springer-Verlag 1995.*

Table 1. *Sex segregation ratios in progeny resulting from open-pollination and controlled crosses of* Actinidia *species*

Species	Cross type[a]	No. of families[b]	Total seedlings n	Flowered seedlings n (%)	Observed M : F	Expected M : F	χ^2_{pool}	χ^2_{heter}	Ref.
Diploids									
A. chinensis	F × M	10	958	947 (99)	0.92	1:1	1.61 ns	15.23 ns	1
A. kolomikta	F o.p.	1	36	34 (94)	1.06	1:1	0.00 ns	–	2
Hexaploids									
A. deliciosa	F × M	4	1176	1126 (96)	0.97	1:1	0.29 ns	4.38 ns	2
	F × I	2	1059	975 (92)	1.10	1:1	2.08 ns	0.01 ns	2
	F × I	7	891	866 (97)	1.50	1:1	17.58 **	35.43 **	3
	I × I	1	487	475 (98)	3.06	3:1	0.03 ns	–	3
	I selfed	1	141	138 (98)	3.06	3:1	0.00 ns	–	2

[a] F = female, M = male, I = inconstant male (see text for explanation), o.p. = open pollinated.
[b] Only families with more than 90% of mature seedlings were considered.
ns, not significant; ** significant at $P = 0.01$
[1] Harvey *et al.*, 1997b; [2] Testolin *et al.*, 1995; [3] McNeilage, 1997

4. The establishment of disomic sex control in polyploids

Given that disomic sex segregation operates in *Actinidia* at different ploidy levels, we have to assume that in polyploid taxa a single chromosome pair carries the allelism for sex determination and that the remaining homologous chromosome pair(s) either do not exist or are homozygous for the female determinant. Zhang and Beuzenberg (1983) estimated a chromosome number of about 170 for *A. deliciosa* (6*x*), which is four fewer than would be expected from tripling the diploid genome ($2x = 58$). They therefore suggested that *A. deliciosa* had only one pair of sex chromosomes, thereby enabling disomic sex segregation to occur in the progeny. However, subsequent counts have reported 116, 174 and 232 chromosomes for tetraploid, hexaploid and octoploid *Actinidia* taxa, respectively (for review, see Ferguson *et al.*, 1996; Ferguson, personal communication). It seems, therefore, that polyploids have a single set of X/Y chromosomes together with a sex-neutral (XX)n set, where n depends on the ploidy level (McNeilage, 1991a, 1997; Testolin *et al.*, 1995).

It is easy to speculate how disomic control of sex could be conserved in polyploids, if it is assumed that the Y chromosome is a strong determinant of sex, as indicated by segregation analysis in hexaploids. Polyploids can result from somatic doubling or from the production of unreduced gametes by means of the mechanisms of parallel spindles or premature cytokinesis. Genotypes producing unreduced gametes have, in fact, been discovered in diploid male and female genotypes of *A. chinensis* (Ferguson *et al.*, 1997). Whatever the route to polyploidy, (X)nXX, (X)nXY and (X)nYY genotypes are produced at the first step from diploids and, since genotypes carrying Y chromosomes are males, irrespective of whether one or two Y chromosomes are present, and since males cannot cross one with another, Y chromosomes cannot accumulate in tetraploids. Higher levels of ploidy can occur by more routes, but the accumulation of Y chromosomes in higher polyploids is basically prevented for the same reasons. Furthermore, selfing of inconstant males will also produce (X)nYY males. At meiosis, such males produce only (X)nY gametes, and on the fertilization of female flowers, the Y chromosomes will be reduced to a single copy.

5. The genes involved

The monofactorial system described above has been classified as being of the active-Y type, where by analogy with the mechanisms operating in mammals and in plants such as *Silene*, the Y is conventionally the chromosome carrying the maleness determinant, (Dellaporta and Calderon-Urrea, 1993; Lebel-Hardenack and Grant, 1997; Westergaard, 1958). To date, however, no dimorphism in a chromosome pair carrying the sex determinants has been observed in *Actinidia*, but this may be due to the small chromosome size, the longest metaphase chromosome measuring only about 0.5 μm (Zhang and Beuzenberg, 1983). In white campion (*Silene latifolia*, syn. *Silene alba* or *Melandrium album*), a dioecious species in which the active Y-system operates, a bipartite gene of the male activator/female suppressor type has been described as being responsible for switching the female developmental programme to male development. A similar model has been postulated for other dioecious species, such as *Asparagus* (see Dellaporta and Calderon-Urrea, 1993 and Ainsworth *et al.*, 1998 for reviews). If we assume that such a system is also operating in *Actinidia*, as suggested by some authors (Fraser *et al.*, 1997; McNeilage, 1991a), the maleness determinant must carry at least two distinct, although closely linked, dominant genes: the first (*O*, ovary abortion) suppresses ovary development, the second (*P*, pollen fertility) maintains stamen maturation. Females must therefore be *op*/*op* at the sex-determining region, with *o* and *p* as recessive alleles, and males must be *op*/*OP* with the dominant alleles *O* and *P* closely linked on the same (Y)-chromosome. If this is the case, rare crossing-over should then result in the appearance of new variants, such as sterile individuals (*op*/*Op*), and hermaphrodites (*op*/*oP*); both variants have been found in kiwi fruit (McNeilage, 1997). In experiments where hermaphrodites were crossed to females and males as well as being selfed, phenotypic gender ratios in the progeny appeared to be those expected (*Table 2*), although segregation ratios could have been biased as families still had a high proportion of juvenile seedlings (McNeilage, 1997). Further observations have been made of sterile progeny in *A. chinensis* (Testolin, unpublished) but we do not know if these individuals are recombinants at the sex-determining regions or are the products of unbalanced chromosome sets as a result of ploidy differences between the parents.

6. Mapping and cloning sex-determining genes

Using bulk segregant analysis and RAPD techniques, Harvey and co-workers have identified two sex-linked markers in diploid *A. chinensis* (Harvey *et al.*, 1997a). One of these markers was found to be linked to the male-determining region at a distance of 2.86 cM (Harvey *et al.*, 1997b). Further investigations on sex markers, based on AFLP, RDA and mRNA differential display techniques are underway (Fraser, 1997 personal communication; Xiao, 1998 personal communication; Testolin, unpublished) and these should facilitate cloning.

An approach to the identification of sex genes in diploid *A. chinensis* based on the screening of male- and female-derived cDNA libraries obtained from flower buds collected at pre- and post-meiotic stages led to the isolation of several putatively sex-specific messengers from the libraries of each sex. Four cDNA clones obtained from the male library are being characterized. The amino acid sequence deduced from one of these clones shows homology with stamen- and/or tapetum-specific proteins found in

Table 2. *Phenotypic sex ratios in crosses involving hermaphrodite mutants of* A. deliciosa

Cross type[a]	No. of families	Total seedlings *n*	Flowered seedlings *n* (%)	Observed M : F[b]	Expected M : F	χ^2_{pool}	χ^2_{heter}	Ref.
H × I	2	501	456 (91)	1.07	1:1	0.56 ns	4.44 ns	1
H selfed	1	232	186 (80)	all F	0:1			1
F × H	1	144	71 (49)	all F	0:1			1

[a] H = hermaphrodite, I = inconstant male (see text for explanation), F = female.
[b] Seedlings were classified provisionally, on the basis of floral phenotype and the presence/absence of fruit. Class M (male) includes therefore both true and inconstant males, and Class F (female) includes both females and hermaphrodites.
ns, not significant; 1, re-arranged from McNeilage (1997).

unrelated species, such as snapdragon, tomato, *Lilium*, *Arabidopsis* and *Brassica* (Fraser *et al.*, 1997).

7. Biochemical aspects and their relationship with sex-controlling genes

Beginning in 1977, Hirsch and co-workers reported that male and female tissues in *A. chinensis* differed in both total peroxidase (PX) activity and the patterns of peroxidase zymogram (Hirsch *et al.*, 1977). Subsequently, either PX and polyphenol oxidase (PPO) isoforms characteristic of male tissues (Auxtová *et al.*, 1994; Hirsch *et al.*, 1977) and PX isoforms characteristic of female tissues (Hirsch and Fortune, 1984; Hirsch *et al.*, 1990) have been described. We do not know whether, or to what extent, such different isoenzymatic patterns are related to the primary sex-determining genes and until such genes are identified and sequenced, the roles of PXs and PPOs in determining gender expression remain speculative.

It is probable that the sex-determining genes affect either the endogenous levels of plant growth substances or their balance, as has recently been demonstrated in maize (reviewed in Lebel-Hardenack and Grant, 1997). Moreover, there is a large body of literature which demonstrates that exogenous application of plant growth substances can affect sex expression in many species (for reviews see, for example, Ainsworth *et al.*, 1998; Dellaporta and Calderon-Urrea, 1993; Irish and Nelson, 1989). In kiwi fruit, application of exogenous gibberellins to male flowers caused the development of ovary and styles, but ovules were not produced and fruit were not set after selfing (Marchetti *et al.*, 1992). As a general rule, treatments which depressed pollen grain viability in kiwi fruit male plants encouraged the development of a variable number of styles, whereas no plant growth substance was found which restored pollen grain development in female flowers (Marchetti *et al.*, 1992). Once again, to understand these and other results obtained by the use of exogenous chemicals we must obtain the sequences of primary sex-determining genes.

References

Ainsworth, C., Parker, J. and Buchanan-Wollaston, V. (1998) Mechanisms of sex determination in plants. *Curr. Top. Develop. Biol.* **38**: 167–223.

Auxtová, O., Šamaj, J., Cholvadová, B. and Khandlová, E. (1994) Isoperoxidase and isopolyphenol oxidase spectra in male and female tissues of *Actinidia deliciosa in vitro*. *Biologia Plantarum* **94**: 535–541.

Beatson, R.A. (1991) Inheritance of fruit characters in *Actinidia* deliciosa. *Acta Horticulturae* **297**: 79–86.

Blanchet, P. and Chartier, J. (1991) Genetic variability among the progeny of 'Hayward' kiwifruit. *Acta Horticulturae* **297**: 87–92.

Dellaporta, S.L. and Calderon-Urrea, A. (1993) Sex determination in flowering plants. *Plant Cell* **5**: 1241–1251.

Ferguson, A.R. (1984) Kiwifruit: a botanical review. *Hort. Rev.* **6**: 1–64.

Ferguson, A.R. (1990) Kiwifruit (*Actinidia*). In: *Genetic Resources of Temperate Fruit and Nut Crops* (eds J.N. Moore and J.R. Ballington Jr). *Acta Horticulturae* **290**: 603–653.

Ferguson, A.R., Seal, A.G., McNeilage, M.A., Fraser, L.G., Harvey, F. and Beatson, R.A. (1996) Kiwifruit. In: *Fruit Breeding* (eds J. Janick and J.N. Moore). John Wiley & Sons, New York, pp. 371–417.

Ferguson, A.R., O'Brien, I.E.W. and Yan, G.J. (1997) Ploidy in *Actinidia*. *Acta Horticulturae* **444**: 67–71.

Fraser, L.G., Harvey, C.F. and Gill, G.P. (1997) Molecular investigations into dioecy in *Actinidia chinensis*. *Acta Horticulturae* **444**: 79–83.

Harvey, C.F. and Fraser, L.G. (1988) Floral biology of two species of *Actinidia* (Actinidiaceae). II. Early embryology. *Bot. Gaz.* **149**: 37–44.

Harvey, C.F., Fraser, L.G. and Gill, G.P. (1997a) Sex determination in *Actinidia*. *Acta Horticulturae* **444**: 85–88.

Harvey, C.F., Gill, G.P., Fraser, L.G., McNeilage, M.A. (1997b) Sex determination in *Actinidia*. 1. Sex-linked markers and progeny sex ratio in diploid *A. chinensis*. *Sex. Plant Reprod.* **10**: 149–154.

Hirsch, A.M. and Fortune, D. (1984) Peroxidase activity and isoperoxidase composition in cultured stem tissue, callus and cell suspensions of *Actinidia chinensis*. *Z. Pflanzenphysiol Bd* **113**: 129–139.

Hirsch, A.M., Bligny, D. and Tripathi, B.K. (1977) Biochemical properties of tissue cultures from different organs of *Actinidia chinensis*. *Acta Horticulturae* **78**: 75–82.

Hirsch, A.M., Fortune, D. and Blanchet, P. (1990) Study of dioecism in kiwi fruit *Actinidia deliciosa* Chevalier. *Acta Horticulturae* **282**: 367–376.

Irish, E.E. and Nelson, T. (1989) Sex determination in monoecious and dioecious plants. *Plant Cell* **1**: 737–744.

Lebel-Hardenack, S. and Grant, S.R. (1997) Genetics of sex determination in flowering plants. *Trends Plant. Sci.* **2**: 130–136.

Liang, C. (1984) *Actinidia*. In: *Flora Reipublicae Popularis Sinicae*, Vol. 49 (2). (ed. K.-M. Feng). Science Press, Beijing, pp. 196–268, 309–324.

Marchetti, S., Zampa, C. and Chiesa, F. (1992) Sex modification in *Actinidia deliciosa var. deliciosa*. *Euphytica* **64**: 205–213.

McNeilage, M.A. (1988) Cytogenetics, dioecism and quantitative variation in *Actinidia*. PhD thesis, University of Auckland, Auckland, New Zealand.

McNeilage, M.A. (1991a) Gender variation in *Actinidia deliciosa*, the kiwifruit. *Sex. Plant Reprod.* **4**: 267–273.

McNeilage, M.A. (1991b) Sex expression in fruiting male vines of kiwifruit. *Sex. Plant Reprod.* **4**: 274–278.

McNeilage, M.A. (1997) Progress in breeding hermaphrodite kiwifruit cultivars and understanding the genetics of sex determination. *Acta Horticulturae* **444**: 73–78.

Messina, R. (1993) Microsporogenesis in male-fertile cv. Matua and male-sterile cv. Hayward of *Actinidia deliciosa* var. *deliciosa* (Kiwifruit). *Adv. Hort. Sci.* **7**: 77–81.

Messina, R., Vischi, M., Marchetti, S., Testolin, R. and Milani, N. (1990) Observations on subdioeciousness and fertilisation in a kiwifruit breeding program. *Acta Horticulturae* **282**: 377–386.

Morton, C.M., Chase, M.W., Kron, K.A. and Swensen, S.M. (1996) A molecular evaluation of the monophyly of the order Ebenales based upon *rbcL* sequence data. *Syst. Bot.* **21**: 567–586.

Seal, A.G. and McNeilage, M.A. (1989) Sex and kiwifruit breeding. *Acta Horticulturae* **240**: 35–38.

Tang, S.-X. and Jiang, S.-F. (1995) Detection and observation on the bud mutation of *Actinidia chinensis* Planch. *Acta Horticulturae* **403**: 71–73.

Testolin, R., Cipriani, G. and Costa, G. (1995) Sex segregation and gender expression in the genus *Actinidia*. *Sex. Plant Reprod.* **8**: 129–132.

Westergaard, M. (1958) The mechanism of sex determination in dioecious flowering plants. *Adv. Genet.* **9**: 217–281.

Zhang, J. and Beuzenberg, E.J. (1983) Chromosome numbers in two varieties of *Actinidia chinensis* Planch. *N.Z. J. Bot.* **21**: 353–355.

12

Maize sex determination

Erin E. Irish

1. Introduction

Zea mays, or maize, belongs to the grass family, a large group of angiosperm species. Most grasses do not exhibit sex determination and, as such, develop perfect flowers. Maize is monoecious, developing male and female flowers in different locations on a single plant. A description of grass flowers and how flowers are arranged in inflorescences (Dahlgren *et al.*, 1985) in the grass family is helpful in understanding sex determination in maize. Grass flowers are simple and inconspicuous, lacking sepals and petals. A typical grass flower consists of a bract-like organ called a palea, a whorl of two or three lodicules (possible petal homologues), three stamens, and a tricarpellate, unilocular pistil, containing a single ovule. The flowers of all grasses are borne multiply in complex inflorescences, the basic unit of which is the spikelet. A spikelet is a short branch that consists of one or more flowers, each subtended by a bract called a lemma, and the entire spikelet is enclosed by a pair of so-called sterile bracts, the glumes. Spikelets can be indeterminate, having many florets (flower plus lemma), a primitive condition in the Gramineae, while determinate spikelets, with only a few florets, are regarded as a more advanced character. Maize belongs to the tribe Andropogoneae, all of whose members have determinate, two-flowered spikelets (Clayton, 1986). The subtribe to which maize belongs, the Maydeae, in addition to having determinate spikelets, also is entirely monoecious (Kellogg and Birchler, 1993). Sex determination in maize is most advanced among the members of the Maydeae, having segregation of male and female flowers into separate inflorescences in distinct locations, the terminal tassel and the lateral ears (Dellaporta and Calderon-Urrea, 1994; Irish, 1996; Veit *et al.*, 1993).

Sex differentiation occurs in maize, as in the majority of angiosperms exhibiting sex determination (Yampolsky and Yampolsky, 1922), by the formation of both stamen and pistil primordia by floral meristems, so that flowers throughout the plant are initially perfect. This is followed by the selective abortion of stamen or pistil primordia, resulting in unisexual flowers (Bonnett, 1948; Cheng *et al.*, 1982; Irish and Nelson, 1993). The identity of the organs that are aborted is determined spatially; the florets in the tassel undergo pistil suppression, resulting in staminate (male) flowers in the terminal inflorescence. In contrast, the stamen primordia of flowers on the ear are suppressed,

Sex Determination in Plants, edited by C.C. Ainsworth.

so that the lateral inflorescences have only pistillate (female) flowers at maturity. The lower floret of each spikelet on ears is also suppressed in most varieties of maize.

The process of sex differentiation in maize is complex: not only is there the primary difference in sex among the flowers in different locations on the plant, but there are also differences in the architecture of the inflorescences and in glume morphology (Irish, 1996). The tassel is thin and bears long branches at its base, while the ear is thick and unbranched. The glumes of the tassel are acute, elongated and photosynthetic, while glumes on the ear are short and broad and remain etiolated. In addition to these differences within the inflorescences, vegetative parts of the plant near the inflorescences develop differently: the internodes subtending the tassel are much longer than those subtending the ear. As a result of this difference, even though there are more nodes between the base of the ear and the point of its attachment to the main stem (*ca.* 10) than there are between the base of the tassel and the point of ear attachment (*ca.* seven), the ear appears to be positioned in the midpoint of the shoot.

2. Genetics of sex determination in maize

The normal process of sex determination in maize requires two different types of organ abortion and their correct spatial regulation within the plant, so that only stamens are suppressed in ear florets and only pistils are suppressed in tassel florets. There are mutations that affect each of these processes (Coe *et al.*, 1988). Mutations that affect stamen suppression result in andromonoecy, as the sex of the tassel florets is unaffected, remaining male, but the ear florets fail to suppress stamens, resulting in perfect flowers. Similarly, feminizing mutations that fail to suppress pistils in the tassel do not alter the sex of the flowers on the ears. Masculinizing and feminizing mutations will be discussed below.

2.1 *Stamen suppression*

The largest class of masculizing mutations, representing genes required for stamen suppression in ear flowers, are the andromonoecious *dwarf* mutations: *d1, d2, d3, d5, D8* and *anther ear* (*an*) (Phinney, 1961, 1984) (in maize genetics the convention is to identify dominant mutations with an upper case first letter and recessive mutations with lower case). All of these mutants form shoots with reduced stature. The recessive mutations *d1, d2, d3, d5* and *an* all affect some specific step in the biosynthesis of the plant hormone gibberellic acid (GA) (Hedden and Phinney, 1979; Phinney and Spray, 1982; Spray *et al.*, 1984). *an* has recently been cloned (Bensen *et al.*, 1995): its product is *ent*-kaurene synthase, an enzyme involved in the cyclization of geranylgeranyl pyrophosphate to form *ent*-kaurene. Plants carrying a deletion spanning the *an* locus nonetheless retain 20% of *ent*-kaurene synthase activity, indicating redundancy in the genome for encoding genes required for this essential hormone. Despite the presence of this residual activity, *an* plants fail to suppress stamen development. The dominant mutations *D8* and *D9* likely affect the production of a GA receptor (Harberd and Freeling, 1989; Winkler and Freeling, 1994), as these mutants, unlike the recessive *dwarf* and *an* mutants, cannot be rescued by the application of exogenous GA (Phinney, 1956). In common with the other *dwarf* mutants, *D8* and *D9* have an anther-eared phenotype. Thus, GA is required to induce stamen suppression, and in plants with significantly reduced levels of GA, stamen suppression processes do not occur. Response to a concentration of GA above some threshold level seems likely to be required to effect stamen suppression, which occurs in ears but not tassels of even normal plants. That there is a minimum concentration of GA required for stamen suppression function is consistent with measurements of endogenous GA levels in different parts of the plant: there is approximately 100 times less GA in developing tassels than there is in developing ears (Rood *et al.*, 1980). Genes that respond to GA concentrations specifically to induce stamen suppression have not yet been identified: such mutants would be expected to have normal stature but an anther-eared phenotype.

Masculinization of the plant is also conditioned by the *silkless* mutation, which like the GA mutants, has normal tassels. Ears on *sk* plants are completely barren as all pistils as well as stamens are aborted (Jones, 1925). Genetic studies have shown that this phenotype is the result of inappropriate expression of pistil-suppression functions that are normally limited to the tassel flowers (see Section 2.2). The *teosinte branched* mutation can also be considered to masculinize the plant, as it converts ears to tassels. *tb* mutant plants retain the ability to develop some female flowers, which form on secondary inflorescences forming laterally on the side branches. *Teosinte branched* is one of five major loci responsible for the morphological differences between maize and its progenitor, teosinte (Doebley *et al.*, 1995). *tb* has recently been cloned and has been shown to have regions of homology to the *Antirrhinum* gene *cycloidea* (Doebley *et al.*, 1997).

2.2 *Pistil suppression*

Feminization of the maize plant by the failure to suppress pistil development in the tassel is conditioned by a large group of mutations, known as *tassel seed* mutations. There are six mapped *ts* loci (Coe *et al.*, 1988), and probably at least as many yet unmapped *ts* genes (E. Irish, unpublished observation). Some of the mutations are dominant or semi-dominant, while others are recessive, and the mutations also vary in their severity. The recessive mutations *ts1* and *ts2* convert all tassel florets from staminate to pistillate (Emerson, 1920). *ts2* has been cloned via transposon tagging (DeLong *et al.*, 1993). The product that it encodes, as deduced from its sequence, is a short-chain alcohol dehydrogenase. How such an enzyme participates in sex determination is not yet known, but it most likely acts directly in the process of pistil abortion, as its mRNA is found in tassel (but not ear) floret pistils at the stage preceding the histological manifestation of sex differentiation (DeLong *et al.*, 1993).

Other *tassel seed* genes are likely to be involved in the patterning of pistil suppression, as deduced from the phenotypes of mutations such as the semi-dominant *Ts5*, which conditions a mixture of staminate and pistillate florets evenly distributed in the tassel and the dominant *Ts3*, which results in tassels with large domains of staminate or pistillate flowers (Nickerson and Dale, 1955). *ramosa 3* mutants have a phenotype similar to that of *Ts5* mutants, but in this case, the mutation is recessive (Coe *et al.*, 1988). Other *tassel seed* genes have multiple functions: both *ts4* (Phipps, 1928) and *Ts6* (Emerson *et al.*, 1935) mutants have highly branched and partially sterile tassels and ears as well as feminized tassels. These genes are thought to play an important role in bestowing determinacy in spikelet pair meristems and spikelet meristems in addition to pistil suppression functions (Irish, 1997a). Lack of determinacy that results in the formation of additional floral meristems in *Ts6* mutant tassels has been shown by ectopic expression of *Knotted* (Irish, 1997b), a gene whose expression is limited to meristems (Jackson *et al.*, 1994). Culturing experiments in which immature tassels were assayed for floral determination showed that although *Ts6* delayed determinacy of certain meristems, the ability to form normal flowers was reached at a wild-type rate (Irish, 1997b).

2.3 *Genetic interactions in sex determination*

The finding that *ts2*, a gene required for pistil suppression, encodes a short-chain alcohol dehydrogenase has led to the suggestion that pistil suppression, like stamen suppression, is regulated by gibberellins, as GA-like molecules might serve as a substrate for such an enzyme (Calderon-Urrea and Dellaporta, 1994). Since relatively high levels of GA are required for stamen suppression, a feminizing effect, it is possible that the masculizing function of pistil suppression could be induced by low levels of GA or by an inability to respond to GA. The tassel seed phenotype would then be the result of either increased levels of GA in the tassel or in a heightened response by tassel flowers to this hormone. This is an attractive model, as a single signalling molecule would be responsible for different patterns of sex differentiation throughout the plant. Results from double mutant studies of *dwarf* and *tassel seed* mutations, however, do not support this model. If the *tassel seed* mutants have a greater sensitivity to GA, one would predict that a *tassel seed, dwarf* double mutant would have a near-normal phenotype in that the deficit in the production of the hormone would be balanced by an enhanced ability to respond to it. Alternatively, if the *tassel seed* mutations alter levels of GA, then epistasis would result and the double mutant would have either a dwarf or a tassel seed phenotype, depending on their relative positions in the genetic pathway. The observed phenotype of *tassel seed, dwarf* double mutants had none of the above phenotypes. Instead, they exhibited a novel phenotype in which flowers on both tassels and ears have both stamens and pistils (Irish *et al.*, 1994). Thus, the additive phenotype of the double mutants provides clear evidence that pistil suppression and stamen suppression occur by the activities of two genetically independent pathways.

Double mutant studies have shown that *ts2* and *sk* genes do interact directly: a *ts2, sk* double mutant has a weak tassel seed phenotype (Irish *et al.*, 1994; Jones, 1932, 1934). This result can be explained by the following: if the function of *ts2* is to suppress pistil development, then *sk* may act to suppress *ts2* activity in the ear. An *sk* mutant would then fail to suppress *ts2*, resulting in pistil suppression throughout the plant. According to this scheme, a *ts2* mutant, regardless of whether the *sk* gene is functional, would not be able to suppress pistils, resulting in the tassel seed phenotype.

3. Tassel seed genes regulate inflorescence meristem fate

Double mutant studies of combinations of different tassel seed mutations have provided new insights into the origins of pistil suppression functions in maize. This has come from the unexpected phenotypes of certain combinations of tassel seed mutations. As described above (Section 2.2) some *tassel seed* mutations affect only pistil suppression while others affect inflorescence development as well. Genetic evidence places these into separate classes of action: Class I (pistil suppression only) includes *ts1, ts2, Ts3, Ts5*, and Class II (pistil suppression and inflorescence branching) is comprised of *ts4* and *Ts6*. Double mutants made with two *tassel seed* mutations within a class had a phenotype that was identical to the more extreme of the two component mutations (epistasis). Double mutants comprised of a Class I plus a Class II *tassel seed* had enormously branched, sterile tassels (Irish *et al.*, 1994). The latter phenotypes provide strong evidence that all *tassel seed* genes, both Class I and Class II, play some role in regulating the extent of branching in normal inflorescences. As the ability to suppress pistil development is a derived characteristic, it is not unreasonable to expect that by mutation analysis it may be possible to determine the origin of the genes that regulate this process in modern maize. It seems likely from these results that pistil suppression functions were gained by genes whose first function was to regulate determinacy of meristems that form spikelets or florets in the inflorescence. It is intriguing to note that all members of the Andropogoneae, which have paired determinate, two-flowered spikelets, also undergo pistil suppression in the lower floret of each spikelet (Clayton, 1986). Thus, in this tribe, homologues of *Ts6* and *ts4* may regulate both meristem determinacy and sex determination. This possibility can be tested by the eventual isolation of their homologues, once the maize tassel seed genes are cloned.

References

Bensen, R.J., Johal, G.S., Crane, V.C., Tossberg, J.T., Schnable, P.S., Meeley, R.B. and Briggs, S.P. (1995) Cloning and characterization of the maize an1 gene. *Plant Cell* 7: 75–84.

Bonnett, O.T. (1948) Ear and tassel development in maize. *Ann. Mo. Bot. Gard.* **35**: 269–287.

Cheng, P.C., Greyson, R.I. and Walden, D.B. (1982) Organ initiation and the development of unisexual flowers in the tassel and ear of *Zea mays. Am. J. Bot.* **70**: 450–462.

Clayton, W.D. (1986) Andropogoneae. In: *Grass Systematics and Evolution* (eds T. Soderstrom, K. Hilu, C. Campbell and M. Barkworth) Smithsonian Institute Press, Washington, pp. 307–309.

Coe E.H., Neuffer, M.G. and Hoisington, D.A. (1988): The genetics of corn. In: *Corn and Corn Improvement*, 2nd Edn. (eds G. Sprague and J. Dudley) American Society of Agronomy, Crop Science Society of America, and Soil Science Society of America, Madison, pp. 81–258.

Dahlgren, R.M.T., Clifford, H.T. and Yeo, P.F. (1985) Order poales. In: *The Families of the Monocotyledons*. Springer-Verlag, New York, pp. 419–460.

Dellaporta, S.L. and Calderon-Urrea, A. (1994) The sex determination process in maize. *Science* **266**: 1501–1505.

DeLong, A., Calderon-Urrea, A. and Dellaporta, S.L. (1993) Sex determination gene Tasselseed2 of maize encodes a short-chain alcohol dehydrogenase required for stage-specific floral organ abortion. *Cell* **74**: 757–768.

Doebley, J., Stec, A. and Gustus C. (1995) *teosinte branched* and the origin of maize: Evidence for epistasis and the evolution of dominance. *Genetics* **141**: 333–346.

Doebley, J., Stec, A. and Hubbard, L. (1997) The evolution of apical dominance in maize. *Nature* **386**: 485–488.

Emerson, R.A. (1920) Heritable characters in maize. II. Pistillate flowered maize plants. *J. Hered.* **11**: 65–76.

Emerson, R.A., Beadle, G.W. and Fraser, A.C. (1935) A summary of linkage studies in maize. *Cornell Univ. Agric. Exp. Sta. Mem.* **180**: 1–83.

Harberd, N.P. and Freeling, M. (1989) Genetics of dominant gibberellin-insensitive dwarfism in maize. *Genetics* **121**: 827–838.

Hedden, P. and Phinney, B.O. (1979) Comparison of ent-kaurene and ent-isokaurene in cell-free systems from etiolated shoots of normal and dwarf-5 maize seedlings. *Phytochemistry* **18**: 1475–1479.

Irish, E.E. and Nelson, T. (1993) Developmental analysis of the inflorescence of the maize mutant Tassel seed 2. *Am. J. Bot.* **80**: 292–299.

Irish, E.E., Langdale, J.A. and Nelson, T. (1994) Interactions between sex determination and inflorescence development loci in maize. *Develop. Genet.* **15**: 155–171.

Irish, E.E. (1996) Sex determination in maize. *BioEssays* **18**: 363–369.

Irish, E.E. (1997a) Class II tassel seed mutations provide evidence for multiple types of inflorescence meristems in maize. *Am. J. Bot.* **84**: 1502–1515.

Irish, E.E. (1997b) Experimental analysis of tassel development in the maize mutant Tassel seed 6. *Plant Physiol.* **114**: 817–825.

Jackson, D., Veit, B. and Hake, S. (1994) Expression of maize *Knotted1* related homeobox genes in the shoot apical meristem predicts patterns of morphogenesis in the vegetative shoot. *Development* **120**: 405–413.

Jones, D.F. (1925) Heritable characters in maize. XXIII. Silkless. *J. Hered.* **16**: 339–341.

Jones, D.F. (1932) The interaction of specific genes determining sex in dioecious maize. *Proc. Sixth Int.Congress Genetics* **2**: 104–107.

Jones, D.F. (1934) Unisexual maize plants and their bearing on sex differentiation in other plants and in animals. *Genetics* **19**: 552–567.

Kellogg, E.A. and Birchler, J.A. (1993) Linking phylogeny and genetics: Zea mays as a tool for phylogenetic studies. *Syst. Biol.* **42**: 415–439.

Nickerson, N.H. and Dale, E.E. (1955) Tassel modification in *Zea mays. Ann. Mo. Bot. Gard.* **42**: 195–212.

Phinney, B.O. (1956) Growth response of single-gene dwarf mutants in maize to gibberellic acid. *Proc. Natl Acad. Sci. USA* **42**: 185–189.

Phinney, B.O. (1961) Dwarfing genes in *Zea mays* and their relation to the gibberellins. In: *Plant Growth Regulation*. (ed. R.M. Klein), Iowa State University Press, Ames, pp. 489–501.

Phinney, B.O. (1984) Gibberellin A1, dwarfism and the control of shoot elongation in higher plants. In: *S.E.B. Seminar Series*, Vol. 3 (eds A. Crozier and J.R. Hillman) Cambridge University Press, Cambridge, pp. 17–41.

Phinney, B.O. and Spray, C. (1982) Chemical genetics and the gibberellin pathway in *Zea mays* L. In: *Plant Growth Regulation* (ed. P. F. Wareing) Academic Press, New York, pp. 101–110.

Phipps, I.F. (1928) Heritable characters in maize. XXXI Tassel-seed 4. *J. Hered.* **19**: 399–404.

Rood, S.B., Pharis, R.P. and Major, D.J. (1980) Changes of endogenous gibberellin-like substances with sex reversal of the apical inflorescence of corn. *Plant Physiol.* **66**: 793–796.

Spray, C., Phinney, B.O., Gaskin, P., Gilmore, S.J. and Macmillan, J. (1984) Internode length in *Zea mays* L. The dwarf-1 mutation controls the 3β-hydroxylation of gibberellin A20 to gibberellin A1. *Planta* **160**: 464–468.

Veit, B., Schmidt, R.J., Hake, S. and Yanofsky, M.F. (1993) Maize floral development: new genes and old mutants. *Plant Cell* **5**: 1205–1215.

Winkler, R.G. and Freeling, M. (1994) Physiological genetics of the dominant gibberellin-nonresponsive maize dwarfs, Dwarf8 and Dwarf9. *Planta* **193**: 431–348.

Yampolsky, C. and Yampolsky, H. (1922) Distribution of sex forms in the phanerogamic flora. *Bibliotheca Genet.* **3**: 1–59.

13

Male to female conversion along the cucumber shoot: approaches to studying sex genes and floral development in *Cucumis sativus*

Rafael Perl-Treves

1. Introduction

The cucurbit family is remarkable for its diversity of sex types. Hermaphroditism is rare among the Cucurbitaceae; monoecy and dioecy are common, and different stages in the presumed evolution of sex-determination mechanisms can be found, from single sex genes (e.g. in cucumber and melon) to heteromorphic sex chromosomes (e.g. in *Coccinia grandis*; Roy and Saran, 1990). *Cucumis sativus*, the cucumber, an important vegetable crop, is undoubtedly the cucurbit species in which plant sex expression has been studied most extensively.

Most traditional cucumber cultivars are monoecious, but the shift of emphasis of the seed industry from open-pollinated varieties to F1 hybrids has stimulated research on the inheritance of sex determination, and the physiological factors that affect it. Pioneer genetic studies on cucumber sex determination were started in Israel in the 1950s, by Galun and co-workers, and in Poland by Kubicki. During the next two decades, hormonal and environmental influences on sex expression were studied in detail. After a long period of quiescence, in which the practical use of these discoveries by cucumber breeders has become routine, but little new was learned about the mechanisms underlying sex expression, the last years have witnessed a renewed interest in this topic. New advances in plant molecular biology and in plant genome analysis may now be utilized to study the mode of action of sex genes, and their interactions with sex-modifying hormones, in a most direct way. In this chapter I review the main findings that emerged from the earlier genetic and physiological studies and the interesting questions raised by them. A discussion of the newer, albeit preliminary results

Sex Determination in Plants, edited by C.C. Ainsworth.

of modern experimental approaches will follow. The interested reader is referred to a previous detailed review by Frankel and Galun (1977), and shorter reviews by Rudich (1990) and Malepszy and Niemirowicz-Szczytt (1991).

2. The cucumber floral bud: transition from bisexual to unisexual development

Cucumber plants may produce male (staminate), female (pistillate; *Figure 1c*) and, less commonly, hermaphrodite (bisexual) flowers. The respective three bud types have, at a young developmental stage, both stamen and carpel primordia, that appear equally developed under the microscope, (*Figure 1d*). Such buds (measuring ~1 mm in length) are, therefore, morphologically bisexual (Atsmon and Galun, 1960). Unisexual male flowers develop from the bisexual buds as a result of inhibition of the carpel primordia as the stamens develop, while female flowers form when stamen primordia are arrested and the carpels differentiate (*Figure 1e* and *f*). A pistillodium (i.e. a rudimentary carpel) and the nectary derived from it, are still present in mature male flowers, and vestigial stamens are still present in mature female flowers (*Figure 1f* and *g*). No evidence for programmed cellular death has been reported, and the inappropriate organ primordia may simply cease to grow upon sexual differentiation of the bud.

Goffinet (1990) applied electron microscopy and allometric methods (measurement of the relative growth of organs), in an attempt to determine the precise time when sexual differentiation of the bud begins. He showed that the first flower meristem formed in the fourth stem node 10 days after sowing, the flower reaching anthesis 22 days later. Seven days after floral meristem formation, the bud is 1 mm long and the first morphological difference between male and female buds is evident. In male buds, growth of carpel primordia slows visibly, and the subsequent buds that form in the same node are not inhibited by the first one. In the female, the ovary begins to enlarge, and the younger buds in the same axil are visibly inhibited; female flowers are usually borne singly on each node, while multiple male flowers form sequentially on the same node (*Figure 1a* and *b*). The allometric study revealed, however, that changes in growth rate have started well before the seventh day: Goffinet claimed that the truly bisexual stage is present, in fact, earlier than visualized, since the determined path of differentiation has been taken already. Galun (1961a) tried to capture the moment of 'physiological bisexuality', i.e. the time at which a bud is still sensitive to sex reversal by hormonal treatment. He treated monoecious plants with gibberellin (GA), that delays the differentiation of the first female flower by several nodes (see Section 6), causing the formation of male flowers instead. This treatment was most effective when given 1 week before the bud reached the morphologically bisexual stage. If this is true, the hormonal stimulus, or developmental information that determines the bud's sex, must be stored in meristematic cells long before visible sex organ primordia have formed. Other studies have cast doubt on the notion that GA can change the specific bud's fate (see Section 6.1). Ethylene treatment given at the first leaf stage induced female flowers on the fourth node instead of the 12th node (Iwahori *et al.*, 1970); treatments delivered later, at the second leaf stage, only affected the seventh node, suggesting that, at the first leaf stage, the sex of the buds in the first four nodes has already been determined – long before the buds reached the 1 mm stage. Another study, however (Robinson *et al.*, 1969), showed that treatments given at the first and third leaf stage affected even the first nodes, that

Figure 1. *(a) Flowering pattern of a gynoecious cucumber genotype. Single pistillate flowers are borne at each node of the main shoot. (b) Flowering pattern of an androecious cucumber genotype. Multiple staminate flowers form sequentially at each node. (c) Male and female flowers at anthesis and on the day preceding anthesis. (d) Morphologically bisexual bud, 1 mm long. Arrows point at stamen and carpel primordia. (e) A day after the stage shown in (d), the ovary of a female-determined bud begins to visibly elongate. (f) Section through intermediate stage (4 mm long) female bud: O, ovary; G, stigma; S, rudimentary stamens. (g) Section through intermediate stage (5 mm long) male bud: A, anthers; C, rudimentary carpel and nectary.*

contained already-formed buds, suggesting that 'physiological bisexuality' with respect to ethylene extends to more advanced bud stages than previously had been suggested.

A different approach, using *in vitro* culture, has been undertaken to address the same question. Cucumber floral buds at different stages of development were detached from the plant and put in organ culture, where the direct effect of hormones on a bud can be separated from whole-plant effects such as bud–leaf interactions. Remarkably, buds detached at very early bisexual stages could grow and differentiate *in vitro* to rather advanced stages. Potentially male buds, 0.7 mm long (i.e. morphologically bisexual; left on the plant, such buds would have become staminate), reverted to female buds upon addition of auxin (Galun *et al.*, 1962, 1963). Physiological bisexuality, as determined by auxin reversion, thus extends up to the 0.7 mm stage. Potentially female buds did not respond *in vitro* to either indole acetic acid (IAA) or GA, so their bisexuality could not be tested in such a system.

3. Sex expression along the cucumber shoot: sex types and sex phases of the whole plant

Having discussed sex expression of the individual bud, we now turn our attention to the whole plant. The distribution of different types of flowers along the plant axis varies in cucumber, giving rise to various sex types. Most genotypes are monoecious and bear separate male and female flowers on the same plant. Clusters of male flowers, or single female flowers, form in the leaf axils. Along the monoecious shoot, sex expression changes, and three phases may be recognized (Currence, 1932): a male phase with only staminate flowers, a mixed phase, and a continuous-female phase. Thus, sex tendency gradually changes from male to female. The mechanism responsible for such an intriguing 'sex gradient' is not yet understood, and it certainly represents one of the interesting questions that are still open. Dax-Fuchs *et al.* (1978) also described an early 'vegetative phase' (a few basal nodes that do not form flowers), and a phase of a few abortive floral-nodes found in some genotypes. They noted a correlation between male tendency and the presence of abortive nodes.

Several genes that affect gene expression have been described (Section 4); such genes can specify sex genotypes other than monoecious, by changing the fate of specific buds, or by affecting the succession of sex phases along the shoot. In addition, sex expression of a given genotype can be strongly influenced by the environment and by the exogenous application of plant growth regulators (Sections 5 and 6 below). The most common deviations from monoecy are:

(i) gynoecious: plants bear only pistillate flowers;
(ii) androecious: plants bear only staminate flowers;
(iii) hermaphrodite, bearing only perfect (bisexual) flowers;
(iv) andromonoecious, with both male and perfect flowers.

Gynoecious and androecious plants are shown in *Figure 1a* and *b.* A schematic representation of the various sex types and their cross-progenies is also given in *Figure 3.*

Since sex patterns change along the plant and during its development, description of a plant's sex may require multiple records at different stages. A descriptor such as the percentage of male or female flowers will change during ontogeny of a monoecious plant. A useful and simpler descriptor of sex tendency, that is genetically stable under a specific set of environmental conditions, is the 'node number' (Shifriss and Galun, 1956), defined as the first node at which a female flower is recorded along the main shoot: the lower the node number of a plant, the stronger its female tendency. For andromonoecious plants, their node number would be the first node where a bisexual flower appears. In addition, the main shoot has usually a stronger male tendency than the lateral branches. When examining the effect of a given treatment on sex expression, one should assess whether the observed effect is a true one, that directly modifies the developmental programme at given nodes – or just an indirect change, such as a general increase or decrease in flowering, bud abortion, or branching, causing an apparent change in the ratio of male/female flowers. The reported effect of increased nitrogen nutrition on cucumber sex expression may represent such an apparent change, where flower sex ratios changed towards femaleness, but actual node-numbers did not change (Atsmon and Galun, 1962).

An interesting attempt to interpret sex-phase shift along the monoecious shoot as a 'physiological gradient' related to leaf age in the vicinity of the bud was made by Atsmon and Galun (1962). They observed that in a gynoecious plant, the buds developing in the axils of the younger leaves are at a more advanced stage of development, as compared to male buds of a monoecious plant, at similar positions on the shoot. This implies that male flowers pass through their critical, bisexual stage while in proximity to relatively older leaves than female buds. The authors hypothesized that the distribution of young and mature leaves may be responsible for a gradient of growth substances along the shoot, influencing the sex ratio. They also monitored leaf and bud age along the shoot of monoecious plants under different environmental regimes. As the plant matures, more advanced bud stages will be found in the axils of younger and younger leaves, because floral development is faster than plastochrone. In addition, under long days and warm temperatures, leaves grow faster, conditions which correlate with increased maleness. Under short days and cold night conditions, the bisexual stage is found in the axils of much younger leaves, correlating with increased femaleness. The authors suggested that the compound coming from younger leaves, building up the leaf-age : femaleness gradient, may be an auxin.

4. Inheritance of sex determination in cucumber

The inheritance of sex expression in cucumber has been elucidated by Galun (1961b) and Kubicki (1969a,b,c,d). The appearance of different flower types is a complex trait, but only three major genes account for most sex phenotypes. The newer gene nomenclature (Pierce and Wehner, 1990) will be used here. *Figure 2* illustrates the interaction between the three genes.

(i) *F/f* (formerly designated *St, Acr*) is a partially dominant gene that controls femaleness. The *F* allele shifts the monoecious sex pattern 'downwards', causing the female phase to start much earlier. Kubicki (1969a,b) described a series of alleles in this locus, that differ in the strength of the phenotype, and the degree of dominance they exhibit over the *f* allele found in monoecious plants.
(ii) *A/a* increases maleness. *A* is epistatic to *F*: in *ff* genotypes, the recessive *a* allele intensifies the male tendency. *aaff* plants are androecious, and will never attain the mixed, or female phases. Rudich *et al.* (1976) tested androecious genotypes under short day (SD) and long day (LD) conditions, and under higher (16–18°C) versus

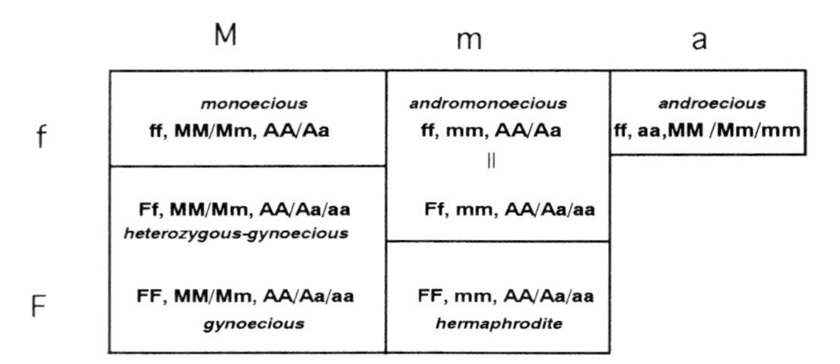

Figure 2. *Schematic representation of the sex phenotypes obtained by different combinations of the genes* F, M *and* A.

lower (10–13°C) night temperatures, showing that stable androecious genotypes never formed female flowers under any of the regimes tested.

(iii) ***M/m*** (former designation: *G*, *A*), is different from the *F* and *A* genes – its primary effect is not on the succession of sex phases along the shoot, but on the floral structure of individual buds – those that are determined to develop an ovary. The dominant allele will only allow the formation of stamen-less female flowers as well as male flowers, while in *mm* plants, bisexual flowers form (in addition to male flowers). The bisexual flowers will occupy, along the shoot of such plants, positions typical of pistillate flowers of *MM* plants (Galun, 1961b; Kubicki, 1969c). We may speculate that in buds determined to develop a carpel, but not in male flowers, *M* may act, either directly or indirectly, as a stamen-suppressor.

The gene *M* apparently has additional, 'secondary' phenotypic effects. Pistillate flowers of *MM* genotypes are usually borne singly on female nodes, but in *mm* genotypes, the bisexual flowers are followed, in the same node, by additional flowers, often displaying increasing male tendency. The same nodes produce also intermediate male–bisexual flowers, with a varying degree of ovary development.

Another apparently secondary effect of the *m* alleles would be to increase the male tendency: *FFmm* genotypes, instead of being full hermaphrodites, display a short male phase (as illustrated in *Figure 3b*). Kubicki (1969c), on the other hand, noted that in andromonoecious genotypes (*ffmm*), *m* seems to increase femaleness, as compared to the monoecious genotypes *ffMM.* He suggested that polygenes affecting femaleness are linked to the *M* locus. The existence of femaleness modifier genes that, in addition to the major gene *F*, can enhance or decrease femaleness, was demonstrated by Galun (1961b). He selected progressively, over eight generations, sister lines derived from one monoecious (*ffMM*) genotype, and obtained lines with an increased, and decreased, node number, respectively; a stabilized node-number difference of four nodes was 'fixed' between the lines.

Additional genes that have been identified include *In-F* (formerly designated *F*; Kubicki, 1969a,b). This gene is unlinked to the major sex gene, *F. In-F* enhances femaleness of monoecious genotypes and, when crossed to *FF* gynoecious cucumbers, it will display an additive effect, rendering their female tendency even stronger; such 'super females' will not respond to GA treatment!

Tr (trimonoecious) is a co-dominant gene, that affects the fate of unisexual flowers, turning them into bisexual ones. Contrary to the *M* gene, *Tr* affects only developing

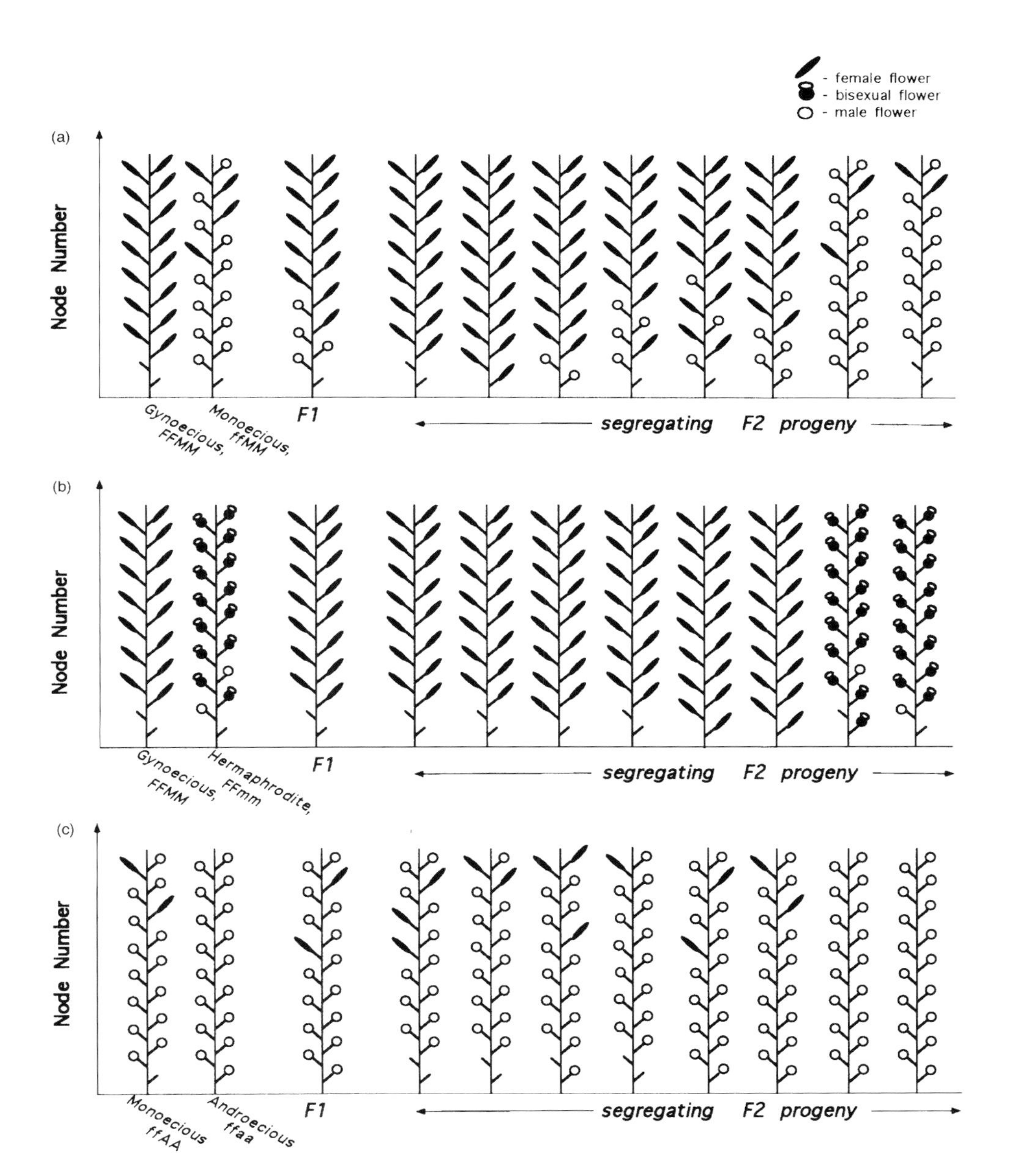

Figure 3. *Graphical representation of the sex type of the parents and the F1 and a sample of the F2 progenies, of three segregating populations. Sex of the flower in each node along the main shoot is indicated. Four breeding lines/cultivars whose sex expression is well characterized were chosen as the parents: (a) Population segregating for the* F *gene. The parents consisted of a gynoecious line, producing only female flowers starting from the first nodes and a monoecious cultivar, displaying typical sex-phases. (b) Population segregating for the* M *gene. The parents were the same gynoecious line as in (a) and a hermaphrodite line having bisexual flowers (occasionally accompanied by male flowers) from the bottom-most nodes. (c) Population segregating for the* A *gene. The parents were the monoecious line [as in (a)] and an androecious line that develops exclusively male flowers even on side branches.*

staminate buds and 'releases' an apparent inhibition of carpel development (Kubicki, 1969d). This results in flowers that are hypogynous (i.e. have a superior ovary, while normal pistillate cucumber flowers have inferior ovaries).

Another, related *Cucumis* species in which sex expression is of economic importance is *C. melo*, the musk melon. Kubicki (1969b) compares the inheritance of femaleness in melon and cucumber. In melon, during ontogeny, the male tendency increases, in contrast to the situation in cucumber. A further difference is that the gene *G*, that controls gynoecy in melon, is recessive, that is, *gg* is gynoecious and *G*– is andromonoecious or monoecious (Kubicki, 1969b). Using different genetic stocks, Kenigsbuch and Cohen (1990) reported that gynoecy was controlled by two recessive genes. Gynoecious melons do not respond to GA, while ethylene acts as a feminizing agent in both species (see below). As for the inheritance of andromonoecy, a gene similar to cucumber's *M* gene, designated *A* (not to be confounded with cucumber's *A* gene), is responsible in melon for the difference between *aa* (andromonoecious) and *A*–(monoecious). The *aa* genotype in melon lacks, however, the 'secondary characteristics' typical of andromonoecious cucumber. Bisexual flowers of andromonoecious melons are borne singly, like female flowers, and no intermediate forms with less-developed ovaries are encountered (Kubicki, 1969b). Taken together, it appears that in these two *Cucumis* species, the genetic mechanisms that regulate sex expression share some similarities, but are different in at least a few important features.

Figure 3 displays the parental phenotypes, and those of the F1 and F2 progenies, of three experimental populations used in our laboratory, each segregating for one of the above three genes. The primary importance of such segregating populations is for studying the inheritance of sex determination. They can also be used to apply modern mapping approaches to the study of cucumber sex expression (Section 9.1).

5. Environmental effects on sex expression

Several studies correlated changes in sex expression with environmental factors. Cucumber flowering is not a photoperiod-sensitive process, but sex expression of the flowers is affected by day length and temperatures (Nitsch *et al.*, 1952). Short days and low night temperatures ('winter conditions') enhanced femaleness in a monoecious cultivar: node number was 10.4, as compared to 18.8 under long days–warm night conditions (Atsmon and Galun, 1962). The photoperiodic response and the temperature response of monoecious cucumbers was characterized further in a series of studies by Matsuo and co-workers, and genetic variability in the type of response, and the number of photoperiods required, was reported (Matsuo, 1968; Matsuo *et al.*, 1969).

6. Chemical treatments that affect sex expression

6.1 *Gibberellins, anti-gibberellins and abscisic acid*

A variety of chemical compounds can alter sex expression in cucumber. During the 1950s and the 1960s plant growth regulators were extensively studied, and cucumber geneticists were excited by the prospect of chemically modifying the plant's sex. Wittwer and Bukovac (1958) were the first to report that GAs enhanced the male tendency of cucumber. Galun (1959a) showed that repeated daily GA treatments (100 ppm) increased the node number, and postponed the continuous-female phase, in heterozygous female plants. Atsmon (1968) noted that GA had a dual effect – it favoured male flower initiation, and also inhibited existing pistillate buds from full development. The assumption that GA can change the fate of a specific bud from female to male was challenged by Fuchs *et al.* (1977): GA appeared to cause the abortion of existing female buds, and the development, in their place, of adventitious male flowers that would otherwise be inhibited. Among different GAs, a mixture of GA4 and GA7 proved more effective than GA3 (Pike and Peterson, 1969).

Compounds that inhibit GA synthesis ('anti-gibberellins') were applied in several studies and were shown to promote femaleness (e.g. Frankel and Galun, 1977; Mitchell and Wittwer, 1962; Rodriguez and Lambeth, 1962; Yin and Quinn, 1995). Results with abscisic acid (ABA) were less clear-cut. Friedlander *et al.* (1977a) reported a variable effect of ABA on different sex genotypes: it enhanced the male tendency of monoecious types, but decreased female bud abortion in gynoecious plants. Another study reported a lack of ABA effect on monoecious plants, but a counter-action of GA effect on gynoecious plants (Rudich and Halevy, 1974).

6.2 *Auxins*

The first report on the effect of auxins on cucumber sex expression was by Laibach and Kribben (1949). Galun (1959a) reported that naphthalene acetic acid (NAA) (100 ppm) promoted female flower formation. Malepszy and Niemirowicz-Szczytt (1991) devised a peculiar system to elevate auxin and cytokinin levels. They infected cucumber seedlings with strains of *Agrobacterium tumefaciens* in which either the auxin or the cytokinin-synthesis genes were inactivated, and observed changes in sex expression in plants recovering from such infection. The results were not easy to explain, because the reaction of different genotypes to the same construct differed: certain hormone over-producing strains enhanced maleness in some genotypes, and femaleness in others. It is possible that in such infected, tumorous plants, hormone levels are much higher than in healthy plants.

6.3 *Ethylene and its inhibitors*

After its re-discovery as an important plant hormone, it was realized that ethylene exerts a strong feminizing effect on cucumbers (McMurray and Miller, 1968; Rudich *et al.*, 1969). Ethylene gas treatment given at the first leaf stage (50–100 ppm) lowered the node number of monoecious plants from 12.7 to 4.2 (Iwahori *et al.*, 1970). The effect of a single treatment lasted for several consecutive nodes. A convenient way to expose plants to ethylene is by application of ethrel, an ethylene-releasing compound. Ethrel increased the number of pistillate flowers in monoecious and andromonoecious lines, and, in addition, decreased the number of male flowers (Robinson *et al.*, 1969; Rudich *et al.*, 1969). It also inhibited stamen formation in the bisexual flowers of andromonoecious plants.

Aminoethoxyvinyl glycine (AVG) and α-aminoxyacetic acid (AOA) inhibit ethylene synthesis at the step catalysed by 1-aminocyclopropane-1-carboxylate (ACC) synthase; both compounds promoted maleness (Rudich, 1990). AVG induced the appearance of male flowers on a strictly gynoecious plant, following a transition phase in which perfect flowers formed (Atsmon and Tabbak, 1979). CO_2 is a competitive inhibitor of ethylene action. Rudich *et al.* (1972a) reported male flower formation on gynoecious plants after a 0.3% CO_2 treatment. On the other hand, Byers *et al.* (1972) observed no effect of 10% CO_2 on gynoecious cucumber.

Beyer (1976) discovered that silver ions act as inhibitors of ethylene action. Gynoecious cucumber plants that were given a single run-off spray of 500 ppm $AgNO_3$ at the third leaf stage developed, as a result, bisexual flowers followed by staminate ones. Ethylene evolution by the treated plants was unaffected. However, in a study by Atsmon and Tabbak (1979), 250 ppm $AgNO_3$ increased ethylene evolution 2.5-fold over the untreated control. The appearance of perfect flowers as an intermediate form on either ethrel- or anti-ethylene-treated plants, both at the onset of sex reversal, and as the effect fades out, indicates that these compounds may affect the fate of the bisexual bud itself. Such intermediates are never induced by GA, and it seems likely that the two compounds, GA and silver ions, act through different pathways. The anti-ethylene compounds also affect the mechanism responsible for the inhibition of subsequent buds in a female node; multiple pistillate/perfect flowers are formed in the boundary of the affected zone. Silver ion treatment was quickly adopted by seed companies; the use of $Ag(S_2O_3)_2^{-3}$, the anionic silver thiosulphate complex, which is less phytotoxic than silver nitrate, is often preferred. Silver-induced staminate flowers are morphologically normal and produce more viable pollen than GA-induced staminate flowers (Den Nijs and Visser, 1980).

Are the endogenous levels of plant hormones responsible for differences in sex expression? Hormone inhibitors have already provided us with indications for such an endogenous role, and in the next section more evidence regarding this important question is presented.

7. Role of endogenous growth regulators in sex expression

Having learned that hormones can exogenously modify sex expression in cucumber, how can we learn more about their mode of action, the interaction between them, and their possible relation with sex-determining genes *F*, *M* and *A*? Several studies have assayed the endogenous hormone level of different sex genotypes, or in a single genotype undergoing a change in sex expression, and tried to correlate the data with sex expression. The development of nearly isogenic lines (NILs) that differ in sex genes but share a similar genetic background, was of fundamental importance for the application of such approach.

7.1 *Non-identified chemical factors implicated in sex expression*

Some of the earlier studies provided indirect support for the notion that endogenous chemical factors regulate sex expression: Atsmon (1968) noted that the *F* allele correlated with decreased node length, hinting that sex genotypes may differ in their endogenous GA content. Galun (1959b) induced increased male tendency in monoecious plants (node number rose from 16 to 19) by removing young leaves; removal of older leaves increased female tendency (node number decreased from 16 to 14). Addition of NAA to the cut petiole counteracted the effect of leaf removal. He suggested that auxin produced by young leaves was responsible for the increase in femaleness along the shoot. The presence of a diffusible, endogenous factor that mediates sex expression was further indicated by grafting experiments (Friedlander *et al.*, 1977b; Takahashi *et al.*, 1982). Friedlander *et al.* (1977b) grafted cucumbers of different sex genotypes on each other, and compared sex expression along the scion and the rootstock shoots to control grafts, in which the two graft partners were of the same genotype. The result was that the sex pattern of the scion changed, becoming more similar to that of the rootstock; in a few cases, an effect of the scion upon the rootstock was observed as well.

7.2 *Measurement of endogenous hormones as related to sex expression*

A few studies correlated endogenous auxin with femaleness. Galun *et al.* (1965) reported that *FFmm* genotypes (hermaphrodites) contained more extractable auxin in their stems than andromonoecious plants (genotype *ffmm*). Rudich *et al.* (1972b) found that monoecious plants contained more auxin under SD than LD conditions. In an earlier study (Galun, 1959b), no difference in auxin activity between gynoecious and monoecious leaf extracts was recorded; a difference in auxin-inhibitory activity was, however, reported, monoecious leaves having more 'inhibitors'. In other studies a less expected, inverse correlation between auxin and femaleness was reported: Saito and Ito (1964) measured lower levels of endogenous auxin under femaleness-enhancing environmental conditions. Retig and Rudich (1972) measured higher auxin-degradative activity in a gynoecious, as compared to a monoecious genotype. Trebitsh *et al.* (1987) determined the content of IAA in cucumber shoot-tips: it was 60% higher in the monoecious genotype, inversely related to the feminizing effect of exogenous IAA. Such discrepancies may suggest that the endogenous auxin level may be irrelevant to sex expression. We may speculate that gynoecious plants would have an increased sensitivity to auxin, resulting in a stronger response of the ethylene-producing apparatus (see below), rather than increased levels of the hormone itself.

Atsmon *et al.* (1968) compared endogenous GA levels in a pair of NILs, one monoecious, the other gynoecious. Diffusates from young seedlings and root exudates from 6-week-old plants were collected and assayed for GA activity by two different bioassays. Monoecious samples had 10-fold more GA activity than gynoecious ones. A similar difference was reported by Rudich *et al.* (1972b). Friedlander *et al.* (1977c) measured a two-fold higher GA content in gynoecious leaves under LD, as compared to SD conditions. Hemphill *et al.* (1972) extracted GA from shoot tips taken at various developmental stages from gynoecious, monoecious and andromonoecious sex types. A peak of GA activity was found 1 week after germination in the monoecious and andromonoecious genotypes, but not in the gynoecious one. Following the peak, GA levels remained constant, and were 50% lower in the gynoecious line.

Friedlander *et al.* (1977c) determined the endogenous content of ABA in monoecious and gynoecious shoot tips by gas liquid chromatography, under LD versus SD conditions. ABA content was higher under SD (feminizing) conditions. On the other hand, monoecious plants contained more ABA than gynoecious ones, while *FFmm* (hermaphrodite) plants had four times more ABA than *ffmm* (andromonoecious) ones. In another study, however, ABA was higher in gynoecious shoot tips, than in monoecious apices (Rudich *et al.*, 1972b), and a female-promoting role, possibly inhibition of GA, was proposed for ABA. Friedlander *et al.* (1977c) suggested that different genotypes respond differentially to exogenous ABA (Section 6.1), because their endogenous hormone levels are situated on the two different 'slopes' of a hypothetical optimum curve for that hormone.

Gas chromatographic measurements of the ethylene produced by various sex-types provided important evidence for the involvement of endogenous ethylene in sex expression. Gynoecious lines produced 50% more ethylene that monoecious ones already at seed germination. A similar difference was measured in intact shoot tips at the fifth leaf stage (Byers *et al.*, 1972).

Rudich *et al.* (1972a) compared a pair of monoecious and gynoecious NILs. At 11 days post-germination, the two lines evolved similar amounts of ethylene from their excised shoot tips, but at 23 days post-germination, a three-fold difference between the genotypes was recorded. Ethylene evolution was doubled under SD, as compared to LD, conditions. It was found that developing male and female buds differ in ethylene evolution: excised female buds released three to four times more ethylene than male buds.

In a subsequent study, Rudich *et al.* (1976) also included androecious genotypes. They found that already at the germination stage, young androecious seedlings produce only 60%, and gynoecious seedlings produced, on average, 132%, of the ethylene made by monoecious ones. At the fourth leaf-stage androecious apices had 80%, and gynoecious apices had 290%, respectively, of the level found in the monoecious line. Expanded leaves did not differ in ethylene production. A prominent peak of ethylene was recorded in the gynoecious line 24 days post-germination, coinciding with the presence of rapidly developing female flowers. Interestingly, a smaller peak appeared in the monoecious plants (that presumably develop, at that stage, their first female flowers), but not in the androecious lines, where no female flowers are produced.

Trebitsh *et al.* (1987) assayed the endogenous levels of ethylene and its precursor ACC, in first leaf-stage apices of monoecious and gynoecious isogenic lines. Both were higher in the gynoecious line (70% more ethylene, 44% more ACC). Ethylene evolution from the gynoecious line was two-fold higher following ACC application. Taken together, these findings suggest that both ACC synthase (ACS) activity (producing ACC) and ACC oxidase (ACO) activity (consuming ACC) are higher in the gynoecious plant.

In conclusion, the studies reviewed in this section do suggest a correlation between GA level and male tendency, and between ethylene and female tendency. Results regarding auxin and ABA are less consistent, or more complex.

7.3 *Evidence for interaction among different hormones*

The next question involves possible interactions between different hormones. Is the level, or action of one hormone, controlled by a second one? Such interactions can be explored by treating a plant with one substance, and monitoring the change in the endogenous content of a second substance. Shannon and De La Guardia (1969) suggested that the feminizing effect of auxin is, in fact, mediated by ethylene, since auxin induced an increase in ethylene evolution, that preceded the sex-reversing effect. Trebitsh *et al.* (1987) reported that incubation of excised apices with 0.1 mM IAA induced higher levels of ACC and ethylene. The induction of ACS activity by exogenous auxin is a well-established phenomenon in several plant systems (references in Rudich, 1990).

Ethylene, on the other hand, may have an inhibitory feedback effect on auxin in cucumber (Rudich *et al.*, 1972b), because auxin activity decreased in cucumber apices after ethrel treatment, while auxin-inhibitory activity increased. Takahashi and Jaffe

(1984) induced femaleness by ACC and auxin treatments. The effect was inhibited by either AVG or $AgNO_3$, clearly suggesting that ethylene acts downstream of auxin. Also, in the study by Trebitsh *et al.* (1987), AVG prevented an auxin-induced rise in ACC and ethylene. Moreover, treatment of gynoecious plants with the anti-auxins β-NAA and PCIB did not alter sex expression, suggesting that the limiting factor in female expression is not endogenous auxin (Trebitsh *et al.*, 1987). It may be concluded, therefore, that the effect of auxin is indirect, ethylene being the true sex hormone of the two.

Atsmon and Tabbak (1979) showed that treatment of young gynoecious apices with GA did not affect their rate of ethylene production, indicating that GA's effect is not mediated by decreasing ethylene levels. It may, therefore, act downstream of ethylene, or through a parallel, unrelated mechanism. Rudich *et al.* (1972a) reported that treatment of monoecious seedlings with ethrel greatly lowered their GA levels – and raised the level of GA-activity inhibitors. This, again, points at GA as being downstream of ethylene, in contrast to the suggestion made by Yin and Quinn (1995; Section 7.4).

7.4 *A model for cucumber sex expression*

Yin and Quinn (1995) proposed a rather elegant theoretical model for sex expression in cucumber, providing suggestions on the possible roles of hormones in sex expression, and the relationships between hormones and sex-determining genes. Based on the opposite effects of the two hormones, GA and ethylene, two alternative hypotheses could be advanced to explain their mechanism of action. (i) A 'two-hormone balance' hypothesis: each of the two hormones promotes the development of one sex type; the relative levels (balance) of the two will determine the sex of a plant, or of floral organs at a particular site on the plant. According to this hypothesis, in a monoecious plant the 'male-hormone' level must always be accompanied, in different zones along the shoot, by low levels of 'female-hormone', and *vice versa*. Such a model will not explain how, in the same monoecious plant, male and female flowers, but no perfect flowers, are produced – even at adjacent sites. (ii) The 'one-hormone' hypothesis suggests that the same hormone regulates both sexes: it inhibits one type of sex organ and, at the same time, promotes the other sex. Variation in the level of an endogenous hormone, and in the sensitivity threshold of its receptors in different regions of the plant, define the regions in which male, female or perfect flowers will form. Yin and Quinn (1995) observed plants treated with either GA, ethylene or their inhibitors, and concluded (in agreement with other reports, Section 6) that each hormone not only promoted one sex, but also inhibited the formation of flowers of the other sex. This fits the 'one-hormone hypothesis' – except that in cucumber, two hormones play a role, each of them exerting such a dual effect. The next question to consider is which of the two hormones, ethylene or GA, is the primary sex hormone, having a more direct effect on the process. To answer this question, andromonoecious and monoecious plants were treated with one effective concentration of GA, paclobutrazol (a GA synthesis inhibitor), ethrel and $AgNO_3$, either singly or in different combinations. The combined GA-and-ethrel treatment had an effect similar to ethrel, and the paclobutrazol plus silver treatment was similar to silver alone. This was interpreted in favour of a more direct effect of ethrel, rather than GA: the authors speculate that GA acts more upstream, and may negatively regulate the levels of endogenous ethylene. Note, however, that other studies (Section 7.3) do not support such a relationship between the two hormones; the overriding effect of ethrel as compared to GA may be due to the specific concentrations chosen for this particular experiment, or to a difference in the persistence of the exogenous compounds absorbed by the plant.

Of greater relevance are the predictions regarding the mode of action of the two sex-determining genes, *F* and *M* (*Figure 4*). According to Yin and Quinn's model, the *F* gene determines the range of sex hormone (i.e. ethylene) concentrations along the plant's shoot. A monoecious plant would have an increasing concentration of ethylene. 'Pushing the range upwards' along the shoot by lowering the ethylene level at given nodes will delay the female phase. Changes in ethylene concentration may be caused by environmental conditions, chemical treatments or by *F* alleles of varying strength. The second element of the model, required to predict which organ type will form at a specific ethylene concentration, consists of the receptors. The female organ receptor would perceive ethrel above a specific threshold concentration, and transduce a promoting signal to carpel primordia. Distinct male organ receptors will transduce an inhibitory signal above their own ethylene concentration threshold. The *M* gene specifies the sensitivity of male organs to the inhibitory effect of ethylene. Its dominant allele presumably encodes a more sensitive variant of the male receptor, compared to the female organ receptors (*Figure 4*). The *m* allele, encoding a less sensitive isoform, will cause an 'overlapping' range of ethylene, that will be both female-promoting and male-permissive. Such 'overlap' allows the formation of perfect flowers in *mm* genotypes, while a non-overlapping range of receptor sensitivities would preclude such a possibility in *M* genotypes.

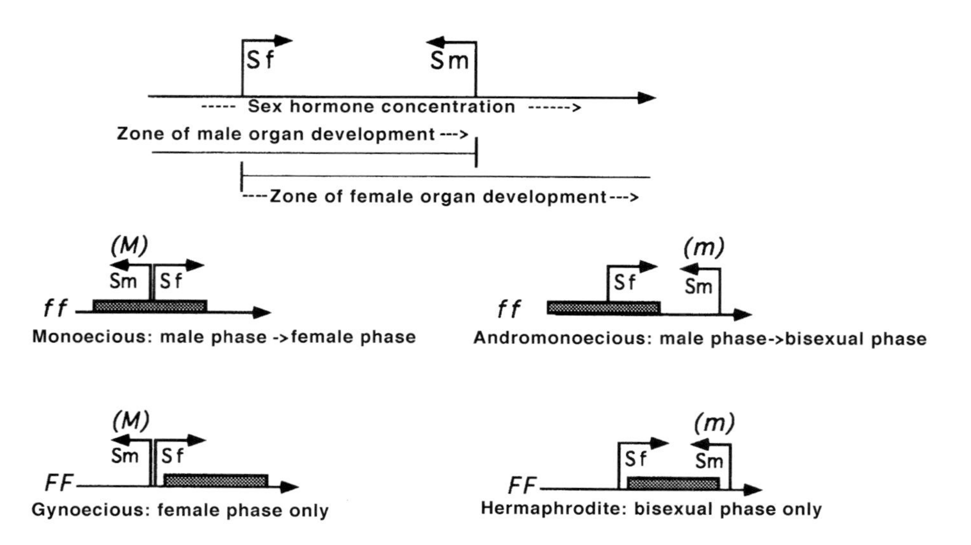

Figure 4. *Theoretical model explaining the interaction of ethylene concentration, ethylene sensitivity and the two sex genes,* F *and* M, *in cucumber sex determination. Horizontal box represents the range of ethylene concentration along the shoot. Sm denotes the threshold-sensitivity of the male-organ receptors, above which male organs are inhibited. Sf denotes the ethylene threshold concentration of the female-organ receptors, above which female organs will develop. Modified from the* American Journal of Botany, *vol. 82, 1995, Yin and Quinn, Tests of a mechanistic model of one hormone regulating both sexes in* Cucumis sativus *(Cucurbitaceae), with permission from the* American Journal of Botany.

We note that the model falls short in explaining some of the features of sex expression in cucumber. If varying male-sensitivity thresholds are possible, the purely monoecious situation would seem a very special case, according to the model, rather than the prevailing rule in cucumbers; it is hard to explain why we never encounter monoecious plants with sterile zones between the male and female phases (i.e. a zone where ethylene is insufficient for female development but still male inhibitory). Moreover, why are trimonoecious patterns, with an intermediate phase of perfect flowers, so rare? One possible explanation is, that the *m* allele is very insensitive (a null allele?), and therefore a pistillate flower phase cannot be attained in *mm* plants. Another aspect that remains unexplained by the model involves the mixed phase, where, after a single female node, the monoecious plant 'regresses' to a male phase for a few nodes. Within *mm* genotypes we also find, in the same node, a 'gradient' of flowers of decreasing femaleness. Such patterns suggest a more dynamic situation along the shoot, and would probably require more elaborate models. For example, it is possible that the ethylene concentration does not increase linearily but fluctuates, resulting in male flowers following female flowers. The action of *M*, being a 'trigger' that allows unisexual flowers, may be better explained if we assume that the putative male-repressor encoded by the *M* allele is expressed, or activated, only in high ethylene regions (female-determined nodes), precluding the formation of a sterile zone, or of occasional bisexual flowers, at intermediate ethylene levels.

Despite its limitations, this model gave proof of predictive power: recent molecular work by Trebitsh *et al.* (1997) indeed mapped an extra copy of an ACC synthase gene to the *F* locus (Section 9.3), suggesting that *F* may indeed determine ethylene levels. It is now tempting to make educated guesses about the molecular nature of the *M* gene. Attempts to isolate and study cucumber ethylene receptors have been recently initiated in ours, as well as in other laboratories.

8. Application of the above studies to agriculture and breeding

In the last few decades, F1 hybrid varieties started to replace open-pollinated cucumber cultivars at an increasing pace. Gynoecious inbred lines have been used as female parents for hybrid seed production, ensuring that the seed harvested from the maternal parent originates from cross-pollination with the paternal inbred line. This practice eliminated the need for emasculation, enclosure of the maternal flowers or even hand pollination, provided that the two parental lines are grown in an isolated plot, in the presence of bees. Monoecious inbreds were first used as pollinators; later, androecious lines were developed for that purpose. The female lines are propagated by self-pollination using GA or silver ion treatments to induce staminate flowers, while androecious lines are maintained by ethrel treatments that induce pistillate flowers (Augustine *et al.*, 1973). The hybrids that result from such crosses are heterozygous-female (*Ff*), with an early, continuous female phase following a short male phase. Such hybrids have higher, earlier and more uniform yields as compared to traditional monoecious varieties, and may be more suitable for mechanical harvest of pickling varieties (Wehner and Miller, 1985). In the last years, the trend in breeding greenhouse varieties is different: hybrids are produced by crossing two gynoecious parents, the paternal one being chemically treated. The resulting F1 is gynoecious (*FF*), with a further increase in uniformity and yield. Parthenocarpic fruit setting is ensured in such genotypes by the dominant gene, *Pc*.

9. Molecular studies on cucumber sex expression

9.1 *Possible approaches to identify genes of interest*

The great recent advances in the fields of molecular plant development and in plant genome analysis have opened new possibilities to elucidate the mechanism of cucumber sex determination, and dissect the process of male and female flower differentiation. A few general approaches may be taken to isolate important genes and study their role in the process.

(i) Mapping strategies: molecular DNA markers generated randomly by techniques such as RFLP, RAPD and AFLP can be applied to mapping populations derived from different sex genotypes, in an attempt to identify markers for the sex loci. *Figure 3* shows three F2 populations used for this purpose in our laboratory. Such markers are of applicative value – they can be used as a breeding aid for early selection of the desired sex genotype. Furthermore, a saturated map with closely linked markers surrounding a locus can be used to clone the gene of interest. The cucumber genome is characterized by a very small size (Arumuganathan and Earle, 1991), but very low levels of DNA polymorphism (Staub *et al.*, 1996). The linkage map developed in cucumber by Kennard *et al.* (1994) guarantees the feasibility of such an approach.
(ii) Molecular study of cucumber genes that may be related to floral differentiation and to hormone metabolism/action. Several genes that have a proven role in floral differentiation and reproductive development have been isolated from model plants. Studying their expression in the cucumber system may shed light on the process of unisexual flower development. Genes that control the metabolism and the perception/transduction of the major sex hormones of cucumber, namely ethylene and GA have also been isolated. Again, one can isolate their cucumber homologues and apply molecular techniques, ranging from Northern and *in situ* analysis of mRNA expression, to modulation of expression in transgenic cucumber plants, and ask whether the expression of such genes is related to sex. Over-expression and silencing of critical genes may also yield novel sex types for research and breeding purposes.
(iii) Isolation of genes of yet-unknown functions that exhibit a differential pattern of expression in buds of different sex, or in buds undergoing sex-reversal. Such genes can be 'fished' using different methods, and their function can then be studied by the techniques mentioned above.

9.2 *Cloning genes that determine floral organ identity*

In the past decade, the developmental process of flower development has been thoroughly investigated at the molecular level. The main model plants that were used were *Arabidopsis thaliana*, *Antirrhinum majus* and *Petunia*. Mutations that alter floral architecture enabled the isolation of genes that determine floral organ identity, floral symmetry and inflorescence determinacy. Studying these genes in dioecious and monoecious plants may shed light on questions related to unisexual flower development, as compared to the more common hermaphrodite flower.

The homeotic gene *AGAMOUS* was the first floral organ identity gene to be cloned from *Arabidopsis*. It encodes a transcription factor essential for reproductive organ development in plants (Yanofsky and Meyerowitz, 1990). *AGAMOUS* is a member of the MADS-box gene superfamily whose representatives are also found in the other eukaryotic kingdoms (Theißen *et al.*, 1996). In plants, most MADS-box genes are expressed in reproductive organs, where they control different stages of inflorescence, flower and floral organ development. They can be classified to distinct sub-families according to their amino acid sequences, their expression domains and their functions (Theißen and Saedler, 1995; Theißen *et al.*, 1996). Genetic and molecular data have led to the formulation, and continued elaboration, of the 'A, B, C model for flower development' (Coen and Meyerowitz, 1991; Weigel and Meyerowitz, 1994), which describes how specific combinations of homeotic gene functions give rise to specific floral organs in the four whorls of a flower. For example, the *AGAMOUS* gene and its functional homologues provide the 'C'-function required for both stamen and carpel formation in the two inner whorls, while B-function genes such as *APETALA3* specify petals and stamens. MADS-box genes often have additional functions. *AGAMOUS* is also responsible for determinacy of the flower, and has a late function in the development of ovules.

Our laboratory set to study the expression of *AGAMOUS* homologues during cucumber flower development, to see whether the expression of such genes could be correlated with sex expression. Floral bud cDNA libraries were thus prepared from two cucumber inbred lines, a gynoecious and an androecious one. The mRNA for the libraries was extracted from 1 mm-size buds (*Figure 1d*), in which the buds are morphologically bisexual; important developmental 'decisions' regarding sex expression may take place at this stage. The libraries were screened with an *AGAMOUS* cDNA probe, and three different *AGAMOUS* homologues were isolated and designated *CAG1* (for *Cucumis AGAMOUS*), *CAG2* and *CAG3* (Perl-Treves *et al.*, 1998a). Other laboratories have also reported the isolation of *AGAMOUS*-like sequences: *Table 1* summarizes the data on clones reviewed in this section and provides necessary cross-reference between them.

The three *CAG* clones appear closely homologous to each other and to the *Arabidopsis* gene, and contain full-length open reading frames. All three contain a MADS-box region that is almost identical to that of *AGAMOUS*. Their homology with representatives of two other major subfamilies of the MADS-box superfamily, *APETALA1* and *APETALA3*, is much lower.

Northern analysis was undertaken, to characterize the stages at which *CAG1*, *CAG2* and *CAG3* are expressed during male and female bud development. Expression of the genes was barely detectable in the earlier stages, and became higher in intermediate-stage buds. Transcript levels were maintained in larger buds and were still present at anthesis. No *CAG2* transcript was detected in staminate buds and flowers, its expression being limited to pistillate buds. The *CAG3* and *CAG1* transcripts, on the other hand, were present in both female and male buds. None of the three transcripts was detected in leaves. In the organs of mature flowers the three genes exhibited distinct patterns of expression. Northern analysis (*Figure 5*) indicated that *CAG2* is a female-specific clone, and is expressed only in the carpels of pistillate flowers. *CAG1* and *CAG3* were expressed in the stamens, carpel and nectaries. The latter two genes are thus expressed in both sexes in the third and fourth floral whorls, which correspond to the expression domain of the original *AGAMOUS* gene. More detailed analysis may reveal, in the future, differences between *CAG1* and *CAG3*. In *Petunia*, two *AGAMOUS*-like genes, *pMADS3* and *FBP6*, are expressed in the two inner

Table 1. *Cloned cucumber genes and cDNAs involved in flower development and ethylene metabolism*

	Clone	Clone description	GenBank accession	Reference
AGAMOUS-like	CAG-1	Stamen- and carpel-specific cDNA; same as *CUM10*	AF022377	Perl-Treves *et al.*,1998a
	CAG-2	Carpel-specific cDNA; same as *CUS1*	AF022378	Perl-Treves *et al.*,1998a
	CAG-3	Stamen- and carpel-specific cDNA; same as *CUM1*	AF022379	Perl-Treves *et al.*,1998a
	CUS-1	Ovary- and embryo-specific cDNA; same as *CAG2*	X97801	Filipecki *et al.*, 1997
	CUM-1	Same as *CAG3*	AF035438	Kater *et al.*, 1998
	CUM-10	Same as *CAG1*	AF035439	Kater *et al.*, 1998
ACC synthases	Cs-ACS-1	Auxin-induced genomic fragment, *F*-linked	U59813	Trebitsh *et al.*, 1997
	Cs-ACS-2	Apex-expressed cDNA	D89732	Kamachi *et al.*, 1997
	Cs-ACS-3	Auxin-treated fruit cDNA, similar to ACS-1	AB003683	Kamachi *et al.*, 1997
	Cs-ACS-4	Wounded fruit cDNA	AB003684	Kamachi *et al.*, 1997
ACC oxidases	Cs-ACO-1	Apex and leaf expressed cDNA	AF033581	Perl-Treves *et al.*,1998b
	Cs-ACO-2	Apex and leaf expressed cDNA	AF033582	Perl-Treves *et al.*,1998b
	Cs-ACO-3	Apex and leaf expressed cDNA	AF033583	Perl-Treves *et al.*,1998b

whorls (Angenent *et al.*, 1995; Kater *et al.*, 1998), and it now seems that the prevailing situation consists of two *AGAMOUS* orthologues providing in concert the 'C-function', *Arabidopsis* being an exception with a single *AGAMOUS* gene. Kater *et al.* (1998) isolated two *AGAMOUS* homologues from cucumber, designated *CUM1* and *CUM10* (*Table 1*); *CUM1* is identical to *CAG3* and *CUM10* to *CAG1*. The two cucumber cDNAs were ectopically expressed in transgenic *Petunia* under the control of the 35S promoter. As a result they were strongly expressed in leaves and in all floral whorls. *CUM1* caused a complete transformation of petals to anthers (in the stronger transgenic phenotypes). Sepals were deformed and, under the scanning electron microscope, carpelloid tissue with stigmatic papillae and ovules was apparent. These effects are consistent with the ABC model of organ identity. The effect of *CUM10* over-expression was more moderate. Corolla limbs in the second whorl were reduced and contained antheroid tissue, but sepals did not exhibit carpelloid features. This work is the first to prove the organ-identity function of the cucumber *AGAMOUS*-like genes, and to show a functional difference between *CAG1* and *CAG3* – albeit in an heterologous system.

The third cDNA isolated, *CAG2*, is not expressed as an *AGAMOUS* orthologue, but has a carpel-specific expression pattern. The same mRNA was cloned in a parallel study from a cucumber embryogenic callus cDNA library by Filipecki *et al.* (1997; *Table 1*), and designated *CUS1*. Expression analysis of *CUS1* complements our findings: the transcript was detected in the transmitting tissue and surrounding the ovules of mature female flowers and developing fruits. It was also localized in heart stage somatic embryos, suggesting a role for this gene in embryo development. Female-specific genes of the *AGAMOUS* subfamily have been isolated also from other plants, for example, maize and *Arabidopsis* (Rounsley *et al.*, 1995; Savidge *et al.*, 1995; Schmidt *et al.*, 1993).

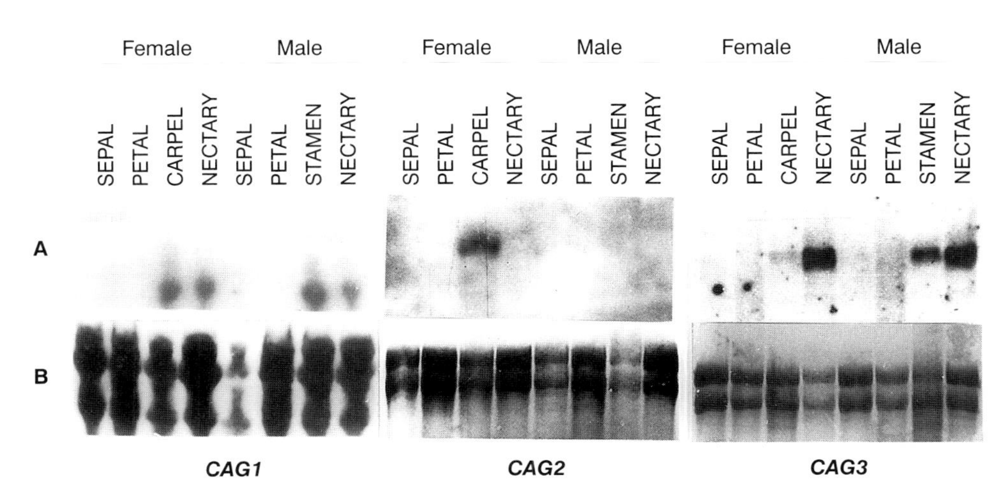

Figure 5. Expression of CAG1, CAG2 and CAG3 in mature flower organs. Lanes contained 10 μg total RNA from sepals (with some fused corolla), petals, stamens and nectary from the androecious genotype ('male') and sepals (with fused corolla), petals, carpel and nectary from the gynoecious genotype ('female') (a) Hybridization to CAG probes. (b) Ribosomal RNA staining by methylene blue. Reprinted from Plant and Cell Physiology, *vol. 39, 1998, Perl-Treves* et al., *Expression of multiple Agamous-like genes in male and female flowers of cucumber (*Cucumis sativus *L.), with permission from the Japanese Society of Plant Physiology.*

Two *Petunia* genes, *FBP7* and *FBP11*, proved to be functionally related to ovule development and determination of ovule identity: when ectopically expressed in whorls 1 and 2, naked ovules formed on the surface of sepals and petals (Angenent *et al.*, 1995; Colombo *et al.*, 1995). *CAG2* may belong to such a group of female-specific *AGAMOUS* homologues.

An important question is, whether cucumber's unisexual flowers are 'designed' by modulating the expression of floral organ-identity genes. For example, lack of *AGAMOUS* expression in whorl 3 or 4 at a critical stage may suppress further development of organ primordia in that whorl. One way to address this question in a future study would be to correlate the precise timing of organ arrest with the expression of MADS-box genes in a developing bud. This would require *in situ* detection of the transcript in the arrested vestigial organs of the unisexual flower.

We also tested whether ethrel and GA treatments that affect sex expression of the plant would induce a change in transcript levels of *AGAMOUS* homologues in cucumber apices. For example, would a feminizing treatment elevate the *CAG2* (female-specific) transcript – or decrease *CAG1* or *CAG3* transcript level in male buds? Shoot apices at the third-leaf stage were treated with ethrel and GA solutions at concentrations used to reverse the sex of breeding lines, and Northern analysis was carried out. We did not observe detectable changes in the levels of *CAG1, CAG2* and *CAG3* transcripts after such treatments in either the male or female genotypes. On the other hand, when we later examined visibly reversed buds that developed from the hormone-treated apices, female-specific expression of *CAG2* became apparent (Perl-Treves *et al.*, 1998a), indicating that such expression is indeed specific to carpel development, and not to the particular genotype. In conclusion, *CAG* gene products are probably required for the initiation and maintenance of stamens and carpels; our study could not, however, implicate the three genes as part of a 'sex-expression cascade' that responds to external (and probably also internal) hormonal stimuli. We may suggest that the mechanism of organ arrest induced by sex hormones may not involve the elevation or repression of a particular *CAG* transcript. This conclusion is, however, tentative, because a transient or very localized change in transcript level following the hormonal treatment would have passed unnoticed in Northern analysis, and would require *in situ* hybridization or promoter–reporter assays to be detected. Moreover, MADS-box gene products may be affected by sex hormones at a post-transcriptional level, for example, by affecting another regulatory protein that interacts with the *CAG* gene product.

9.3 *Cloning of ethylene-synthesis genes*

ACC synthase. Due to its importance as a 'sex hormone' in cucumber, molecular characterization of cucumber genes that control ethylene metabolism has been started. Trebitsh *et al.* (1997) applied PCR primers that matched conserved sequence domains of plant ACSs, and isolated a genomic DNA fragment, designated *CS-ACS-1*, from a gynoecious cucumber line (*Table 1*). When tested on Southern blots, it hybridized to a single band in monoecious cultivars, and to a pair of homologous bands in gynoecious cultivars (*Figure 6*). The authors interpreted such a pattern as a duplication of the ACS gene in the gynoecious genotypes. Expression of the *ACS-1* transcript was studied by Northern blot analysis in a pair of monoecious and gynoecious lines. The two did not differ in the levels of *ACS* transcript in their shoot apices and leaves. Treatment of excised

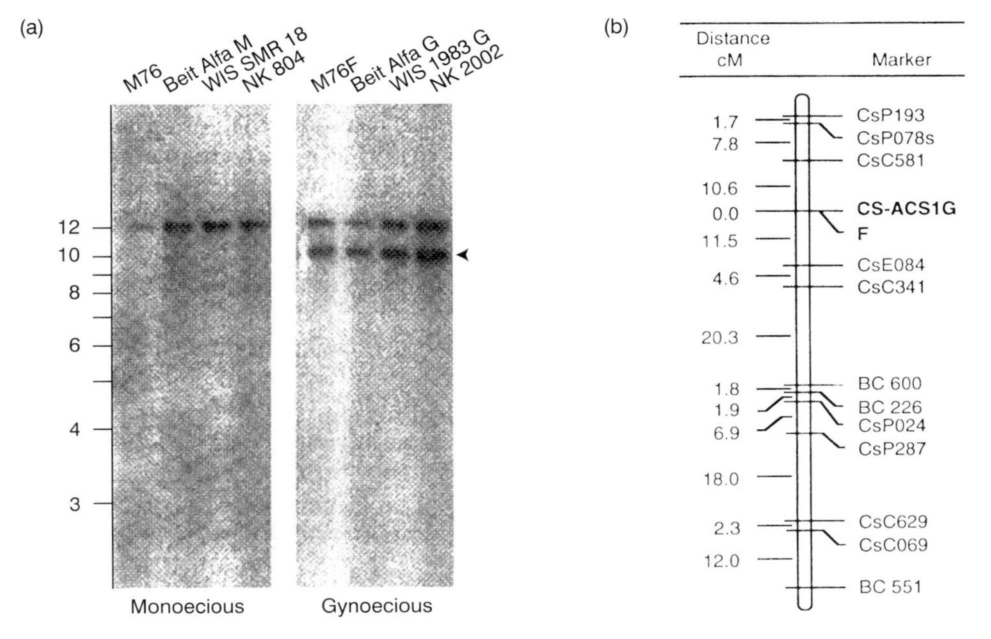

Figure 6. *Mapping of ACS1 to the F locus. (a) Southern blot analysis of cucumber DNA with the ACS-1 probe. Gynoecious cultivars contained an additional ACS-RFLP band, indicated by an arrow, as compared to monoecious near-isogenic cultivars. Reprinted from* Plant Physiology, *vol. 113, 1997, Trebitsh* et al., *Identification of a 1-aminocyclopropane-1-carboxylic acid synthase gene linked to the Female (F) locus that enables female sex expression in cucumber, with permission from the American Society of Plant Physiologists. (b) Such RFLP was mapped using the mapping population developed by Kennard* et al. *(1994) to the F locus at 100% linkage. Modified with the permission of Springer-Verlag from* Theoretical and Applied Genetics, *vol. 89, 1994, Kennard* et al., *Linkages among RFLP, RAPD, isozyme, disease resistance, and morphological markers in narrow and wide crosses of cucumber, pp. 42–48, Figure 4B. © Springer-Verlag, 1994.*

apices with auxin caused a strong, transient elevation of transcript level that was similar in both genotypes. Interestingly, the sequence of *ACS-1* is highly homologous to a subfamily of auxin-inducible ACSs from various plant species. A mapping population, segregating for the *F* locus, used by Kennard *et al.* (1994) to construct a cucumber linkage map, was used to map *ACS-1* with respect to *F*. A second population consisting of the F2 progeny of a gynoecious and a monoecious NILs was analysed as well. A 100% linkage (not a single recombinant) was found between the *F* locus and the *ACS-1* RFLP! Such tight linkage strongly suggests that the putative extra copy of *ACS-1* is the *F* gene itself. Additional examples of linkage between plant hormone synthesis and sex-determination include *Mercurialis annua*, where sex-determination genes co-segregated with the level of zeatin, a sex-determining hormone in this plant (Durand and Durand, 1991). In maize, *anther-ear1*, a mutation that affects sex expression, was identified at the molecular level as the gene encoding ent-kaurene synthase, a primary step in GA synthesis (Bensen *et al.*, 1995). It remains, however, to explain the apparent lack of correlation between *ACS-1* expression and the sex phenotype. It could be argued that the difference in transcript level underlying the sex-determination event is a transient, or very localized one, undetected in whole shoot-apices by Northern analysis. Note that it is still impossible to know which of the two duplicated copies, *ACS-1* or *ACS-1G* (the latter is the putative female-specific copy), was actually cloned. A thorough analysis of the *F*-locus genomic structure, the development of gene-specific probes and isolation of the two respective promoters may shed light on the function of the *ACS* genes in this locus. Conclusive proof identifying *F* as *ACS-1* may come from transgenic-complementation experiments.

The molecular approach is often complicated by the fact that the genes investigated are part of gene families, and it is rather difficult to distinguish between the activity of individual gene members. In another recent study, Kamachi *et al.* (1997) have used reverse transcriptase-PCR with degenerate primers to isolate ACS sequences from cucumber. Three cDNA fragments were amplified from different tissues: *Cs-ACS-2* from shoot apices of a monoecious cultivar; *Cs-ACS3* from auxin-treated immature fruits; and *Cs-ACS-4* from wounded immature fruits (*Table 1*). Full-length clones were then obtained by the rapid amplification of cDNA ends (RACE) technique. Expression of the three genes was followed by reverse transcriptase-PCR: only *ACS-2* was expressed in the shoot apex. *ACS-3* was auxin-inducible, and its sequence was very similar to the *ACS-1* gene isolated by Trebitsh *et al.* (1997). The authors went on to study the expression of *ACS-2* in shoot tips of three genotypes. Transcript levels in apices harvested at different developmental stages were compared, by Northern analysis, and found to correlate with the timing and intensity of female flowering. The transcript was first detectable in 15-day-old apices of the gynoecious cultivar, later attaining very high levels; in a monoecious line with a relatively strong female tendency, transcripts were detected after 20 days and reached moderate levels; in the third cultivar, which had the most delayed female phase, transcripts were hardly detectable after 25 days. Further studies would be required to elucidate the relationship between the expression of the different *ACS* genes and sex expression.

ACC oxidase. ACO ('ethylene-forming enzyme') catalyses the second, final step in ethylene formation, following ACC production by ACS, and its expression is tightly regulated in plants (e.g. Barry *et al.*, 1996; Kim *et al.*, 1997). We set out to study the expression of ACO genes in cucumber plants, in order to try to relate it to sex-expression patterns in various sex genotypes. Cucumber bud cDNA libraries were screened at moderate stringency with a tomato ACO probe (Hamilton *et al.*, 1991). Three distinct clones (*ACO-1, ACO-2* and *ACO-3*) were isolated and sequenced (Perl-Treves *et al.*, 1998b). Preliminary characterization of their expression by Northern analysis indicated that all three mRNAs are expressed in leaves and shoot apices of cucumber plants. *ACO-2* was analysed in more detail. Four cucumber genotypes displaying contrasting sex phenotypes were compared. These included a gynoecious line, a hermaphrodite, an androecious and a monoecious cultivar. The first two are *FF* homozygotes, the latter two are homozygous-recessive (*ff*). Northern analysis was carried out on plantlets at the third unfolded leaf stage, when the shoot apex contains unfolded leaves and developing floral meristems and buds at early stages of differentiation.

In the first unfolded leaf, *ACO2* transcript levels were higher in gynoecious and hermaphrodite plants, in concordance with the presumably higher ethylene evolution in female genotypes (we were, however unable to reproduce in our laboratory, the genotypic differences in ethylene evolution reported in the literature). Transcript levels in shoot apices displayed, on the other hand, an inverse correlation: the two *ff* genotypes gave higher *ACO2* signals in their apices (Kahana and Perl-Treves, unpublished results). While we lack a conclusive picture on the relationship between sex and

ACO expression, such inverse correlation is intriguing and warrants further investigation. A possible explanation would be that increased ethylene production is critical for female development, but its most important site of production is not the whole shoot tip. Alternatively, it may be that *ACO* transcript level is irrelevant to final ethylene level, and other genes (*ACS*?) may be responsible for the higher ethylene produced by female genotypes. In that case, we may be observing a feedback inhibition of *ACO* expression in the apex by ethylene; both positive and negative feedback control of ethylene over its own synthesis have been reported (e.g. Liu *et al.*, 1997).

9.4 *Genes that are differentially expressed in floral buds*

A possible strategy to identify genes that are important in a developmental process, is to screen for mRNAs that are present in one sample and absent (or rare) in a second one. To clone transcripts related to sex expression in cucumber, several experiments may be designed, for example, a screen of transcripts induced by a sex-reverting hormonal treatment. We have chosen to compare untreated male and female young buds at the morphologically bisexual stage (*Figure 1d*). These were collected from two contrasting genotypes, an androecious and a gynoecious one (*Figure 1a* and *b*).

We employed two techniques to isolate differentially expressed male and female clones. One was differential display (Liang *et al.*, 1993), in which two total RNA samples are 'fingerprinted' by a combination of reverse transcriptase and PCR amplification with short random primers. Patterns of amplification products from a pair of samples are displayed on sequencing gels, and bands representing differentially expressed mRNA species can be excised and cloned. We have run pairs of reactions, each pair consisting of male bud versus female bud RNA from different bud stages and young leaves (Kahana, Silberstein and Perl-Treves, unpublished results). We have run 49 pairs of 1 mm bud displays and a smaller number of leaf and older bud displays. In order to verify the result, putative- differential bands were labelled and applied as probes on to Northern blots, containing RNA from male and female buds.

The second technique employed to obtain male/female-specific genes was differential screening of cDNA libraries. Aliquots of a male, and a female 1 mm bud library, were plated, and two replicate filters were prepared. These were hybridized to total cDNA probes of male and female small buds, respectively. Despite the small scale of the experiment run, and the fact that detection is restricted to the more abundant mRNAs, differentially expressed clones have been selected and tested on Northern blots.

The results of these two experiments have yet to be finalized. We noted that many of the putative-differential bands from the differential display experiment turned out to be 'false positive', exhibiting non-differential expression on Northern blots. In addition, since differential display often amplifies rare mRNAs, many of its products had too weak expression to be detected on Northern blots. This was the case with about 30% of the putative-differential bands, and confirming their differential nature would require more sensitive techniques. We have cloned so far six cDNAs, all of which are male-specific. For a few of them, the result of a sequence homology search would suggest cell wall-related functions. The role of cell wall proteins in morphogenesis is an area of active research (Keller and Lamb, 1989; Showalter, 1993). Several reports exist on differential expression of such proteins in floral organs (e.g. Goldman *et al.*, 1992). Two of the clones that are preferentially expressed in more mature male buds, as compared to female buds, encoded a ubiquitin homologue and a ribosomal protein, respectively. It remains to be seen whether such genes have a differential role in male flower development, or whether their higher expression is due to a less specific cause, such as more rapid protein metabolism in developing male buds.

10. Concluding remarks

The molecular elucidation of cucumber sex expression has only begun, and the results published so far are few. Only one sex-determining gene, *F*, has been placed on a linkage map, and a 'strong' suggestion regarding its molecular nature has been advanced. The other two major genes (as well as others described in the literature) have not been cloned, or even mapped. Correlations between the expression of specific cucumber genes with sex phenotypes are still scant. The involvement of ethylene and GA synthesis in sex expression, their mode of action and the interaction between them, are still poorly understood. The cellular mechanisms that operate downstream of the hormonal/sex-determining signal, causing the arrest of stamen or carpel primordia, have not been unravelled. Nevertheless, a few genes that participate in different steps of sex expression, from hormone metabolism and perception to sex organ determination and differentiation, have been cloned recently, and their molecular analysis has begun. In the coming years we shall hopefully obtain a better-resolved description, and an improved understanding, of this fascinating but complex system. Such knowledge may then be expanded to other important cucurbit species (melon, pumpkin, squash and watermelon), where sex expression is less understood.

References

Angenent, G.C., Franken, J., Busscher, M., van Dijken, A., van Went, J.L., Dons, H.J.M. and van Tunen, A.J. (1995) A novel class of MADS box genes is involved in ovule development in *Petunia*. *Plant Cell* **7:** 1569–1582.

Arumuganathan, K. and Earle, E.D. (1991) Nuclear DNA content of some important plant species. *Plant Mol. Biol. Reporter* **9:** 208–218.

Atsmon, D. (1968) The interaction of genetic, environmental and hormonal factors in stem elongation and floral development of cucumber plants. *Ann. Bot.* **32:** 877–882.

Atsmon, D. and Galun, E. (1960) A morphogenetic study of staminate, pistillate and hermaphrodite flowers in *Cucumis sativus* L. *Phytomorphology* **10:** 110–115.

Atsmon, D. and Galun, E. (1962) Physiology of sex in *Cucumis sativus* L. Leaf age patterns and sexual differentiation of floral buds. *Ann. Bot.* **26:** 137–146.

Atsmon, D. and Tabbak, C. (1979) Comparative effects of gibberellin, silver nitrate and aminoethoxyvinyl glycine on sexual tendency and ethylene evolution in the cucumber plant (*Cucumis sativus* L.). *Plant Cell Physiol.* **20:** 1547–1555.

Atsmon, D., Lang, A. and Light, E.N. (1968) Contents and recovery of gibberellins in monoecious and gynoecious cucumber plants. *Plant Physiol.* **43:** 806–810.

Augustine, J.J., Baker, L.R. and Sell, H.M. (1973) Female flower induction in an androecious cucumber, *Cucumis sativus* L. *J. Am. Soc. Hortic. Sci.* **98:** 197–199.

Barry, C.S., Blume, B., Bouzayen, M., Cooper, W., Hamilton, A.J. and Grierson, D. (1996). Differential expression of the 1-amino-cyclopropane-1-carboxylate oxidase gene family of tomato. *Plant J.* **9:** 525–535.

Bensen, R.J., Johal, G.S., Crane, V.C., Tossberg, J.T., Schnable, P.S., Meeley, R.B. and Briggs, S.P. (1995) Cloning and characterization of the maize *An1* gene. *Plant Cell* **7:** 75–84.

Beyer, E. (1976) Silver ion: a potent antiethylene agent in cucumber and tomato. *HortScience* **11:** 195–196.

Byers, R.E., Baker, L.R., Sell, H.M., Herner, R.C. and Dilley, D.R. (1972) Ethylene: a natural regulator of sex expression of *Cucumis melo* L. *Proc. Natl Acad. Sci. USA* **69:** 717–720.

Coen, E.S. and Meyerowitz, E. (1991) The war of whorls: genetic interactions controlling flower development. *Nature* **353:** 31–37.

Colombo, L., Franken, J., Koetje, E., van Went, J., Dons, H.J.M., Angenent, G.C. and van Tunen, A.J. (1995) The petunia MADS Box gene *Fbp11* determines ovule identity. *Plant Cell* **7:** 1859–1868.

Currence, T.M. (1932) Nodal sequence of flower types in the cucumber. *Proc. Am. Soc. Hort. Sci.* **29:** 477–479.

Dax-Fuchs, E., Atsmon, D. and Halevy, A.H. (1978) Vegetative and floral bud abortion in cucumber plants: hormonal and environmental effects. *Sci. Hortic.* **9:** 317–327.

Den Nijs, A.P.M. and Visser, D.L. (1980) Induction of male flowers in gynoecious cucumber (*Cucumis sativus* L.) by silver ions. *Euphytica* **29:** 273–280.

Durand, B. and Durand, R. (1991) Sex determination and reproductive organ differentiation in *Mercurialis*. *Plant Sci.* **80:** 49–65.

Filipecki, M.K., Sommer, H. and Malepszy, S. (1997) The mads-box gene *CUS1* is expressed during cucumber somatic embryogenesis. *Plant Sci.* **125:** 63–74.

Frankel, R. and Galun, E. (1977) *Pollination Mechanisms, Reproduction and Plant Breeding*. Springer-Verlag, Berlin.

Friedlander, M., Atsmon, D. and Galun, E. (1977a) Sexual differentiation in cucumber: the effect of AbA and other growth regulators on various sex genotypes. *Plant Cell Physiol.* **18:** 261–269.

Friedlander, M., Atsmon, D. and Galun, E. (1977b) The effect of grafting on sex expression in cucumber. *Plant Cell Physiol.* **18:** 1343–1350.

Friedlander, M., Atsmon, D. and Galun, E. (1977c) Sexual differentiation in cucumber: abscisic acid and gibberellic acid contents of various sex genotypes. *Plant Cell Physiol.* **18:** 681–691.

Fuchs, E., Atsmon, D. and Halevy, A.H. (1977) Adventitious staminate flower formation in gibberellin treated gynoecious cucumber plants. *Plant Cell Physiol.* **18:** 1193–1201.

Galun, E. (1959a) Effect of gibberellic acid and naphthaleneacetic acid in sex expression and some morphological characters in the cucumber plant. *Phyton* **13:** 1–8.

Galun, E. (1959b) The role of auxin in the sex expression of the cucumber. *Physiol. Plant.* **12:** 48–61.

Galun, E. (1961a) Gibberellic acid as a tool for the estimation of the time interval between physiological and morphological bisexuality of cucumber floral buds. *Phyton* **16:** 57–62.

Galun, E. (1961b) Study of the inheritance of sex expression in the cucumber, the interactions of major genes with modifying genetic and non-genetic factors. *Genetica* **32:** 134–163.

Galun, E., Jung, Y. and Lang, A. (1962) Culture and sex modification of male cucumber buds in vitro. *Nature* **194:** 596–598.

Galun, E., Jung, Y. and Lang, A. (1963) Morphogenesis of floral buds of cucumber cultured in vitro. *Develop. Biol.* **6:** 370–387.

Galun, E., Izhar, S. and Atsmon, D. (1965). Determination of relative auxin content in hermaphrodite and andromonecious *Cucumis sativus* L. *Plant Physiol.* **40:** 321–326.

Goffinet, M.C. (1990) Comparative ontogeny of male and female flowers of *Cucumis sativus*. In: *Biology and Utilization of the Cucurbitaceae* (eds D.M. Bates, R.W. Robinson and C. Jeffrey). Cornell University Press, New York, pp. 288–304.

Goldman, M.H., Pezzotti, M., Seurinck, J. and Mariani, C. (1992) Developmental expression of tobacco pistil-specific genes encoding novel extensin-like proteins. *Plant Cell* **4:** 1041–1051.

Hamilton, A.J., Bouzayen, M. and Grierson, D. (1991) Identification of a tomato gene for the ethylene-forming enzyme by expression in yeast. *Proc. Natl Acad. Sci. USA* **88:** 7434–7437.

Hemphill, D.D., Baker, L.R. and Sell, H.M. (1972) Different sex phenotypes of *Cucumis sativus* L. and *C. melo* L. and their endogenous gibberellin activity. *Euphytica* **21:** 285–291.

Iwahori, S., Lynos J.M. and Smith O.E. (1970) Sex expression in cucumber as affected by 2-chloroethylphosphosphonic acid, ethylene and growth regulators. *Plant Physiol.* **46:** 412–415.

Kamachi, S., Sekimoto, H., Kondo, N. and Sakai, S. (1997) Cloning of a cDNA for a 1-aminocyclopropane-1-carboxylate synthase that is expressed during development of female flowers at the apices of *Cucumis sativus* L. *Plant Cell Physiol.* **38:** 1197–1206.

Kater, M.M., Colombo, L., Franken, J., Busscher, M., Masiero, S., Van Lookeren Campagne, M.M. and Angenent, G.C. (1998) Multiple *AGAMOUS* homologs from cucumber and petunia differ in their ability to induce reproductive organ fate. *Plant Cell* **10:** 171–182.

Keller, B. and Lamb, C.J. (1989) Specific expression of a novel cell wall hydroxyproline-rich glycoprotein gene in lateral root initiation. *Genes Develop.* **3:** 1639–1646.

Kenigsbuch, D. and Cohen, Y. (1990) The inheritance of gynoecy in muskmelon. *Genome* **33:** 317–320.

Kennard, W.C., Poetter, K., Dijkhuizen, A., Meglic, V., Staub, J. and Havey, M. (1994) Linkages among RFLP, RAPD, isozyme, disease resistance, and morphological markers in narrow and wide crosses of cucumber. *Theor. Appl. Genet.* **89:** 42–48.

Kim, J.H., Kim, W.T., Kang, B.G. and Yang, S.F. (1997) Induction of 1-aminocyclopropane 1-carboxylate oxidase messenger RNA by ethylene in mung bean hypocotyls – involvement of both protein phosphorylation and dephosphorylation in ethylene signalling. *Plant J.* **11:** 399–405.

Kubicki, B. (1969a) Investigations on sex determination in cucumber (*Cucumis sativus* L.). V. Genes controlling intensity of femaleness. *Genet. Pol.* **10:** 69–85.

Kubicki, B. (1969b) Comparative studies on sex determination in cucumber (*Cucumis sativus* L.) and muskmelon (*Cucumis melo* L.). *Genet. Pol.* **10:** 167–183.

Kubicki, B. (1969c) Investigations on sex determination in cucumber (*Cucumis sativus* L.). VII. Andromonoecism and hermaphroditism. *Genet. Pol.* **10:** 101–120.

Kubicki, B. (1969d) Investigations on sex determination in cucumber (*Cucumis sativus* L.). VIII. Trimonoecism. *Genet. Pol.* **10:** 123–142.

Laibach, F. and Kribben, F.J. (1949) Der Einfluss von Wuchsstoff auf die bildung männlicher und weiblicher Blüten bei einer monözischen Pflanze (*Cucumis sativus*). *Ber. Det. Botan. Gesell.* **62:** 53–55.

Liang, P., Averboukh, L. and Pardee, A.B. (1993). Distribution and cloning of eukaryotic mRNA by means of differential display: refinements and optimization. *Nucleic Acids Res.* **21:** 3269–3279.

Liu, J.H., Lee-Tamon, S.H. and Reid, D.M. (1997). Differential and wound-inducible expression of 1-aminocylopane-1-carboxylate oxidase genes in sunflower seedlings. *Plant Mol. Biol.* **34:** 923–933.

Malepszy, S. and Niemirowicz-Szczytt, K. (1991) Sex determination in cucumber (*Cucumis sativus*) as a model system for molecular biology. *Plant Sci.* **80:** 39–47.

Matsuo, E. (1968) Studies on the photoperiodic sex differentiation in cucumber, *Cucumis sativus* L. I. Effect of temperature and photoperiod upon the sex differentiation. *J. Fac. Agric. Kyushu Univ.* **14:** 483–506.

Matsuo, E., Uemoto, S. and Fukushima, E. (1969) Studies on the photoperiodic sex differentiation in cucumber, *Cucumis sativus* L. II. Aging effect on the photoperiodic dependency of sex differentiation. *J. Fac. Agric. Kyushu Univ.* **15:** 287–303.

McMurray, A.L. and Miller, C.H. (1968) Cucumber sex expression modified by 2-chloroethane phosphonic acid. *Science* **162:** 1397–1398.

Mitchell, W.D. and Wittwer, S.H. (1962) Chemical regulation of sex expression and vegetative growth in *Cucumis sativus*. *Science* **136:** 880–881.

Nitsch, J., Kurtz, E.B., Livermann, J.L. and Went, F.W. (1952) The development of sex expression in cucurbit flowers. *Am. J. Bot.* **39:** 32–43.

Perl-Treves, R., Kahana, A., Rosenmann, N., Yu, X. and Silberstein, L. (1998a) Expression of multiple Agamous-like genes in male and female flowers of cucumber (*Cucumis sativus* L.). *Plant Cell Physiol.* **39:** 701–710.

Perl-Treves, R., Kahana, A., Korach, T. and Kessler, N. (1998b) Cloning of three cDNAs encoding 1-aminocycloproane-1-carboxylate oxidases from cucumber floral buds (accession Nos. AF033581, AF033582 and AF033583) (PGR98–037). *Plant Physiol.* **116:** 1192.

Pierce, L.K. and Wehner, T.C. (1990) Review of genes and linkage groups in cucumber. *HortScience* **25:** 605–615.

Pike, L.M. and Peterson, C.E. (1969) Gibberellin A4/A7, for induction of staminate flowers on the gynoecious cucumber (*Cucumis sativus* L.). *Euphytica* **18:** 106–109.

Retig, N. and Rudich, J. (1972) Peroxidase and IAA oxidase activity and isoenzyme patterns in cucumber plants, as affected by sex expression and ethephon. *Physiol. Plant* **27:** 156–160.

Robinson, R.W., Shannon, S. and De La Guardia, M.D. (1969) Regulation of sex expression in the cucumber. *BioScience* **19:** 141–142.

Rodriguez, B.P. and Lambeth, V.N. (1962) Synergism and antagonism of GA and growth inhibitors on growth and sex expression in cucumber. *J. Am. Soc. Hortic. Sci.* **97:** 90–92.

Rounsley, S.D., Ditta, G.S. and Yanofsky, M.F. (1995) Diverse roles for MADS box genes in *Arabidopsis* development. *Plant Cell* **7:** 1259–1269.

Roy, R.P. (1990) Sex expression in the Cucurbitaceae. In: *Biology and Utilization of the Cucurbitaceae* (eds D.M. Bates, R.W. Robinson and C. Jeffrey). Cornell University Press, New York, pp. 251–268.

Rudich, J. (1990) Biochemical aspects of hormonal regulation of sex expression in cucurbits. In: *Biology and Utilization of the Cucurbitaceae* (eds D.M. Bates, R.W. Robinson and C. Jeffrey). Cornell University Press, New York, pp. 269–279.

Rudich, J. and Halevy, A.H. (1974) Involvement of abscisic acid in the regulation of sex expression in the cucumber. *Plant Cell Physiol.* **15:** 635–642.

Rudich, J., Halevy, A.H. and Kedar, N. (1969) Increase in femaleness of three cucurbits by treatment with Ethrel, an ethylene-releasing compound. *Planta* **86:** 69–76.

Rudich, J., Halevy, A.H. and Kedar, N. (1972a) Ethylene evolution from cucumber plants as related to sex expression. *Plant Physiol.* **49:** 998–999.

Rudich, J., Halevy, A.H. and Kedar, N. (1972b) The level of phytohormones in monoecious and gynoecious cucumbers as affected by photoperiod and ethephon. *Plant Physiol.* **50:** 585–590.

Rudich, J., Baker, L.R., Scott, J.W. and Sell, H.M. (1976) Phenotypic stability and ethylene evolution in androecious cucumber. *J. Am. Soc. Hortic. Sci.* **101:** 48–51.

Saito, T. and Ito, H. (1964) Factors responsible for the sex expression of the cucumber plant. XIV. Auxin and gibberellin content in the stem apex and the sex pattern of flowers. *Tohoku J. Agric. Res.* **14:** 227–239.

Savidge, B., Rounsley, S.D. and Yanofsky, M.F. (1995) Temporal relationship between the transcription of two *Arabidopsis* MADS-box genes and the floral organ identity genes. *Plant Cell* **7:** 721–733.

Schmidt, R.J., Veit, B., Mandel, M.A., Mena, M., Hake, S. and Yanofsky, M.F. (1993) Identification and molecular characterization of *ZAG1*, the maize homolog of the *Arabidopsis* floral homeotic gene *AGAMOUS*. *Plant Cell* **5:** 729–737.

Shannon, S., and De La Guardia, M.D. (1969). Sex expression and the production of ethylene induced by auxin in the cucumber (*Cucumis sativus* L.). *Nature* **223:** 186.

Shifriss, O. and Galun, E. (1956) Sex expression in cucumber. *Proc. Am. Soc. Hortic. Sci.* **67:** 479–486.

Showalter, A.M. (1993) Structure and function of plant cell wall proteins. *Plant Cell* 5: 9–23.

Staub, J.E., Bacher, J. and Poetter, K. (1996) Factors affecting the application of random amplified polymorphic DNAs in cucumber (*Cucumis sativus* L.). *HortScience* **31:** 262–266.

Takahashi, H. and Jaffe, M.J. (1984) Further studies on auxin and ACC induced feminization of the cucumber plant using ethylene inhibitors. *Phyton* **44:** 81–86.

Takahashi, H., Saito, T. and Suge, H. (1982) Intergeneric translocation of floral stimulus across a graft in monoecious Cucurbitaceae with special reference to the sex expression of flowers. *Plant Cell Physiol.* **23:** 1–9.

Theißen, G. and Saedler, H. (1995) MADS box genes in plant ontogeny and phylogeny: 'Haeckel's biogenetic law' revisited. *Curr. Opinions. Genet. Develop.* **5:** 628–639.

Theißen, G., Kim, J.T. and Saedler, H. (1996) Classification and phylogeny of the MADS-Box multigene family suggest defined roles of MADS-Box gene subfamilies in the morphological evolution of eukaryotes. *J. Mol. Evol.* **43:** 484–516.

Trebitsh, T., Rudich, J. and Riov, J. (1987) Auxin, biosynthesis of ethylene and sex expression in cucumber (*Cucumis sativus*). *Plant Growth Regul.* **5:** 105–113.

Trebitsh, T., Staub, J.E. and O'Neill, S.D. (1997) Identification of a 1-aminocyclopropane-1-carboxylic acid synthase gene linked to the *Female (F)* locus that enhances female sex expression in cucumber. *Plant Physiol.* **113:** 987–995.

Wehner, T.C. and Miller, C.H. (1985) Effect of gynoecious expression on yield and earliness of a fresh market cucumber hybrid. *J. Am. Soc. Hortic. Sci.* **110:** 464–466.

Weigel, D. and Meyerowitz, E.M. (1994) The ABCs of floral homeotic genes. *Cell* **78:** 203–209.

Wittwer, S.H. and Bukovac, M.J. (1958) The effect of gibberellins on economic crops. *Econ. Bot.* **12:** 213–255.

Yanofsky, M.F. and Meyerowitz, E.M. (1990) The protein encoded by the *Arabidopsis* homeotic gene *AGAMOUS* resembles transcription factors. *Nature* **346:** 35–39.

Yin, T. and Quinn, J.A. (1995) Tests of a mechanistic model of one hormone regulating both sexes in *Cucumis sativus* (Cucurbitaceae). *Am. J. Bot.* **82:** 1537–1546.

Index